Heat and Mass Transfer in the Thermal Control System of Technical and Technological Energy Equipment (HMTTSC-2018)

MATEC Web of Conferences Volume 194 (2018)

Tomsk, Russia
24 - 26 April 2018

Editors:

G.V. Kuznetsov
E. Bulba
D. Feoktistov

ISBN: 978-1-5108-7004-8

Printed from e-media with permission by:

Curran Associates, Inc.
57 Morehouse Lane
Red Hook, NY 12571

proceedings.com

Some format issues inherent in the e-media version may also appear in this print version.

Printed by Curran Associates, Inc. (2018)

For additional information, please contact EDP Sciences – Web of Conferences
at the address below.

EDP Sciences – Web of Conferences
17, Avenue du Hoggar
Parc d'Activité de Courtabœuf
BP 112
F-91944 Les Ulis Cedex A
France

Phone: +33 (0) 1 69 18 75 75
Fax: +33 (0) 1 69 28 84 91

contact-edps@webofconferences.org

Additional copies of this publication are available from:

Curran Associates, Inc.
57 Morehouse Lane
Red Hook, NY 12571 USA
Phone: 845-758-0400
Fax: 845-758-2633
Email: curran@proceedings.com
Web: www.proceedings.com

TABLE OF CONTENTS

Statement of Peer review

In submitting conference proceedings to *Web of Conferences*, the editors of the proceedings certify to the Publisher that

1. They adhere to its **Policy on Publishing Integrity** in order to safeguard good scientific practice in publishing.
2. All articles have been subjected to peer review administered by the proceedings editors.
3. Reviews have been conducted by expert referees, who have been requested to provide unbiased and constructive comments aimed, whenever possible, at improving the work.
4. Proceedings editors have taken all reasonable steps to ensure the quality of the materials they publish and their decision to accept or reject a paper for publication has been based only on the merits of the work and the relevance to the journal.

Title, date and place of the conference

Heat and Mass transfer in the Thermal Control System of technical and technological energy equipment - 2018
24-26 April 2018
Tomsk, Russia

Proceedings editor(s):

/Genii Kuznetsov
/Elena Bulba
/Dmitrii Feoktistov

Date and editor's signature

25/07/2018

17, avenue du Hoggar - PA de Courtabœuf – BP 112 - 91944 Les Ulis Cedex A (France)
Tél. : 33 (0)1 69 18 75 75 - Fax : 33(0)1 69 07 45 17 – www.edpsciences.org

Mathematical modeling of ignition of coal-water slurries particles coated with water film

V. G. Afanaseva[1,*]

[1]National Research Tomsk Polytechnic University, 634050 Tomsk, Russia

Abstract. The results of the mathematical modeling of heat and mass transfer processes occurring jointly with ignition of coal-water slurries particles covered with a water film are presented. It is shown that water film thickness can have a significant effect on the ignition dynamics. In addition, it was found out that the main fuel layer is heated to the evaporation temperature during water film evaporation. As a result, following processes of thermal preparation (evaporation of moisture, thermal decomposition) occur faster.

1 Introduction

Presently, climate change problem is particularly acute. The main cause of the current global warming trend is increasing concentration of carbon dioxide in the atmosphere, resulting in the greenhouse effect. In Russia, about 70% of the carbon dioxide evolution is caused by the fossil fuel combustion in thermal power plants. The highest level of emissions is observed during coal combustion, therefore, the first thing we have to do is develop clean coal technologies to minimize the negative impact on the environment.

One of such technological solutions that ensure safety and efficiency of coal use is the technology of combustion it in the form of coal water slurry (CWS) [1-4].

The large-scale inmplementation of CWS in the energy sector is hampered by many factors, one of which is insufficient study of the processes occurring during the CWS thermal preparation for ignition. Experimental studies of CWS combustion in boiler units require theoretical studies of the ignition and burning regularities, which are necessary for solving the organization of suspension combustion problems.

At the time of CWS spraying in furnace volume of a boiler, two classes of CWS drops form: the first class includes partially dried CWS particles, the second class contains more watered drops with a near-surface water film [5]. The ignition characteristics of the latter class are currently insufficiently studied, but it is clear that the presence of a water film can have an impact on the ignition of CWS particles due to the high endothermic evaporation and heat capacity of water [6].

Thus, the purpose of the article is mathematical modeling of the heat and mass transfer processes occurring jointly with ignition of coal-water slurries particles covered with a water film.

* Corresponding author: ssyrodoy@yandex.ru

2 Formulation of the problem

The mathematical formulation of the problem, the corresponding stated higher of physical model is a non-stationary system of differential equations in partial derivatives:

CWS particle has a polyhedral shape that depends on many factors and can vary randomly, therefore, the particle was observed in the spherical form for simplification [7].

At the initial time moment (t=0) the CWS particle enters the gas environment, the temperature of which far exceeds the particle temperature. The particle absorbs the gas environment heat through convection and radiation, as a result the evaporation of the water film begins. It is taken that thermal radiation is absorbed only by the water film [6].

After the water film evaporation, further warming of the main fuel layer begins, the warming is followed by pore moisture evaporation and thermal decomposition of the fuel combustible matter with volatile content. The processes run before the volatiles igniting near the particle. It is taken that combustion of volatiles includes the following reactions:

$$2H_2 + O_2 \rightarrow 2H_2O + 115.5\ {}^{kJ}/_{mole},$$

$$CH_4 + 2O_2 \rightarrow CO_2 + 2H_2O + 892\ {}^{kJ}/_{mole},$$

$$2CO + O_2 \rightarrow 2CO_2 + 285\ {}^{kJ}/_{mole}.$$

The mathematical formulation of the problem includes non-stationary differential equations in partial derivatives, which are shown below.

The energy equation for the CWS particle:

$$c_p(T)\rho_p(T)\frac{\partial T}{\partial t} = \frac{1}{r^2}\frac{\partial}{\partial r}\left(r^2\lambda_p(T)\frac{\partial T}{\partial r}\right) + \frac{1}{r^2\sin\varphi}\frac{\partial}{\partial\varphi}\left(\sin\varphi\,\lambda_p(T)\frac{\partial T}{\partial\varphi}\right) - q_{th}W_{th} - Q_{ev}W_{ev} \quad (1)$$

The thermo-physical properties of CWS particle (c_p, ρ_p, λ_p) were evaluated by position of evaporation front and were calculated taking into account volume ratio of CWS particle components.

There is the boundary condition of the fourth kind at the interface of the "CWS particle-water film", taking into account the radiation heat transfer:

$$r = r_p, 0 < \varphi < \pi: \begin{cases} \lambda_p(T)\dfrac{\partial T_p}{\partial r} = \lambda_\Sigma\dfrac{\partial T_w}{\partial r} \\ T_p = T_w \end{cases} \quad (2)$$

The energy equation for the water film ($t > 0, r_p < r < r_w, 0 < \varphi < \pi$):

$$c_w\rho_w\frac{\partial T}{\partial t} = \frac{1}{r^2}\frac{\partial}{\partial r}\left(r^2\lambda_\Sigma\frac{\partial T}{\partial r}\right) + \frac{1}{r^2\sin\varphi}\frac{\partial}{\partial\varphi}\left(\sin\varphi\,\lambda_\Sigma\frac{\partial T}{\partial\varphi}\right) \quad (3)$$

The radiation absorption by water film was adopted in accordance with the Rosseland radiation model [8]:

$$\lambda_\Sigma = \lambda_w + \frac{16\sigma T^3}{3a_R}$$

There is the boundary condition of the fourth kind at the interface of the "water film-gas environment", taking into account the radiation heat transfer and the thermal effect of vaporization:

$$r = r_w, 0 < \varphi < \pi: \begin{cases} \lambda_\Sigma \dfrac{\partial T_w}{\partial r} = \lambda_g \dfrac{\partial T_g}{\partial r} - Q_{ev} W_{ev}, \\ T_w = T_g \end{cases} \quad (4)$$

In the equations (1) and (4) the mass evaporation rate was calculated from the expression:

$$W_{ev} = W_0 \cdot exp\left[\frac{Q_{ev}\mu(T - T_f)}{RTT_f}\right]$$

Thermal decomposition of the fuel combustible matter was described by chemical kinetics equation:

$$\begin{cases} \dfrac{\partial \eta}{\partial t} = (1-\eta) k_{th} exp\left(\dfrac{-E_{th}}{RT}\right) \\ W_{th} = (1-\eta) \cdot k_{th} \cdot \rho(T) \cdot exp\left(\dfrac{-E_{th}}{RT}\right) \end{cases}.$$

The energy equation for the gas environment:

$$c_g \rho_g \frac{\partial T}{\partial t} = \frac{1}{r^2} \frac{\partial}{\partial r}\left(r^2 \lambda_g \frac{\partial T}{\partial r}\right) + \frac{1}{r^2 \sin\varphi} \frac{\partial}{\partial \varphi}\left(\sin\varphi\, \lambda_g \frac{\partial T}{\partial \varphi}\right) + Q_i W_i \quad (5)$$

In the equation (5) the rate of chemical reactions was calculated from the mathematical expression of the Arrhenius law:

$$W_i = k_i C_i \left(1 - \sum_{i=1}^{4} C_i\right) \rho_g \cdot exp\left(\frac{-E_i}{RT}\right)$$

The boundary condition at the gas frontier line:

$$r = r_g, 0 < \varphi < \pi: \frac{\partial^2 T}{\partial r^2} = 0$$

The volatile substances concentration dependence of coordinates and time is described by the diffusion equation ($t > 0, r_p < r < r_g, 0 < \varphi < \pi$):

$$\frac{\partial C_i}{\partial t} = \frac{D_i}{r^2}\left[\frac{\partial}{\partial r}\left(r^2 \frac{\partial C_i}{\partial r}\right) + \frac{1}{\sin\varphi} \frac{\partial}{\partial \varphi}\left(\sin\varphi\, \lambda_g \frac{\partial C_i}{\partial \varphi}\right)\right] \quad (6)$$

The equations (1-6) were solved with the following boundary conditions:

$$\varphi = 0, 0 < r < r_g: \frac{\partial T}{\partial \varphi} = 0$$

$$\varphi = \pi, 0 < r < r_g: \frac{\partial T}{\partial \varphi} = 0$$

$$r = r_p, 0 < \varphi < \pi: \begin{cases} \dfrac{\partial C_i}{\partial r} = -\dfrac{W_i}{\rho_g D_i} \\ W_i = V_i (1-\eta) \rho(T) \cdot exp\left(\dfrac{-E_{th}}{RT}\right) \end{cases}$$

$$r = r_g, 0 < \varphi < \pi: \frac{\partial^2 C_i}{\partial r^2} = 0$$

$$\varphi = 0, r_p < r < r_g: \frac{\partial C_i}{\partial \varphi} = 0$$

$$\varphi = \pi, r_p < r < r_g: \frac{\partial c_i}{\partial \varphi} = 0$$

The designations: $T0$ – initial temperature of the particle, K; rp – radius of the CWS particle, m; rw – radius of the CWS particle with the water film, m; rg – radius from center of the CWS particle to boundary of the gas environment, m; λp – coefficient of thermal conductivity of the CWS particle, W/(m·K); cp – heat capacity of the CWS particle, J/(kg·K); ρp – density of the CWS particle, kg/m^3; λw – coefficient of thermal conductivity of water, W/(m·K); cw – heat capacity of water, J/(kg·K); ρw – density of water, kg/m^3; λg – coefficient of thermal conductivity of the gas, W/(m·K); cg – heat capacity of the gas, J/(kg·K); ρg – density of the gas, kg/m^3; Wev – mass rate of water evaporation, kg/(m^2·s); Qev – thermal effect of the water evaporation , J/kg; Tf – temperature of ice point, K; $W0$ – mass rate of water evaporation for temperature Tf, kg(m^2·s); μ – molar mass of water, kg/mole; R – universal gas constant, J/(mole·K); Qi – thermal effect of a chemical reaction, J/kg; ki – preexponent of chemical reaction 1/s; Ei – activation energy of a chemical reaction, J/mole; Di – binary diffusion coefficient, m^2/s; σ – Stefan-Boltzmann constant, W·m^{-2}·K^{-4}; aR – Rosseland mean absorption coefficient, 1/m, Vi – volatile yield.

3 The results

At the time of CWS spraying in furnace volume of a boiler, a lot of particles are formed, the sizes of which differ from each other. The thickness of the water film which is coated the CWS particles can also assume a different value. Accordingly, it can be said that the water film thickness is a significant factor affecting on the characteristics of the CWS particles ignition. At the present time the degree of such influence is poorly known.

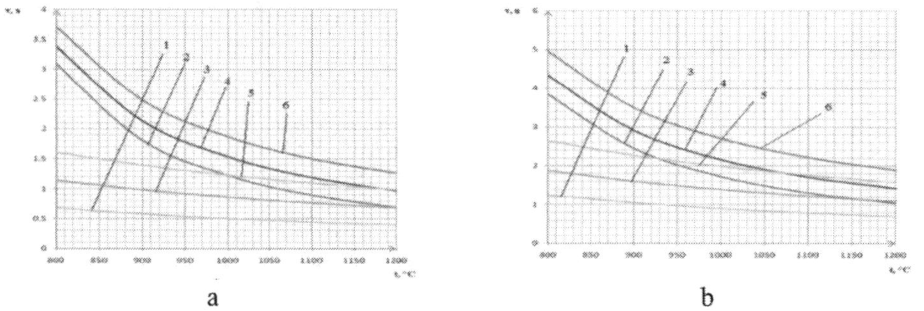

Fig. 1. The dependence of the water film evaporation times and the ignition delay times of the ambient temperature with the water film thickness of 0.12·10^{-3}m (a) and 0.12·10^{-3}m (b) at a rate (d) particles: 1, 2 – d=2·10^{-3}m; 3, 4 – d=3·10^{-3}m; 5, 6 – d=4·10^{-3}m.

Fig. 1 illustrates the time dependence of the CWS particle ignition delay and the water film evaporation, with variation of particle diameter and film thickness. Analysis of dependences revealed that the furnace volume temperature increase reduces the ignition delay time. It can be noted that the diameter increase leads to the induction period decrease. So the diameter increase from 2 to 4 mm results in the ignition delay time increase by 30% (from 3 to 5 s), at T_e = 1073K.

The film thickness increase (Fig. 1, b) leads to an elongation of the induction period due to the high endothermic evaporation and heat capacity of water.

Fig. 2 shows the time dependence of the evaporation time of the water film and the ignition delay time of CWS particles coated with a water film, which has different thicknesses. It can be noted that increasing the film thickness from 0.12 to 0.4 mm leads to the ignition process slowing down by 50%.

Analysis of the dependences in Fig. 2 shows that the evaporation of a water film has a significant effect on the thermal preparation for the CWS particles ignition.

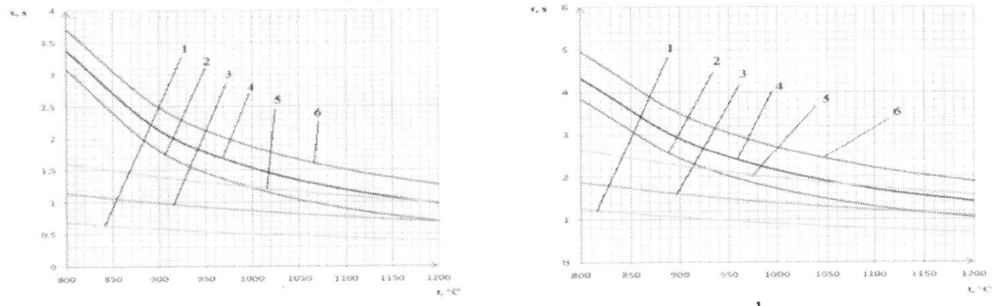

a b

Fig. 2. The dependence of the water film evaporation times and the ignition delay times of the ambient temperature with the particle diameter of $2\cdot10^{-3}$m (a) and $3\cdot10^{-3}$m (b) at a film thickness (Δ):1, 2 – Δ=$0.12\cdot10^{-3}$m; 3, 4 – Δ=$0.25\cdot10^{-3}$m; 5, 6 – Δ=$0.4\cdot10^{-3}$m.

It can be noted that the film evaporation time for almost the entire temperature range is up to 60% of the induction period, it is due to the fact that during the evaporation of the water film the main fuel layer is heated to the evaporation temperature. For that reason, the following stages of the thermal preparation of the CWS particle (without a film) and ignition proceed much faster.

4 Conclusion

According to the results of the computational experiments it was established that the near-surface water film can have a significant effect on the characteristics and conditions of the CWS particles ignition. It is shown that the film thickness increase leads to an elongation of the induction period. Increase of the ambient temperature and film thickness leads to increase of the evaporation process role on the integral characteristics of the ignition process. When the water film evaporates, the main fuel layer is heated to the evaporation temperature. So, the thermal preparation processes (evaporation of moisture in the CWS particle, thermal decomposition) proceed more quickly.

The research has been carried out at the expense of the grant of the Russian Scientific Foundation (project No 15-19-10003).

References

1. O. Erenkov, I. Galanina, E. Urazova Misyura, S. L. Elistratov, R. A. Dekhtyar, Perspective development of science, engineering and technology: Materials of the Third International scientifically-practical conference **1**, 397 (2013)
2. O. Kalanikhin, Ecological safety **2**, 78 (2013)
3. A. Kuznetsov, I. Tokarev, Modern science **1**, 32 (2010)
4. A. Kuznetsov, I. Tokarev, Modern science **1**, 21 (2011)
5. J. Senchurova, V. Murko, V. Fedyaev, D. Dziuba, E. Pusirev, The Tomsk Polytechnic University news **312**, 38 (2008)
6. S. Syrodoy, N. Gutareva, R. Taburchinov, MATEC Web Conf. **72**, 01109 (2016)
7. E. Flek, The Journal Omsk Scientific Bulletin **6**, 102 (2017)
8. R. Siegel, *Thermal radiation heat transfer* (Mir, Moscow, 1975) [in Russian]

MATEC Web of Conferences **194**, 01002 (2018) https://doi.org/10.1051/matecconf/201819401002
HMTTSC-2018

An experimental study of the dynamics of ascent of the bubble system in the presence of a surfactant

Alexandra Antonnikova[1], *Sergey* Basalaev[1,*], *Anna* Usanina[1], and *Eugene* Maslov[1]

[1]National Research Tomsk State University, 634050 Tomsk, Russia

Abstract. This paper presents investigations on the new experimental setup for obtaining a compact cluster of monodisperse bubbles of a given diameter is presented. Also we provided the results of experimental study of the bubble cluster floating-up in the presence of a surfactant in a wide range of Reynolds numbers. There was held a comparison of the dynamics of the floating-up of a monodisperse bubble cluster in a glycerol medium and in the medium glycerin supplemented with a surfactant.

1 Introduction

The behavior of fluid containing bubbles is distinctly different from behavior of homogeneous fluids under different physical and physical-chemical influence. These differences are abundantly used in industry – boiling, heat transfer in the two-phase mediums, cavitation, foaming, flotation. There is a question of the bubble cluster of given size generation in a number of problems, in particular, while researching an ignition of fluid electric discharge in fluids by way of specially created cavitational bubbles [1], while researching surfactants and acoustic waves on dynamics of bubble clusters [2, 3].

When two-phase flows with deformable particles of the disperse phase (liquid-drop, bubble flows) move, an important role is played by the properties of the interface, in particular, the surface tension coefficient. One of the physical factors affecting the boundary conditions is the presence of surfactants, which can change the criterial dependencies for the motion of the particle of the disperse phase by an order of magnitude. On the free boundary of a bubble or a droplet moving in a liquid medium with a surfactant, Marangoni tangential capillary forces arise, contributing to an increase in the coefficient of resistance of the particle. The problems arising in studying the patterns of motion and deformation of bubbles (or droplets) in a liquid containing a surfactant are formulated in the works [4, 5].

The regularities of the motion of bubbles in the presence of surfactants were investigated mainly for a single bubble [6, 7]. In contrast to the known papers, the results of an experimental study of the dynamics of the ascent of a set of spherical air bubbles in the presence of surface active substances in the Reynolds number range are presented $Re=0.001\div30$. There was held a generalizing the data on the coefficient of resistance of a

* Corresponding author: tarm@niipmm.tsu.ru

set of bubbles. Were used as a liquid: distilled water, water-glycerol solutions; as a surfactant: sodium dodecyl sulfate, potassium stearate.

2 The scheme of the experimental setup

An experimental setup for studying the dynamics of the emergence of a cluster of monodisperse bubbles in the presence of surfactants is shown in Fig. 1. The experimental setup comprises a reservoir 1 with a liquid located in the lower part, where air bubbles were produced by device 2, connected by a tube 4 to a compressed air reservoir 5, the pressure of which is supplied to the device 2 through a microreducer 6 with a control manometer 7 and a reducer 8. In addition, there is a separate container 9 connected to the tube 4 through a reducer 10 and a tee 11 designed to create a positive pressure preventing the working fluid from flowing into the device 2.

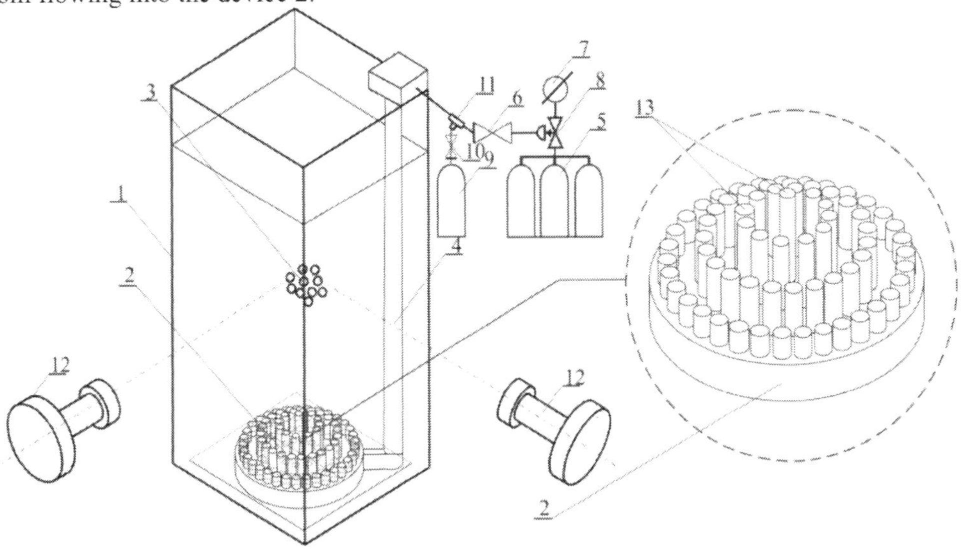

Fig. 1. The scheme of the experimental setup.

Receiver 1 is executed in the form of a cuvette with plane-parallel walls from optical glass of 0.3×0.3×0.6 m in size for visualization possibility of upfloat process of bubbles cluster. To visualize the dynamics of the bubble cluster 3 emergence in a liquid medium with the addition of surfactants, two high-speed video cameras 12 are used arranged to registration a bubble cluster in perpendicular planes.

The device for obtaining a monodisperse cluster of bubbles 2 is made in the form of a collector with perforations in the top cover starting from the center and along equidistant concentric circles, which makes it possible to obtain an axisymmetric bubble cluster. The use of microtubules of identical diameter 13, installed in perforations, provides the formation of monodisperse bubbles. Microtubes of the same height, located along each of the concentric circles, provide simultaneous formation of a "ring" of bubbles for each of the circles. The linear decrease in the height of the microtubes located on the circles, with an increase in the radius of the circle, ensures the sequential formation of each "ring" of bubbles with the same delay in time as they move away from the center of the collector cover, which makes it possible to obtain a compact cluster with a uniform spatial distribution of bubbles.

3 Results of the experimental study

During the experimental studies, the number of bubbles $n = (1 \div 69)$ and their diameters $D = (3 \div 7)$ mm in a monodisperse cluster varied, as well as the spatial arrangement of the bubbles among themselves [8]. Were used as a liquid: distilled water, water-glycerol solutions; as a surfactant: sodium dodecyl sulfate, potassium stearate.

The results of the visualization were given in a coordinate format for further analysis. In Fig. 2 shows a typical view of the spatial arrangement of bubbles at different times of ascent (n=5, D=3.9 mm).

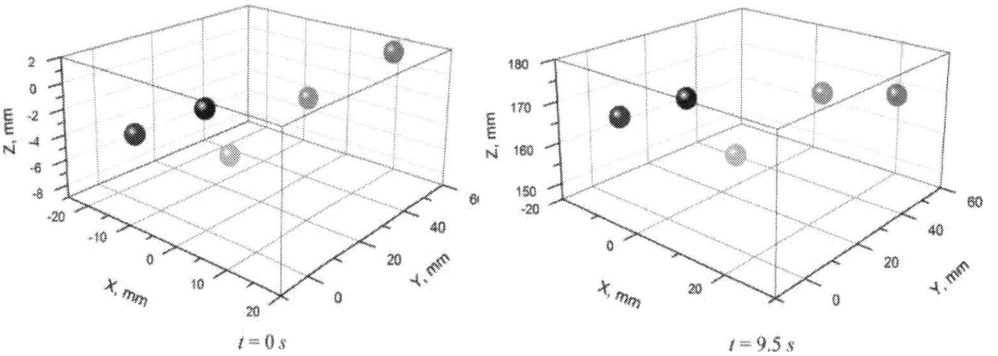

Fig. 2. Coordinates of the location of bubbles at different times (t = 0, 9.5 s).

The results of the experimental investigation for the emergence of a single bubble (n = 1) in the presence of surfactant are consistent with the results of known works on this topic and show that the bubble regime changes. The new bubble rise regime corresponds to the motion of a solid particle - the Stokes law at Reynolds numbers Re < 1 and the Klyachko law for Re > 1.

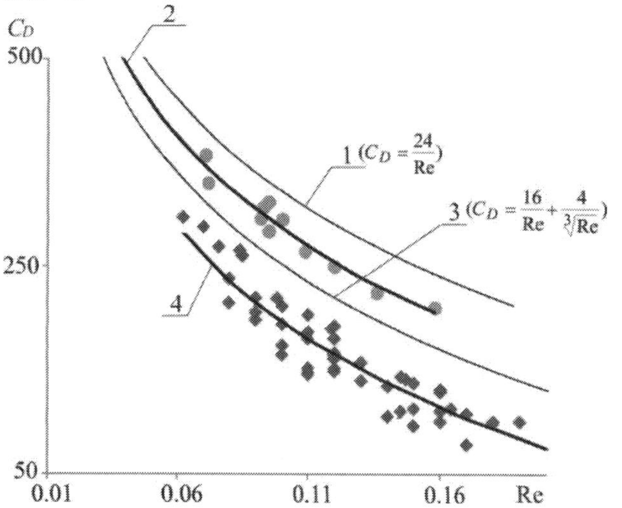

Fig. 3. Dependence of the resistance coefficient of a group of bubbles as a function of the Reynolds number (1 is the Stokes dependence; 2 is the empirical dependence, with the surfactant; 3 is the Hadamard-Rybczynski dependence; 4 is the empirical dependence, without surfactant).

To assess the effect of the presence of surfactant in a liquid on the dynamics of the ascent of a group of bubbles, two series of experiments were carried out: in a pure liquid (without surfactant) and in the presence of a surfactant (Fig. 3).

In experiments on the ascent of a group of bubbles in glycerin without surfactant additives, a value of a volume concentration C > 0.002 corresponding to the onset of the "partially blown" cloud. An empirical dependence is obtained for the coefficient of resistance of a group of bubbles C_D, moving in glycerine without surfactant additives in the Re <1 region (curve 4 in Fig. 3):

$$C_D = \frac{11.3}{\mathrm{Re}} \tag{1}$$

The empirical dependence for the coefficient of resistance of a group of bubbles popping up in glycerin with the addition of surfactant has the form (curve 2 in Fig. 3):

$$C_D = \frac{18.8}{\mathrm{Re}} \tag{2}$$

It was found that the effect of surfactants on the dynamics of bubble emergence is determined by the Reynolds number. When Re <1, the bubble surfacing regime change occurs at a lower concentration than at Re> 1.

4 Conclusion

- There is a certain critical value of the surfactant concentration in the liquid, at which the ascent bubble regime changes. In this case, the critical concentration value is a function of the bubble size (with increasing bubble size, the critical concentration value increases).
- The presence of surfactant leads to an increase in the coefficient of resistance by 33% and, accordingly, to a decrease in the ascent rate of a single bubble. In this case, the law of resistance changes from the Hadamard-Rybczynski formula to the Stokes formula.
- Empirical relationships for the coefficient of resistance of a group of bubbles ascent up in a liquid with a surfactant and without a surfactant are received.
- A decrease in the ascent rate of a group of bubbles in a liquid containing a surfactant has been found, compared with the velocity of a group of bubbles in a pure liquid.

This study was supported by the Russian Science Foundation (Project No. 15-19-10014).

References

1. A. P. Drozhzhin, S. M. Korobeynikov, B. C. Teslenko, Scientific bulletin of NGTU **2** (2003) [in Russian]
2. V. G. Levich, *Physical and chemical hydrodynamics* (Fizmatgiz, Moscow, 1959) [in Russian]
3. V. A Arkhipov, I. M. Vasenin, A. S. Usanina, Fl. Dyn. **51** (2), (2016)
4. R. Clift, J.R. Grace, M.E. Weber, *Bubble. Drops and particles* (Academic Press., London, 1978)
5. R. Griffith, Chem. Eng. Sc. **17** (1962).
6. J. Harper, App. Sci. Res. **38** (1982).
7. E. Almatroushi., A. Borhan, Ann. N.Y. Acad. Sci. **1027** (2004).
8. S. Goldaev, S. Basalaev, K. Afanasyev, MATEC Web Conf. **110**, 01033 (2017)

Experimental and Numerical Studies of Suppression of Forest Combustible Material Pyrolysis under Influence of Steam-Water Curtain

Dmitry Antonov[1,*], *Konstantin* Osipov[2], and *Irek* Khasanov[3]

[1]National Research Tomsk Polytechnic University, 634050 Tomsk, Russia
[2]V.E. Zuev Institute of Atmospheric Optics SB RAS, 634055 Tomsk, Russia
[3]All-Russian Research Institute for Fire Protection of Russian Emergency Ministry, 143903 Moscow Region, Russia

Abstract. Forest fires inflict enormous damage to the environment and the economy of many countries around the world. This paper presents the results of experimental and numerical studies of the suppression of the pyrolysis reaction of forest combustible materials (birch leaves, spruce needles) under the influence of water vapor. The dependences of the characteristic times of suppression of the thermal decomposition and the complete decomposition of forest combustible material on the thickness of its layer were established. We used typical model of fire source with fixed height 0.04 m and varied diameter 0.02–0.06 m. The processes of suppressing the pyrolysis reaction of forest combustible materials were recorded using a high-speed video camera (recording frequency $6 \cdot 10^5$ fps). Numerical studies were conducted using Ansys Fluent and Matlab. The obtained results are important for the development of technologies of extinguishing forest bottom fires.

1 Introduction

Wildfire fighting is an urgent problem for the entire world community [1–4]. Only in Russia the number of forest fires on average is about 30 thousand each year [5, 6]. There are various ways to extinguish forest fires using water [1–4]. The use of water vapor is the most interesting from the standpoint of extinguishing forest bottom fires. The principle of operation is based on the fact that the water vapor creates a vapor shell above the combustion zone and separates it from the oxidant of the external medium, thereby stopping combustion. However, this method of fire extinguishing is poorly studied both experimentally and theoretically and requires detailed investigation.

The aim of this work is experimental and numerical studies of the suppression of the pyrolysis reaction of forest combustible materials under the influence of water vapor.

* Corresponding author: 1aadsdfn@gmail.com

2 Experimental setup and procedure

Fig. 1 shows the scheme of experimental setup used in carrying out researches. As the main recording equipment we chose a high-speed digital CMOS video camera Phantom V411 *1* with a recording frequency is of $6 \cdot 10^6$ fps and maximum resolution of 1280×1280 pixels) and high-speed (the minimum polling rate is 0.1 s) analog input module "National Instruments NI 9216" *2* for temperature sensors connection.

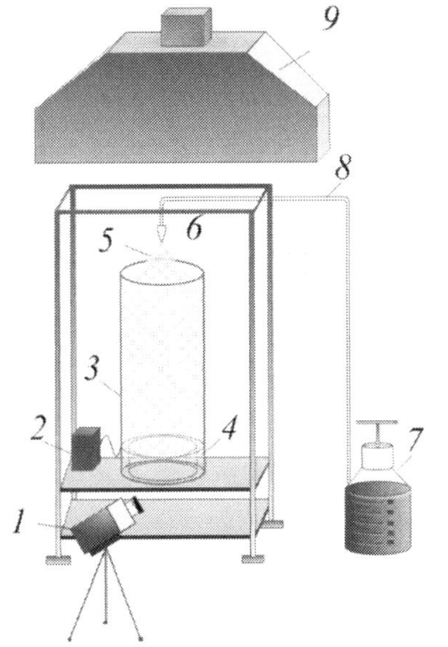

Fig. 1. A scheme of experimental setup: *1* – high-speed video camera; *2* – high-speed analog input module; *3* – cylinder with a forest material; *4* – metal tray; *5* – liquid aerosol; *6* – spraying nozzle; *7* – bulb with quenching liquid; *8* – quenching liquid supply channel; *9* – exhaust system.

The experiments were carried out in two stages: the first one was used to determine the times of complete burning out of model sources of forest material burning – τ_h; the second one included recording the extinguishing times (suppression of flame burning and thermal decomposition) of model sources under water vapor extinguishing – τ_d.

Cylinder *3* with a sample of forest material was installed on a non-combustible metal tray *4* (Fig. 1). Next, we ignited the model fire seat uniformly over the entire area of the open surface of the forest material with the simultaneous use of three piezoelectric gas burners. The time recording began via the electronic stopwatch (the time step was 0.01 s). The time of complete burnout (τ_h) of the total volume of forest combustible material (the period of time from the onset of combustion to the moment of burning out) was recorded. At the second stage of the experiments, the extinguishing of the model fire seat was carried out.

During the experimental studies, a vapor-drop mixture (vapor-water cloud) with a predetermined temperature over the decomposing forest combustible material was created (T_m=300–450 K). We recorded the time during which the temperature of the heated forest combustible material became lower than the temperature of its thermal decomposition beginning. This value corresponded to the conditions for stopping the pyrolysis reaction of forest combustible material. The tested sample was a layer (filling the cylindrical channel *3*) of a forest combustible material with a thickness from 0.02 m to 0.06 m.

The systematic errors in determining the times τ_h and τ_d are 0.5 s. The maximum random errors in determining the temperatures T_m during the thermal decomposition of forest combustible material did not exceed 30 K.

3 Mathematical model

Two software packages Ansys Fluent and Matlab were used when carrying out the numerical studies. These software packages differ primarily in the difference in the solution method. Matlab uses the finite difference method; Ansys Fluent uses the finite element method. In addition, our own code was created in Matlab with the refinement of all the terms of the system of differential equations of heat and mass transfer in partial derivatives.

The solved problem of determining the burning times and the extinguishing times of forest combustible material was two-dimensional. The thickness of the near-surface layer of forest combustible material was varied in the range L_f=35–80 mm. The dimensions of the solution region were varied depending on the values of L_f in the ranges H=0.1–0.5 m and L=0.1–0.5 m. The solution region of the problem under consideration for the system "gas-vapor mixture – forest combustible material" is presented in Fig. 2.

Numerical analysis of the investigated physical and chemical processes is performed under the following assumptions, which do not impose significant restrictions on the generality of the problem formulation:

1. It was assumed that in a small neighborhood of a trace of a typical water mass, the concentration of the oxidizer is close to zero due to sufficiently intense vaporization.

2. It was believed that the vapor-gas mixture consists of water vapor (90%) and combustion products (10%).

3. We did not take into account possible oxidation reactions of pyrolysis products of forest combustible material when modeling the processes of heat transfer in the gas phase.

4. It was assumed that the thermophysical characteristics of forest material and water do not depend on temperature when setting the heat transfer problem.

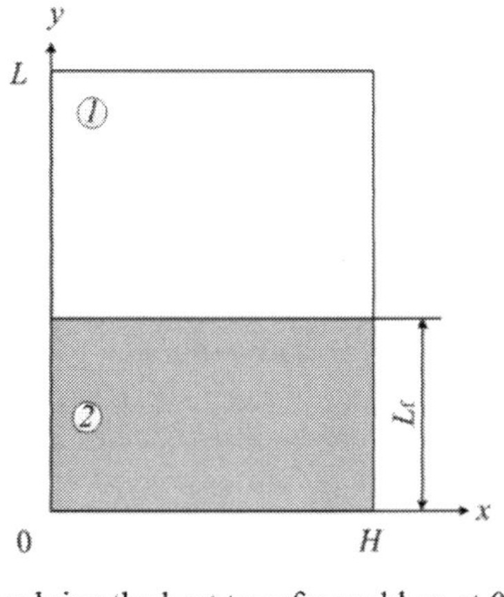

Fig. 2. Scheme of the region for solving the heat transfer problem at $0 < t < \tau_d(\tau_h)$ for the system "vapor-gas mixture – forest combustible material"; *1* – vapor-gas mixture, *2* – forest combustible material.

The solution of the problem of heat and mass transfer using own code in Matlab is considered in detail in [7]. In this paper, we describe a system of partial differential equations for the considered system "vapor-gas mixture-water vapor – forest combustible material".

In the software complex Ansys Fluent, the solution of the heat and mass transfer problem consisted of several stages. The first stage was the creation of a grid model, which is an integral part of the process of mathematical modeling. The quality of the computational grid affects the accuracy, convergence and speed of the solution. In addition, the time to create a grid, often takes a significant proportion of the total execution time. In this paper, the time steps Δt=0.1 s and the coordinate steps Δx=Δy=0.05 mm were used to solve the heat and mass transfer task. Near the boundaries of the phase transition, the grid was thickened to improve the calculation accuracy (up to $\Delta x=\Delta y=\Delta r=0.001$ mm. The second stage was the description of the main used models (Energy Model, Specific Transport Model, Turbulence Model and e.t.).

To ensure high accuracy of calculations, the turbulence model corresponding to the formulated problem k–ε was chosen [8]. It is a two-parameter model in which two transfer equations are solved to determine the turbulent velocity and the time of the vortex dissipation.

4 Results and discussion

Typical temperature fields at extinguishing forest combustible material by water vapor are presented in Fig. 3. The moments before the suppression of the forest combustible material thermal decomposition (τ=50 s), in the process (τ=200 s) and after completion (τ=500 s) of thermal decomposition are shown.

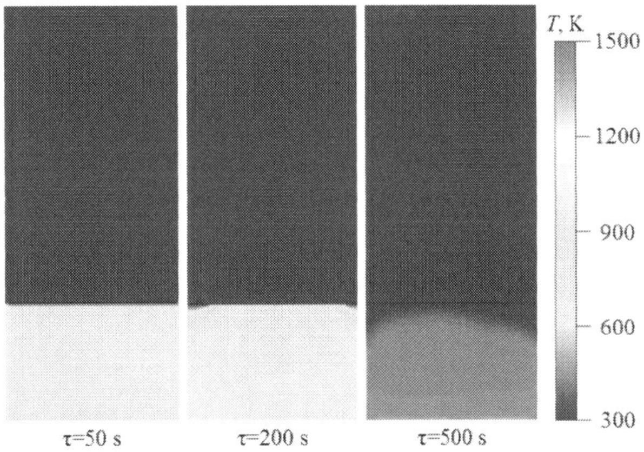

Fig. 3. Temperature fields during thermal decomposition of LCM at different instants of time.

Fig. 4 shows the dependence of the characteristic times of suppression of the thermal decomposition of forest combustible material (τ_d) and the times of total forest combustible material thermal decomposition (τ_h) as a function of the thickness of the forest combustible material layer. It should be noted the non-linear nature of the time variation as a function of L_f. This is due to the nonlinear dependence of the thermal decomposition rate in the near-surface layer of forest combustible material [3, 5, 9–12].

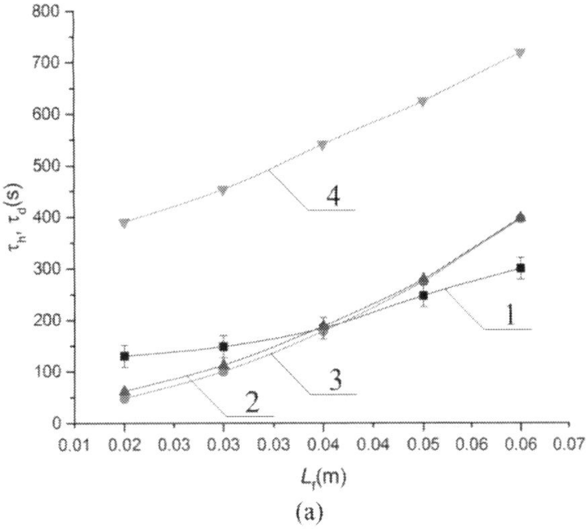

(a)

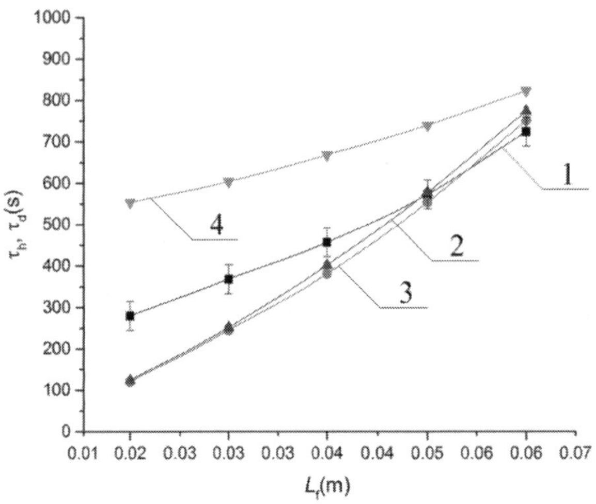

(b)

Fig. 4. Characteristic times of suppression of the thermal decomposition reaction of forest combustible material (birch leaves (a), spruce needles (b)): 1– experimental data, 2 – results of numerical modeling in Matlab, 3 – results of numerical modeling in Ansys Fluent, 4 – time of complete burning of LCM.

We have established a satisfactory agreement between the results of numerical calculations performed in software complexes (Matlab and Ansys Fluent) and experimental studies (Fig. 4). Deviations of the theoretical curves from the experimental ones can be explained by the fact that the constants (λ, C and ρ) were used as the values of thermophysical parameters when modeling. To approximate theoretical models to real applications, it is useful to consider the dependence of λ, C and ρ on moisture content and temperature.

In addition, it should be noted that the initial parameters of solving the heat transfer problem have a significant effect on the deviations of theoretical results. This is, first of all, the thermal decomposition kinetics (activation energy, pre-exponential factor, heat of combustion, initial temperature of water vapor).

We used the kinetics of thermal decomposition for deciduous and coniferous tree species [1, 3, 5, 9] in numerical simulation. From the analysis [1, 3, 5, 9] it can be concluded that forest materials of different regions differ in their thermophysical properties and kinetic parameters.

The effect of the change in the fragments of the samples from experiment to experiment is enhanced when carrying out experiments with large thicknesses of the reacting layer. Most likely, the position of the theoretical curve relative to the experimental curve is due precisely to this factor when the thickness of the sample of birch leaves is more than 0.05 m. Such an effect is not observed with spruce needles, since the structure of the fragments is more uniform in comparison with the leaves. However, it was not possible to ensure the absolute identity of spruce needles from experiment to experiment.

5 Conclusion

As a result of the experimental and theoretical studies, we established the values of times of forest material (birch leaves, spruce needles) extinguishing and burning out depending on the thickness of the layer of forest combustible material. The obtained results can be used to predict the times of extinguishing forest bottom fires.

The research was funded by the Russian Science Foundation (project 18-19-00056).

References

1. O. P. Korobeinichev, A. G. Shmakov, V. M. Shvartsberg, A. A. Chernov, S. A. Yakimov, K. P. Koutsenogii, V. I. Makarov, Fire Saf. J. **51**, 102 (2012), DOI: 10.1016/j.firesaf.2012.04.003
2. C. C. Remy, M. Lavoie, M. P. Girardin, H. Asselin, A. A. Ali, J. Biogeogr. **1** (2017)
3. R. S. Volkov, A. O. Zhdanova, G. V. Kuznetsov, P. A. Strizhak, Tech. Phys. **60**, 1443 (2015), DOI: 10.1134/S1063784215100291
4. D. Morvan, Fire Saf. J. **71**, 34 (2015), DOI: 10.1016/j.firesaf.2014.11.012
5. A. O. Zhdanova, G. V. Kuznetsov, J. C. Legros, P. A. Strizhak, Thermal Sci. **21** (6A), 2559 (2017), DOI: 10.2298/TSCI151006121Z
6. O. Ulybina, Forest Policy and Economics, **38**, 143 (2014), DOI: 10.1016/j.forpol.2013.06.019
7. A. O. Zhdanova, G. V. Kuznetsov, P. A. Strizhak, Applied Mechanics and Materials, **692**, 267 (2014), DOI: 10.4028/www.scientific.net/AMM.692.267
8. A. D. Kliukvin, Science & Education of Bauman MSTU, **8**, 256 (2014), DOI: 10.7463/0814.0725648
9. A. I. Karpov, V. B. Novozhilov, A. A. Galat, V. K. Bulgakov, Fire safety science: proceeding of eight international symposium **27**, 753 (2005), DOI: 10.3801/IAFSS.FSS.8-753
10. D. V. Barovik, Math. Modell. Anal. **15**, 161 (2010), DOI: 10.3846/1392-6292.2010.15.161-174
11. E. Pastor, L. Zarate, E. Planas, J. Arnaldos, Prog. Energy Combust. Sci. **29**, 139 (2003), DOI: 10.1016/S0360-1285(03)00017-0
12. J. Qie, L. Yang, Y. Wang, J. Dai, X. Zhou, J. Fire Sci. **29**, 243 (2011), DOI: 10.1177/0734904110392961

Icing control model and algorithm for wasteheat exchangers of ventilation systems

Andrew Arbatskiy[1,*], *Andrew* Garyaev[1], and *Vasiliy* Glasov[1]

[1]Moscow Power Engineering Institute, 111250 Moscow, Russia

Abstract. Currently, ice control on various heat exchangers to be used for waste heat recovery of discharge air is a rather topical subject because ice building is a factor to reduce efficiency of heat recovery. In such systems, icing always takes place on the side of humid discharge air, with a background of heat exchange between inlet air at temperatures of -30°C and below, and discharge air. To effectively prevent this, it is required to find solutions to problems as follows: study a mechanism of ice building on heat exchange surfaces when interacting with humid air; determine engineering methods to prevent icing, for each type of heat-exchange apparatus, and check efficiency of their operation; develop a mathematical model of ice growth on heat exchange surfaces to enable to vary key parameters (both geometrical and algorithmical ones), determined beforehand, and engineering means aimed at icing prevention.

1 Introduction

These issues were raised in the paper [1]. Within the framework of this paper, one:

- developed a generalized mathematical model to control icing on a heat exchanger on the warm side of a heat recovery system with an intermediate heat carrier;
- proposed an engineering concept to regulate mode parameters of a heat exchanger to provide for its maximum efficiency at any point of time;
- determined, at a first approximation, an algorithm of improvement of intermediate heat carrier parameters, at various parameters of heating and heated media.

2 Generalized mathematical model to control icing on a heat exchanger

A rationale for development of such a mathematical model for operating impact upon icing on heat exchangers within a heat recovery system with intermediate heat carrier (Fig.1) is presented in the paper [1]:

* Corresponding author: arbatsky1985@mail.ru

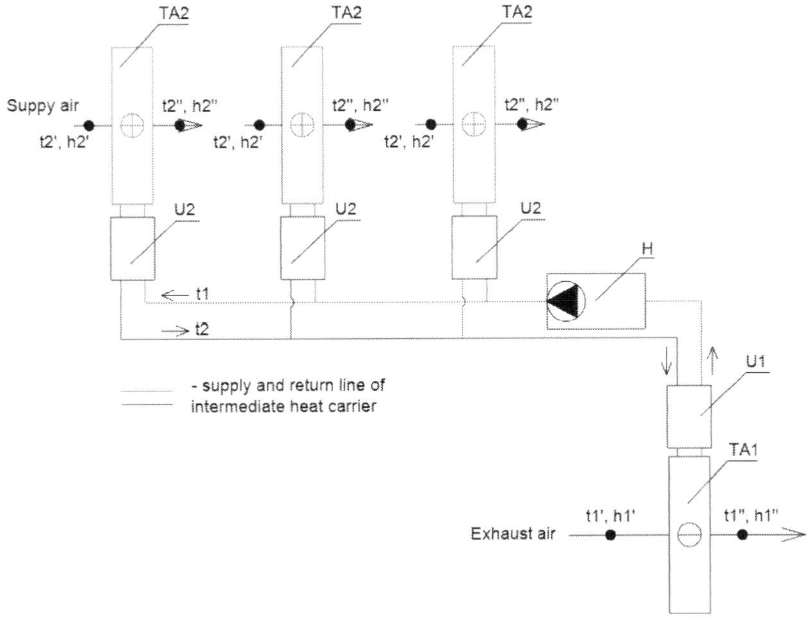

Fig. 1. Principal diagram of a heat recovery system with intermediate heat carrier: TA1 – wasteheat exchanger on the heated side; TA2 - wasteheat exchanger on the heating side; U1 – TA1 regulating unit; U2 – TA2 regulating unit; N – pump group; t2', t2'' – in- and outlet temperatures of heat carrier heated; h2', h2'' – in- and outlet total heat of heat carrier heated; t1, t2 – intermediate heat carrier supply and return line temperatures; t1', t1'' – in- and outlet temperatures of the heat carrier heating; h1', h1'' – in- and outlet total heat of the heat carrier heating.

This type of waste heat exchangers is notable for the lowest thermal effectiveness among other methods to save energy ressources in ventilation systems. This is related to factors as follows:

1) significant losses during heat transmission by means of intermediate heat carrier (IHC);

2) an IHC temperature chart can be improved by equalizing inline parameters of heat exchangers, both on the inlet and discharge side. This, in most cases, results in negative temperatures of IHC, and, consequently, occasional icing on the heat exchanger on the discharge side, especially at higher temperatures of discharge air;

3) Efficiency of the system as a whole depends upon efficiency of an air heater (both inlet and discharge sides). At that, condensate formation in the heat exchanger (discharge side) contributes to, on the one hand, higher efficiency, and on the other hand, when subject to icing, the time-averaged heat exchanger efficiency, on the inlet side, remains unchanged. However, heat current transferred to air by the heat carrier on the inlet side, significantly changes because of changes to temperature head resulting from changes to the temperature chart of the heat carrier.

Before we proceed to creation of a mathematical model to represent ice growth on the TA1 heat exchanger, we have to develop a generalized mathematical algorithm to regulate currents of heat carrier via this apparatus by means of the U1 regulating unit.

Figure 2 shows a principal diagram of the U1 regulating unit:

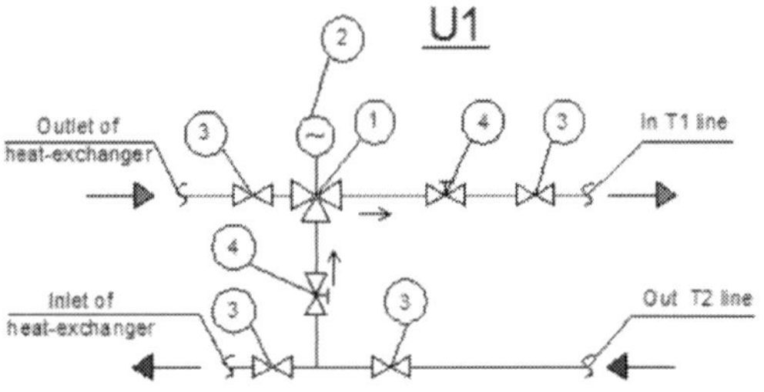

equipment specification			
1	3-way valve	N.	1
2	Drive of 3-way valve	N.	1
3	Valve	N.	4
4	Balancing valve	N.	1

Fig. 2. Principal diagram of the U1 regulating unit.

The diagram enables to control parameters t1 and t2, as well as control release of cold IHC into TA1 heat exchanger for removal of external ice. Regulation accuracy is provided due to settings of the balancing valve – 4, because resistance of a bypass where this valve is installed must be considerably higher than that of the heat exchanger to ensure accurate measurement of consumption on the basis of pressure differences on the three-way valve – 2. A problem to create an accurate regulating algorithm of this unit is related to generation of the mathematical model of TA 1 heat exchanger icing.

Development of a generalized mathematical model to control U1 is, in the first place, related to qualitative nature of changes to temperature t_1 during regulations on the three-way valve 2 (Fig. 2). An engineering arrangement of control and regulating mechanisms is presented in Fig. 3.

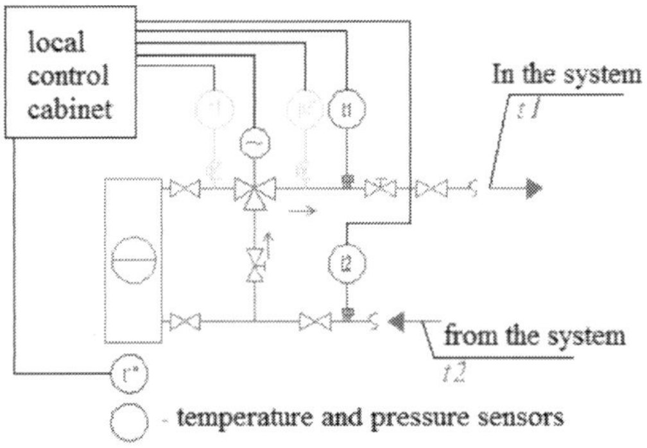

Fig. 3. Principal diagram of the U1 regulating unit.

A position of the three-way valve is determined on the basis of pressure difference:

$$\Delta P_m = p_1 - p_2 \tag{1}$$

The regulating scheme of that kind makes it possible to achieve higher accuracy as compared to valve position regulations using an external signal 0-10V.

Correspondingly, we can write a sequence of the regulating algorithm in the following way:

- When started, all three-way valves are opened; one has to measure pressure difference with the valves opened and write it as an initial constant. Further calculations follow on the basis of this value of ΔP_m;

- One has to record the following variable at each point of time: t_1, t_2, t_1'';

- A value of temperature t_1 at each point of time shall conform to the condition as follows:

$$t_1(t_2) \geq t_2 + k_1 \tag{2}$$

where: k1 is relationship of heat exchanger specific heat power to a coefficient of thermal transmission, °C (actual temperature difference by a liquid provided by an air heater in the design mode).

- Resulting from ice building on the heat exchanger, temperature t_l takes a value that does not conform to the condition (2). In this case one begins regulating the three-way valve on the basis of recurrence dependence:

$$\Delta P_n(\Delta P_{n-1}) = 1.013\Delta P_{n-1} \tag{3}$$

At that, $\Delta P_l = \Delta P_m$. The coefficient of 1.013 is related to an averaged empirical characteristic of the three-way valves, diameters of which range from 15 to 50 mm [2].

- Upon each regulating action of the three-way valve, changes to temperature t1 can be written as follows:

$$t_1(\tau) = \mathrm{k}_2\tau^{\varepsilon} \tag{4}$$

where: k2 and ε are constants to be determined experimentally for each type of heat exchangers; τ is time to be counted from the beginning of each regulating action upon the three-way valve, c;

One has to regulate in such a way so that inequations are always attained:

Table 1. Values of referenced in equations.

λ	τ
$\frac{dt_1(\tau)}{d\tau} = \left\|\mathrm{k}_2\varepsilon\tau^{\varepsilon-1}\right\| \geq \left\|\lambda_1\right\|$	$0.1\tau_n$
$\frac{dt_1(\tau)}{d\tau} = \left\|\mathrm{k}_2\varepsilon\tau^{\varepsilon-1}\right\| \geq \left\|\lambda_2\right\|$	$\tau_n/2$
$\frac{dt_1(\tau)}{d\tau} = \left\|\mathrm{k}_2\varepsilon\tau^{\varepsilon-1}\right\| \geq \left\|\lambda_3\right\|$	τ_n

Where: λ_n are reference values of a principal derivative for each point of time upon a regulating action.

The principal derivative $t_1(\tau)$ in that case characterizes a graph slope angle at each point of time upon regulating actions. With ice on a surface, the rate-of-change of temperature t1 at any movement of the three-way valve will be less than that with no ice, and, consequently, the graph will have a lesser angle of slope;

The algorithm described makes it possible to avoid a need for referencing to specific values of discharge air parameters and practically establish reverence values of the derivative for each type of heat exchangers.

3 IHC parameters optimization

To ensure optimal parameters of heat recovery, a temperature chart of IHC must be in the middle between temperature charts of heating and heated media. It means that if a heated medium is air used for ventilation systems, then, IHC parameters can be determined from the equal averaged logarithmical temperature head values on each side, with allowance for efficiency of heat exchangers, due to phase transition to TA1 [3]:

$$\frac{\varepsilon_{TA1}}{\varepsilon_{TA2}}\overline{\Delta t_{TA1}} = \overline{\Delta t_{TA2}} \tag{5}$$

where: $\overline{\Delta t_{TA1}}$ - averaged logarithmical temperature head for TA1, °C; $\overline{\Delta t_{TA2}}$ - averaged logarithmical temperature head for TA2, °C; ε_{TA1} - efficiency of TA1; ε_{TA2} - efficiency of TA2.

To solve this problem, in the first place one has to analyze ratio of heat exchangers efficiencies at the same temperature head, but under differing air humidity values. One shall carry out calculations on the basis of averaged values of dry coefficients of transmission, typical for ribbed air heaters at gas velocities making 3-6 m/s k=60 W/m²/K [4]. This method of calculation will be appropriate, subject to equal flow conditions of heat carriers in TA1 and TA2. In such a way, we can write an equation as follows:

$$Q_{TA1} = kF_1\overline{\Delta t_{TA1}} + Q_\kappa \tag{6}$$

where: k is a dry coefficient of heat transfer (with no humidity taken into account), W/m²/K; F_1 – a heat exchange surface area TA1, m²; Q_κ – heat of condensation, W; Q_{TA1} – total amount of heat transferred within a heat exchanger per unit time, W.

As $Q_{TA1} = \Sigma Q_{TA2}$, subject to equation (6) we obtain as follows:

$$kF_1\overline{\Delta t_{TA1}} + Q_\kappa = kF_2\frac{\varepsilon_{TA1}}{\varepsilon_{TA2}}\overline{\Delta t_{TA1}} \tag{7}$$

where: F_2 – total surface area of heat exchangers TA2, m².

From (7) we obtain an equation (8) as follows:

$$\frac{\varepsilon_{TA1}}{\varepsilon_{TA2}} = \frac{F_1}{F_2} + \frac{Q_\kappa}{kF_2\overline{\Delta t_{TA1}}} \tag{8}$$

It is evident that the second addend on the right is nothing else but ratio of heat of condensation to sensible heat transferred within TA2 that can be written as:

$$\frac{Q_\kappa}{kF_2\overline{\Delta t_{TA1}}} = \frac{\varepsilon_{TA1}G_1((h_1^{'} - h_1^{''}) - c_{p1}(t_1^{'} - t_1^{''}))}{\varepsilon_{TA2}G_2c_{p2}(t_2^{''} - t_2^{'})} \tag{9}$$

where: c_{p1} and c_{p2} are mean specific heat values of air, J/kg/K; G_1 and G_2 represent consumption values of heating and heated heat carriers, kg/s.

Inserting (9) into (8), we obtain the final equation as follows:

$$\frac{\varepsilon_{TA1}}{\varepsilon_{TA2}} = \frac{F_1}{F_2} + \frac{\varepsilon_{TA1}}{\varepsilon_{TA2}}\frac{G_1((h_1^{'} - h_1^{''}) - c_{p1}(t_1^{'} - t_1^{''}))}{G_2c_{p2}(t_2^{''} - t_2^{'})} \tag{10}$$

It may be affirmed that we always will obtain equation of averaged logarithmical temperature heads, subject to equation of averaged values of temperature heads, basing upon which we can write a ratio as follows:

$$\frac{F_1}{F_2} + \frac{t_1 - t_2^{''} + t_2 - t_2^{'}}{t_1^{''} - t_2 + t_1^{'} - t_1}\frac{G_1((h_1^{'} - h_1^{''}) - c_{p1}(t_1^{'} - t_1^{''}))}{G_2c_{p2}(t_2^{''} - t_2^{'})} = \frac{t_1 - t_2^{''} + t_2 - t_2^{'}}{t_1^{''} - t_2 + t_1^{'} - t_1} \tag{11}$$

From the expression (11), a formula can be derived to calculate an average temperature of IHC:

$$\frac{t_1+t_2}{2}=0.5\frac{\dfrac{F_1}{F_2}(t_1^{''}+t_1^{'})-\dfrac{G_1((h_1^{'}-h_1^{''})-c_{p1}(t_1^{'}-t_1^{''}))}{G_2c_{p2}(t_2^{''}-t_2^{'})}(t_2^{''}-t_2^{'})+t_2^{''}+t_2^{'}}{\dfrac{F_1}{F_2}-\dfrac{G_1((h_1^{'}-h_1^{''})-c_{p1}(t_1^{'}-t_1^{''}))}{G_2c_{p2}(t_2^{''}-t_2^{'})}+1} \quad (12)$$

Using dependence (12) for various schemes of heat recovery, we can always achieve the best IHC temperature values.

References

1. A. A Arbatskiy, V. S. Glasov, Status and prospects of the development of electro- and heat technology **2**, 320 (2017) [in Russian]
2. Esbe. Catalogue of 3G three-way valves characteristics [in Russian]
3. V. P. Isachenko, V. A. Osipova, A. S. Sukomel, *Heat transmissio* (Energy, Moscow, 1975) [in Russian]
4. A. M. Baklastov, V. A. Gorbenko, O. L. Danilov, *Industrial-scale heat-mass-exchange apparatus and processes* (Energoatomizdat, Moscow, 1986) [in Russian]

Experimental setup for investigating the dynamics of the ascent of a cluster of bubbles in a liquid

Vladimir Arkhipov[1], *Sergey* Basalaev[1,*], *Sergey* Orlov[1], and *Sergey* Polenchuk[1]

[1]National Research Tomsk State University, 634050 Tomsk, Russia

Abstract. There is a scheme of experimental setup providing preparation of the compact cluster of monodispersed bubbles of the given diameter presented in the work. Further, in this research we have calculated setup parameters that are necessary for preparation of the compact air bubble cluster in glycerin at room temperature.

1 Introduction

The behavior of fluid containing bubbles is distinctly different from behavior of homogeneous fluids under different physical and physical-chemical influence. These differences are abundantly used in industry – boiling, heat transfer in the two-phase mediums, cavitation, foaming, flotation. There is a question of the bubble cluster of given size generation in a number of problems, in particular, while researching an ignition of fluid electric discharge in fluids by way of specially created cavitational bubbles [1], while researching surfactants and acoustic waves on dynamics of bubble clusters [2-3].

In practice of the experimental studies there are setup relating to aeration devices, aim to gas introduction to the liquid fluid. The carried out analysis devices work [4-5] showed that setups are intended for creation of the continuous flow of bubbles in fluid. The [6] disadvantage is impossibility of cluster generation from monodispersed bubbles of millimetric sizes and also complexity of the setups realizing these methods. The method [7] does not allow to generate a bubble cluster of controlled form.

In this paper, the experimental scheme of setup allowing to provide a compact cluster of monodispersed bubbles of the given diameter was described.

2 The scheme of the experimental setup

The experimental setup for generation of a compact cluster of monodispersed bubbles [8] is presented in Fig. 1. The setup includes collector 3 that placed in a lower part of receiver 1 with fluid 2 with gas-permeable upper cover 4, connected by branch pipe 5 with a source of compressed gas. Receiver 1 is executed in the form of a cuvette with plane-parallel walls

* Corresponding author: tarm@niipmm.tsu.ru

from optical glass of 0.3×0.3×0.6 m in size for visualization possibility of upfloat process of bubbles cluster.

In upper cover 4 of collector 3 are executed in its center and for equidistant concentric circles of perforation in which central 6 and peripheral tubes 7 of identical diameter are established which height that is identical to the tubes located on each circle and linearly decreases with increase of circle radius. Injection medical needles are used as tubes.

Cylinder 8 with the compressed gas, connected through low pressure reducer 9 and also through high pressure reducer 11 and electropneumatic valve 13 with branch pipe 5 of collector 3 is used as a source of gas.

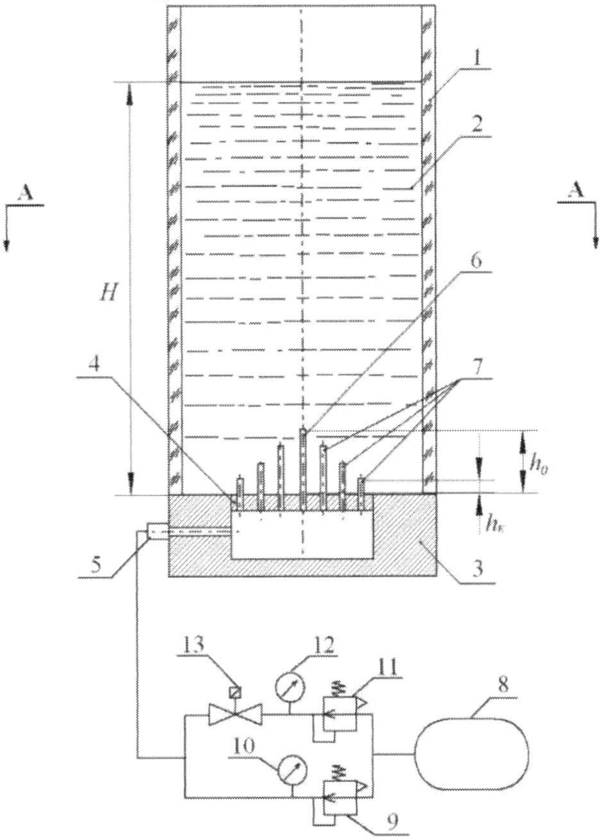

Fig. 1. The scheme of the experimental setup for creation of a compact cluster of monodispersed bubbles.

Work of the setup is carried out as follows. The constant pressure p_{min} controlled by a pressure gage 10 is settled by means of the reducer 9, it hinders fluid 2 flowing through tubes 6, 7 into the collector 3. By means of high pressure reducer 11 and the electropneumatic valve 13 the compressed gas is pulsewise supplied under the pressure p_{max} through the branch pipe 5 into the collector 3. Gas from the collector 3 through tubes 6, 7 in the form of bubbles comes to ambient liquid 2. After bubbles departure from tubes 6, 7 in liquid 2 the compact cluster of spherical shape bubbles floating up is formed.

3 Results of the experimental study

Let`s consider the results of the compact cluster preparation of air monodispersed bubbles in glycerin at room temperature. Parameters of the device, necessary for calculations, are specified in table 1, where D – the required diameter of the formed bubble; d – inside diameter of a tube; h_0 – height of the central tube; R – radius of the upper cover of a collector; H – height of fluid column in the receiver over the collector upper cover.

Table 1. Parameters of a setup.

D (mm)	d (mm)	h_0 (mm)	R, (mm)	H, (mm)
5	0.26	30	64	500

Physical characteristics of air and glycerol [9] at a temperature of 20°C, are shown in table 2, where ρ – density; p_{atm} – atmospheric pressure; σ – coefficient of superficial liquid tension.

Table 2. Physical characteristics of air and glycerin.

Parameter	Air	Glycerin
ρ, kg/m^3	1.205	1260
μ, Pa·c	$1.808 \cdot 10^{-5}$	–
σ, H/m	–	$63 \cdot 10^{-3}$

Now we are calculating parameters of the setup. For determination of the tubes height h_i located on circle radius r_i, we consider similar triangles of ABC and AB_1C_1 (Fig. 2).

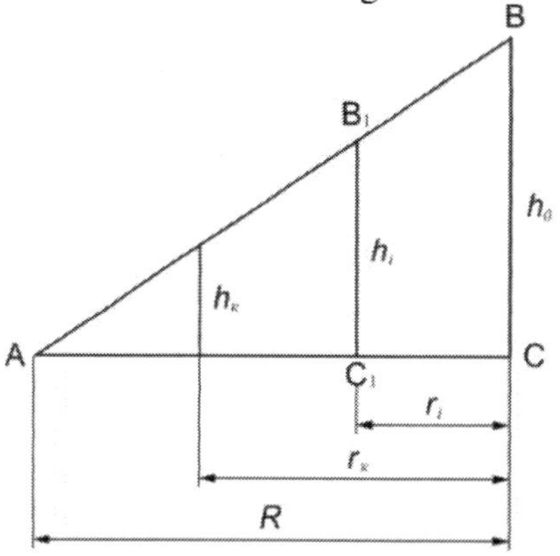

Fig. 2. Similar triangles of ABC and AB_1C_1.

From the similarity condition follows:

$$\frac{BC}{AC} = \frac{B_1C_1}{AC_1}. \tag{1}$$

From Fig. 1 follows:

$$BC=h_0,\ AC=R,\ B_1C_1=h_i,\ AC_1=R-r_i, \tag{2}$$

where h_i, tube height of located on radius r_i (i = 1, 2,...,κ).

Substituting (2) in (1), we are getting: $h_0/R = h_i/(R - r_i)$,
from here follows the ratio:

$$h_i = h_0\left(1 - \frac{r_i}{R}\right). \tag{3}$$

The ratio (3) provides linear reduction of tubes height from h_0 to h_k with increase in radius of circle r_i. The results of calculations of tubes height located on a circle with radius r_i are provided in table 3.

Table 3. Height of the tubes located on a circle radius r_i.

i	1	2	3
r_i (mm)	16	32	48
h_i (mm)	22.5	15	7.5

Under a pressure in the collector equal to the minimum pressure quantity of gas p_{min}, gas shouldn't come into liquid through tubes. It is provided under the following condition:

$$p_{\min} < p_{\text{atm}} + p_h , \quad (4)$$

where $p_h = \rho g h$, hydrostatic pressure; ρ, liquid density; g, acceleration of gravity; h, liquid column height over discharge ring of the tube.

The smallest hydrostatic pressure is implemented for the central tube with height h_0 for which $h = H - h_0$. From (4) follows:

$$p_{\min} < p_{\text{atm}} + \rho g \left(H - h_0 \right) . \quad (5)$$

During the development of the setup we experimentally get more specific condition (5):

$$p_{\min} = p_{\text{atm}} + 0.8 \rho g \left(H - h_0 \right) = 105954 \text{ Pa} . \quad (6)$$

While performing ratio (6) gas doesn't come into liquid through the central tube and also through the tubes located on concentric circles, because hydrostatic pressure is higher for them, than for the central tube.

Under a pressure in the collector equal to the maximum pressure quantity of gas p_{max}, gas should come through the tubes into liquid. It is provided under the following condition:

$$p_{\max} > p_{\text{atm}} + p_h . \quad (7)$$

The largest hydrostatic pressure is implemented for peripheral tubes with height h_k for which $h = H - h_k$. From (7) follows:

$$p_{\max} > p_{\text{atm}} + \rho g \left(H - h_k \right) . \quad (8)$$

During the development of the setup we experimentally get more specific condition (8):

$$p_{\max} = p_{\text{atm}} + 1.2 \rho g \left(H - h_k \right) = 108611 \text{ Pa} . \quad (9)$$

While performing ratio (9) gas bubbles come into liquid through the tubes located on peripheral circle radius r_k and also through other tubes, because hydrostatic pressure is less for them, than for peripheral tubes.

For determination of impulse duration τ of the electropneumatic valve openning we will consider the formula of gas consumption through the tube [10]:

$$Q = \varphi S \sqrt{\frac{2 \Delta p}{\rho_g}} , \quad (10)$$

where Q, volume consumption of gas; φ – consumption coefficient; $S = \pi d^3 / 4$, tube cross-sectional area; $\Delta p = 0.2 \rho g (H - h_k)$, pressure difference on a tube; ρ_g, gas density.

The volume of the gas coming to liquid during time τ, is defined by a formula:

$$V_g = Q \cdot \tau = \varphi S \tau \sqrt{\frac{2\Delta p}{\rho_g}} . \tag{11}$$

The input of gas portion with volume V_g forms a bubble which volume is equal to the volume of the entered gas:

$$V = \frac{\pi D^3}{6} = V_g . \tag{12}$$

From (11), (12) the ratio for definition follows τ:

$$\tau = \frac{\pi D^3}{6\varphi S}\sqrt{\frac{\rho_g}{2\Delta p}} = \frac{2}{3}\frac{D^3}{\varphi d^2}\sqrt{\frac{\rho_g}{0.4\rho g(H-h_k)}} = 0.055\ \text{s} = 55\ \text{ms} \tag{13}$$

When calculating τ the value of the consumption coefficient $\varphi = 0.5$ is defined according to [10, 11].

For the calculated parameters of the setup (p_{min}=105954 Pa; p_{max}=108611 Pa; $\tau = 0.055$ s) there was a series of experiments conducted, for preparation of the monodispersed bubbles compact cluster and the analysis of dynamics of its upfloat in nonperfect fluid. The video frames of upfloating of the monodispersed bubbles compact cluster received in two perpendicular planes are provided on Fig. 3. Diameter of bubbles of $D \approx 5$ mm received experimentally.

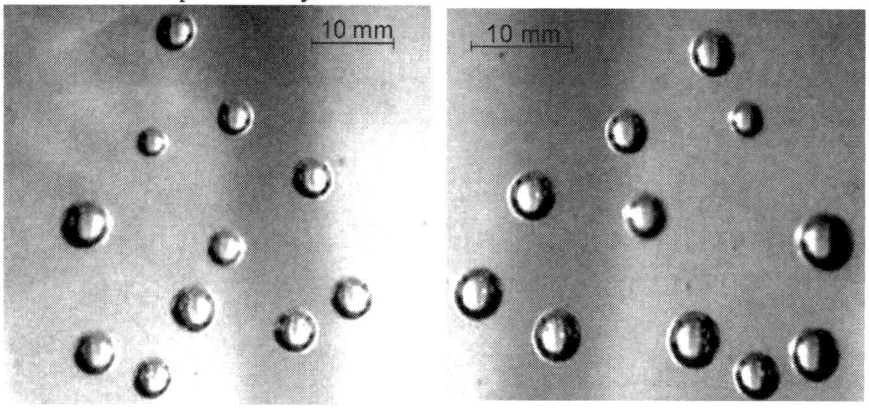

Fig. 3. The video frames of upfloating of the monodispersed bubbles compact cluster.

4 Conclusion

- The modified scheme of experimental setup providing the compact cluster preparation of monodispersed bubbles of the given diameter is presented in the work.
- An increase in the ascent rate of a cluster of bubbles was found, compared to the velocity of a single bubble in a pure liquid.
- Performing perforations in the top cover of the collector at its center and along equidistant concentric circles makes it possible to obtain an axisymmetric bubble cluster.
- The use of microtubules of the same diameter, installed in perforations, provides the formation of monodisperse bubbles.
- The use of microtubules of equal height located along each of the concentric circles ensures the simultaneous formation of a "ring" of bubbles for each of the circles.

• The linear decrease in the height of the tubes located on the circles, with an increase in the radius of the circle, ensures the sequential formation of each "ring" of bubbles with the same time lag as they move away from the center of the collector cover. This allows us to obtain a compact cluster with a uniform spatial distribution of bubbles.

This study was supported by the Russian Science Foundation (Project No. 15-19-10014).

References

1. A. P. Drozhzhin, S. M. Korobeynikov, B. C. Teslenko, Scientific bulletin of NGTU **2** (2003) [in Russian]
2. V. G. Levich, *Physical and chemical hydrodynamics* (Fizmatgiz, Moscow, 1959) [in Russian]
3. V. A. Arkhipov, I. M. Vasenin, A. S. Usanina, Fl. Dyn. **51** (2), (2016)
4. M. Hanok, Aeration device for introduction of gas bubbles on liquid medium, Patent RU 2491116, (Int.Cl.: B01F 3/04, B01F 13/00, 2013) [in Russian]
5. II. Sheaffer, R. John Aeration device for aquatic environments Patent WO 2016003926 (Int.Cl.: B01F 3/04262, C02F 1/74, 2016)
6. S. Naohiro, A. Keita, S. Toshihiko, Ultras. **77** (2017)
7. E. O. Oksman. Bubble-generating aerator, Patent US 3970731 (Int.Cl.: B01F 3/04, C02F 3/20, 1976)
8. V. A. Arkhipov, A. S. Usanina, K. G. Perfilieva, S. A. Basalaev, S. N. Polenchuk, V. I. Romandin. The setup for generation of a compact cluster of monodispersed bubbles. The application on patent RU 2017142492 (Int.Cl.: B01F 3/04, C02F 3/20, 2017) [in Russian]
9. F. V. Nevolin. *Chemistry and technology of glycerin production* (Chemistry, Moscow, 1954) [in Russian]
10. P. P. Kremlin. *Flowmeters and counters of quantity Reference book 4th edition, revised and enlarged* (Machine manufacturing, Leningrad, 1989) [in Russian].
11. S. Goldaev, S. Basalaev, K. Afanasyev, MATEC Web Conf. **110**, 01033 (2017)

Approach to detecting latent defects of the printed circuit boards thin-filmed coatings

Alexey Balastov[1,*], *Boris* Kovrigin[1], and *Valery* Lavrinovich[2]

[1]National Research Tomsk State University, 634050 Tomsk, Russia
[2]All-Russian Scientific Research Institute of Technical Physics, 456770 Snezhinsk, Russia

Abstract. One of the main reasons for the space vehicles (SV) power supply systems failure is a secondary arc. In the statistics of fatal failures that can lead to the SV loss, the failure rate of the power supply systems is 9%, but despite this, there are no standard methods for diagnosing the on-board equipment for the arc stability, due to the relative novelty of the problem. The problem we encountered is associated with the need to switch power supply systems to the voltage levels that are much higher than the arc voltage in the space vacuum. From the point of view of the defects downing that can lead to arcing, the most vulnerable are the insulation coatings. Therefore, a significant part of the problem solution is concentrated in the development of the specific methods for these coatings diagnostics. In space technology, such diagnostics significance is increased due to the inability to provide necessary repairs or services in the outer space.

1 Introduction

The transition to the new operating voltage standard in satellite electronics has led to the problem of arcing between the printed circuit board elements. In order to keep the mass of the satellite as small as possible, the space inside the circuit housing is not filled with an insulator or is not sealed. It contributes to the emergence of electric discharges, which could potentially extend to the satellite on-board equipment. The ignition of the electric arc in the onboard satellite system always leads to failure. In most cases, it leads to the satellite termination. This research main aim is to improve the quality of the satellites on-board electronic equipment. The final result of the research is a proven and patented installation and method for determining the printed circuit boards thin-filmed coatings defects for the satellites on-board electronic equipment. The obtained results will be used to develop documentation and prototypes testing.

2 How it works?

Insulation coatings are the most vulnerable to arcing. Therefore, a significant part of the problem's solution is concentrated in the development of a new insulation coatings inspection

* Corresponding author: lexiusb@gmail.com

method. The corona discharge phenomenon is known for a long time. There are many practical and theoretical materials explaining the nature and properties of the corona discharge [1]. We have proposed the possibility of finding latent defects in the printed circuit boards (PCB) coatings, based on the corona discharge properties, using a reverse corona, that can be registered visually or via ampermeter. Each defect in the insulation coating is a good conductive channel [2]. Our installation can register this conductivity due to the corona discharge properties.

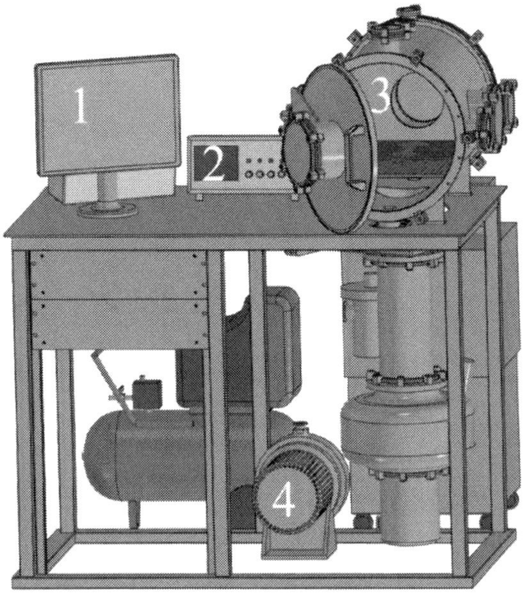

Fig. 1. T-Flex CAD model of the developed installation: 1 – Computer; 2 – Oscilloscope; 3 – Vacuum chamber; 4 – Vacuum pump.

PCB (4) is placed between point-plane system with different potentials (3 and 5). Thermal electron emission gives sufficient number of electrons to generate a little electrical current. If insulation coating has a defect, we can register it via the ampermeter (6).

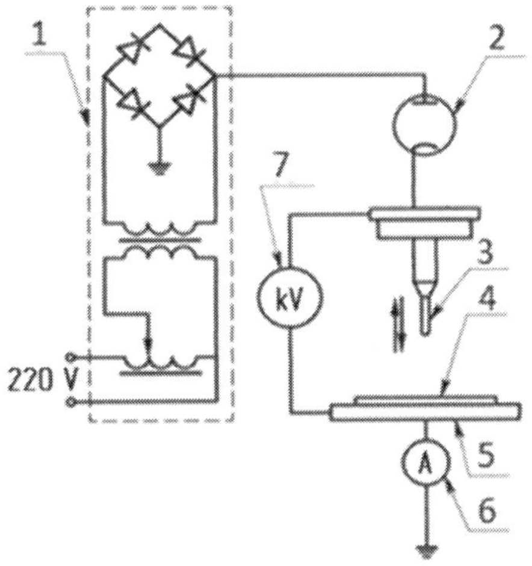

Fig. 2. Diagnostic electric scheme.

There are given types of defects that can be detected by corona discharge testing:

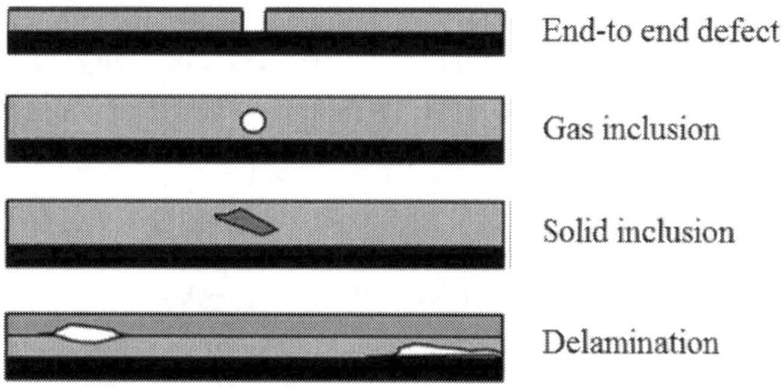

Fig. 3. Type of defects.

3 Glowing effect

High voltage corona discharge can glow. When it happens, this glowing area covers the defected zone and the way from the point to the zone. This simple property provides good PCB coating defects visualization.

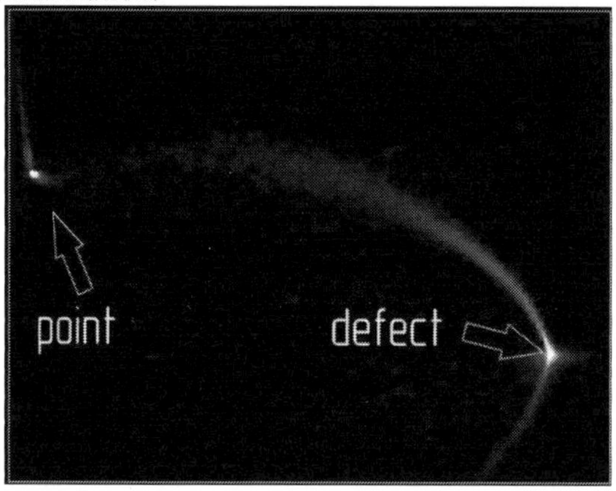

Fig. 4. Glowing area for 1 defect.

The glow effect will work even if the printed circuit board has more than one defect. This feature allows us to re-evaluate the potential of this testing method. Using the corona discharge it is possible to determine not only the defect presence, but also to localize it with high accuracy [3].

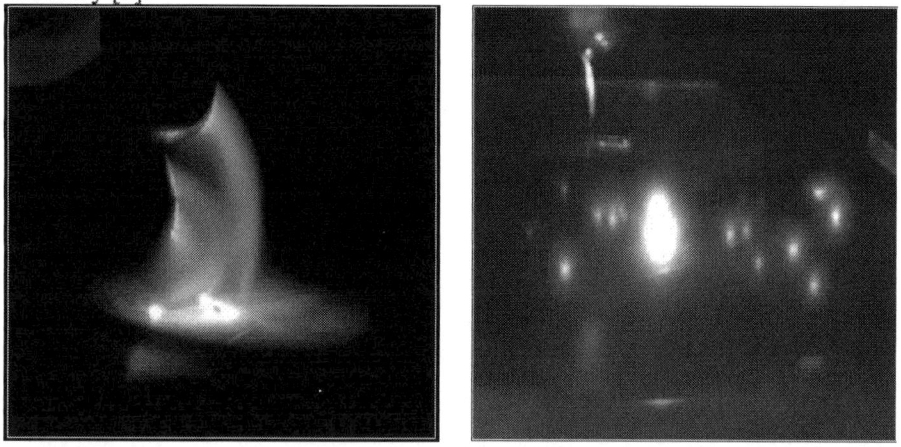

Fig. 5. Glowing area for 3 defects (left) and for 14 defects (right).

4 The research preliminary results

There can't be used tests that can decrease PCB service life because of the satellite electronics high complexity and cost. Limiting the choice of diagnostic methods and providing testing only via the nondestructive one, significantly reduces the range of possible diagnostic methods application. An additional condition is the method's ability to determine various types of insulation defects and to provide testing object entire surface diagnostics, including when the method is applied to the printed circuit boards consisting of a large number of elements and complex topology.

The purpose of the first experiment was the detection of a specially made defect in the test sample insulation coating.

There were obtained the following results:

- The defect registration occurs before reaching the breakdown voltage;
- There is enough 1 kV voltage in order to detect a defect;
- There is recorded current from 10 up to 50 μA if the insulating coating has got a defect;
- There is a clear possibility to detect multi-defects and to limit the breakdown current.

5 Conclusion

The use of a controlled corona discharge as a diagnostic tool, in our opinion, will make it possible to: exclude a contact with the dielectric coating for detecting latent defects; reduce testing voltage to 1 kV and lower; reduce testing current in the circuit up to microamps. It is necessary to note that the defect determination always occurs before the breakdown. Therefore, the corona discharge testing can be referred to the nondestructive testing methods.

There are some plans for the further studies:

1. Automate the process of detection and lower the test voltage;
2. Determine the experiment optimal parameters;
3. Upgrade the installation;
4. Conduct further experiments.

References

1. R. Islam, P. D. Pedrov, Shuzheg Xie, K. R. Englund, Trans. Plasma Sci. **43**, 3695 (2015)
2. J.J. Lowke, R. Morrow, Pure & Appl. Chem. **66**, 1287 (1994)
3. V.A. Lavrinovich, Bulletin of the Tomsk Polytechnic University. Geo Assets Engineering **327**, 58 (2016)

Modeling of conjugated heat and mass transfer in the entrance region of falling liquid film at various values of the Froude number

Maria V. Bartashevich[1,*]

[1]Kutateladze Institute of Thermophysics, 630090 Novosibirsk, Russia

Abstract. Mathematical model of conjugated heat and mass transfer in absorption on the entrance region of the semi-infinite liquid film of lithium bromide water solution is investigated for different values of Froude number. The calculations shown that larger values of Froude number corresponds to a smaller thickness of the falling film. It was demonstrated that for large values of the Froude number the heat transfer from the surface is greater than for smaller values.

1 Problem statement

In absorption chillers and in the spray-type devices the heat and mass transfer processes take place and proceed with a pressure-driven liquid film flow through holes or slits. In other types of devices, including film-like devices, such flows are possible at the exit of the distributors. The liquid flow with a free boundaries in the real devices are often conjugated by the various heat and mass transfer processes such as absorption.

The solutions for simple problems of the coupled heat and mass transfer during absorption at the liquid/gas interfaces of different geometry types (films, jets, drops, sheets) are given in [1, 2]. Modelling of the non-isothermal absorption of vapour into expanding liquid sheets was performed by Acosta-Iborra and co-workers [3]. In [4] it has been shown that a significant change of the salt concentration leads to different absorption regimes.

In present work we have been investigated the coupled heat and mass transfer during absorption in the entrance region of the falling axisymmetric semi-infinite film surrounded by the pure vapour or gas at various values of the Froude number.

We assume, that the liquid film falls under the action of gravity with the uniform vertical velocity u_i. Downstream, the velocity is defined as $u = \sqrt{u_i^2 + 2gx}$. The initial thickness of the film is h_i. The Froude number $Fr = (gh_i)/u_i^2$.

The process of heat and mass transfer at film absorption is described by the equations of heat conductivity and diffusion:

$$u\,\partial T/\partial x = a\,\partial^2 T/\partial y^2 \,, \tag{1}$$

* Corresponding author: bartashevichmv@gmail.com

$$u\,\partial C/\partial x = D\partial^2 C/\partial y^2\,, \qquad (2)$$

a is thermal diffusivity, m²/s; D is diffusion coefficient, m²/s. The boundary conditions at the inlet at $x = 0$ are:

$$T = T_i, C = C_i, h = h_i\,. \qquad (3)$$

Similarly to [1], the equilibrium of solution-vapor system occurs on the interface at $y = h$:

$$C_d = f(T_d)\,,\ \lambda\partial T/\partial y = \chi\partial C/\partial y\,. \qquad (4)$$

where $\chi = L\rho D/(1 - C_i)$, L is latent heat of phase transformation, J/kg; ρ is solution density, kg/m³; λ is heat conductivity, W/(mK). This equilibrium condition joins equilibrium temperature T_d with concentration C_d. We assume the asymptotic boundary condition at $y' \to \infty$:

$$T = T_i, C = C_i\,, \qquad (5)$$

where $y' = y - h$ and directed from the interface to the film depth . The conditions of absorption are $T_i < T_s$ and $C_i < C_s$. Here, $C_s = f(T_i)$ is equilibrium concentration, corresponding to initial temperature of solution T_i; T_s is equilibrium temperature corresponding to initial concentration C_i ($C_i = f(T_s)$), similarly to [1]. We have the temperature gradient at the free liquid solution interface:

$$\frac{\partial T}{\partial y} = \Delta T T_i \left(Fr^{1/2} Pe^{1/2} \frac{1}{h_i\sqrt{\pi}\left(2Fr\dfrac{x}{h_i} + 1\right)^{1/4}} \right), \qquad (6)$$

where $\Delta T = T_s - T_i$ is temperature difference between initial T_i and equilibrium T_s temperatures, $Pe = \dfrac{u_i h_i}{a}$ is a Peclet number. Also, Nusselt number dependence from longitudinal coordinate can be obtained as:

$$Nu(x) = T_i Fr^{1/2} Pe^{1/2} \frac{1}{\sqrt{\pi}\left(2Fr\dfrac{x}{h_i} + 1\right)^{1/4}}\,. \qquad (7)$$

2 Calculations

Numerical calculations were carried out for different values of Froude number. The results of calculations of absorption on the entrance region of the semi-infinite liquid film of lithium bromide water solution are given in Fig.1 – Fig.2. The calculations were performed at T_i =20 °C, T_s = 46 °C, C_i = 0.4, C_s = 0.56, Fr =0.25, 0.5, 1.0, Pe =200, 400.

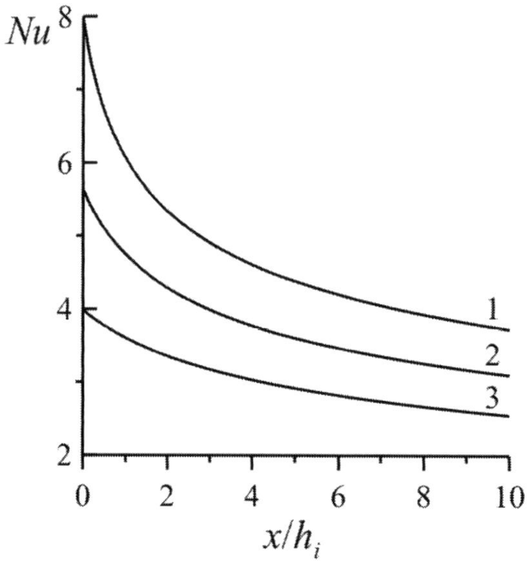

Fig. 1. The Nusselt number along the film-flow direction at the entrance region for different values of Froude number, Pe = 200. 1 – Fr = 0.25; 2 – Fr = 0.5; 3 – Fr = 1.0.

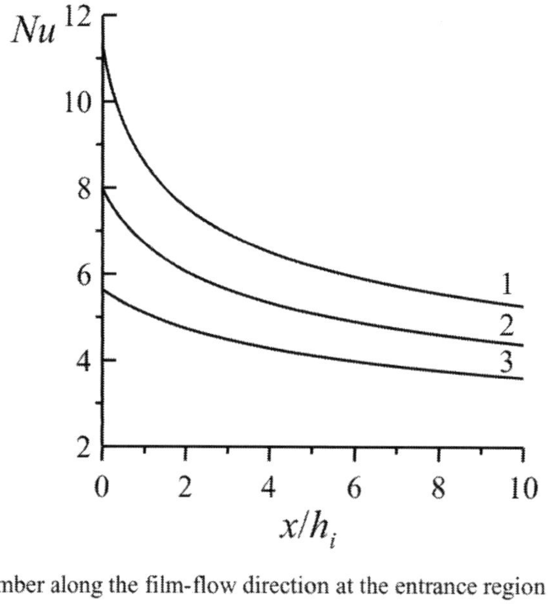

Fig. 2. The Nusselt number along the film-flow direction at the entrance region for different values of Froude number, Pe = 400. 1 – Fr = 0.25; 2 – Fr = 0.5; 3 – Fr = 1.0.

Fig. 1 and Fig. 2 show the calculated distributions of the Nusselt number along the film-flow direction for different values of the Froude and Peclet numbers. For Pe = 400 the Nusselt number is almost 1.5 times more than for Pe = 200 at the inlet. However, with the distance from the entrance region, the difference decreases.

3 Conclusions

In the absorption heat pumps, four heat and mass transfer processes can occur simultaneously: absorption, desorption, condensation, evaporation. In this paper, the Nusselt number (*Nu*), for different values of the Froude (*Fr*) and Peclet (*Pe*) numbers, was calculated for the absorption element of the heat pump in the entrance region of the free-flowing film. The dependence of the Nusselt number on the dimensionless longitudinal coordinate is constructed. It was shown in [2] that a larger thickness of the Froude number corresponds to a smaller thickness of the flowing film. In the calculations presented here it was demonstrated that for the large values of the Froude number, the heat transfer from the surface is greater than for smaller values and the heat transfer more intensive for large Peclet numbers for such regime of the falling film.

This work was carried out at the Kutateladze Institute of Thermophysics SB RAS and financially supported by the Russian Science Foundation (project number 15-19-10025).

References

1. V.E. Nakoryakov, N.I. Grigorieva, *Non-Isothermal Absorption in Thermal Transformers* (Nauka, Novosibirsk, 2010)
2. V.E. Nakoryakov, N.I. Grigoryeva, M.V. Bartashevich, Int. J. Heat Mass Transfer **54** (21-22) (2011)
3. A. Acosta-Iborra, N. Garcia-Hernando and D. Santana, Int. J. Heat Mass Transfer **52** (2009)
4. V.E. Nakoryakov, S.Y. Misyura, J. Eng. Thermophys. **25** (2016)

Numerical simulation of non-isothermal absorption in a liquid film moving over a cooled horizontal surface under the action of a gas flow

Maria V. Bartashevich[1,*]

[1]Kutateladze Institute of Thermophysics, 630090 Novosibirsk, Russia

Abstract. Mathematical model of absorption on a horizontal film of lithium bromide water solution flowing on a cooled wall under the action of shear stress is numerically investigated in this paper. The shear stress on the film surface is set by the motion of co-current or counter-current surrounding saturated water vapor flow. It is known that the external shear stress is caused by the motion of surrounding saturated water vapor intensificate the process of non-isothermal absorption in comparison with the fixed-vapor regime. Our calculations have shown that at low values of heat flux the film temperature and vapor concentration in the solution downstream increases due to absorption.

1 Problem statement

The necessity of heat and mass transfer intensification in the absorption chillers is caused by the limitations towards their size and consumption of materials. The use of thin liquid films just solves these problems. Our numerical modelling of the conjugated heat and mass transfer in absorption chillers focuses on the film flow process on cooling wall. The liquid film moving on a cooled horizontal wall under the action of co-current or counter-current vapor flow have been investigated. The liquid film of lithium-bromide water solution has been considered. The lithium bromide–water solution widely used in absorption refrigeration systems due to its effectivity especially for low heat temperatures [1]. The non-isothermal absorption takes place in the various energy devices [2,3]. The investigation of thin liquid film flowing down a vertical wall in the roll-wave regime in presence of heat and mass transfer through the free surface was presented in [4,5]. The absorption and desorption processes under the conditions, corresponding to the operating modes of thermal transformers are described in [6]. The analytical solution to the problem of conjugate heat and mass transfer in a laminar falling liquid film with a linear velocity profile is presented in [7]. The results of the numerical simulation of absorption and their comparison with the experimental data are presented in [8]. The results of experimental studies concerning the development of critical phenomena and structure formation in the process of boiling in falling films and during liquid cavitation were presented in [9]. Also, in [10,11] it was

* Corresponding author: bartashevichmv@gmail.com

shown that free convection in the gas significantly affects the desorption (adsorption) of the aqueous solution of the salt.

We assume, that the lithium bromide–water solution liquid film flowing under the action of the external tangential stress on the film surface. The external shear stress is caused by the motion of surrounding saturated water vapor. It is assumed that the shear stress applied to the interface will allow intensification of the process of non-isothermal absorption in comparison with the fixed-vapor regime. We consider the two-dimensional stationary flow of a laminar liquid film of thickness h over a horizontal plate. The surface of liquid is in contact with moving vapor. The shear stress on the film surface is τ. Absorption is considered in the framework of ordinary assumptions [2,3]. The thermal-physical properties of solution are assumed constant.

Let us introduce the Cartesian coordinate system with Ox axis in the direction of the flow and Oy axis directed along the normal to the plate. The velocity profile in the film takes form

$$u = \frac{y\tau}{\mu}.$$

The process of heat and mass transfer at film absorption is described by the equations of heat conductivity and diffusion:

$$u\frac{\partial T}{\partial x} = a\frac{\partial^2 T}{\partial y^2}, \tag{1}$$

$$u\frac{\partial C}{\partial x} = D\frac{\partial^2 C}{\partial y^2}, \tag{2}$$

a is thermal diffusivity, m²/s; D is diffusion coefficient, m²/s. The boundary conditions at the inlet at $x = 0$ are:

$$T = T_i, C = C_i.$$

Similarly to [1], the equilibrium of solution-vapor system occurs on the interface at $y = h$:

$$C_d = f(T_d),\ \lambda\frac{\partial T}{\partial y} = \chi\frac{\partial C}{\partial y}.$$

where $\chi = L\rho D/(1 - C_i)$, L is latent heat of phase transformation, J/kg; ρ is solution density, kg/m³; λ is heat conductivity, W/(mK). This equilibrium condition joins equilibrium temperature T_d with concentration C_d. We set on the wall at $y = 0$:

$$\lambda\frac{\partial T}{\partial y}\Big|_{y=0} = q_w,\ \frac{\partial C}{\partial y}\Big|_{y=0} = 0,$$

where q_w is heat flux on the wall. The conditions of absorption are $T_i < T_s$ and $C_i < C_s$. Here, $C_s = f(T_i)$ is equilibrium concentration, corresponding to initial temperature of solution T_i; T_s is equilibrium temperature corresponding to initial concentration C_i (

$C_i = f(T_s)$), similarly to [1]. Then we turn to dimensionless variables $\xi = x/l$, $l = \tau h^3/(\mu a)$, $\eta = y/h$, $\theta = (T - T_i)/(T_s - T_i)$.

2 Numerical calculations

Numerical calculations were carried out by the finite difference method. The results of calculations of absorption on a film moving under the action of a tangential stress over a horizontal cooled wall are given in Fig. 1 – Fig. 3. The calculations were performed at T_i = 20 °C, T_s = 46 °C, C_i = 0.4, C_s = 0.56, $\bar{q}_w$ is a dimensionless heat flux. Fig. 1 shows the calculated distributions of the dimensionless temperature along the film surface for different heat fluxes on the wall. Corresponding concentration distributions vs. longitudinal coordinate are shown in Fig. 2. Fig. 3 shows the calculated distributions of the dimensionless concentration at different distances from the inlet.

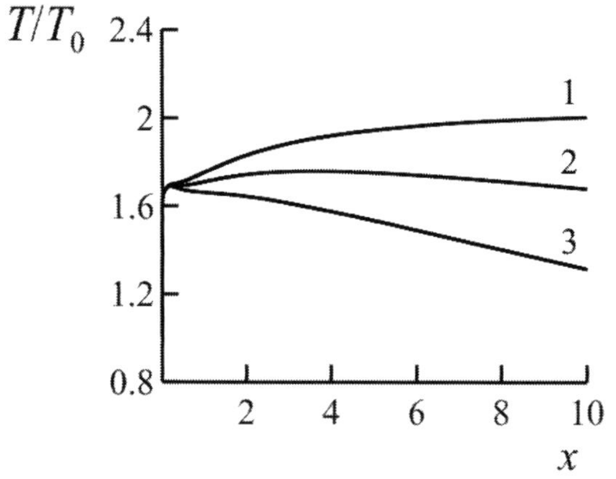

Fig. 1. Profiles of dimensionless temperature of the film surface at different heat fluxes: 1 – $\bar{q}_w = 0.01$; 2 – $\bar{q}_w = 0.1$; 3 – $\bar{q}_w = 0.2$.

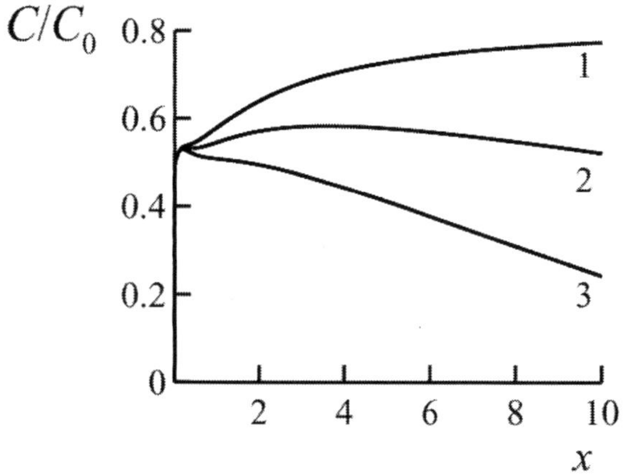

Fig. 2. Profiles of dimensionless concentration on the film surface at different heat fluxes: 1 – $\bar{q}_w = 0.01$; 2 – $\bar{q}_w = 0.1$; 3 – $\bar{q}_w = 0.2$.

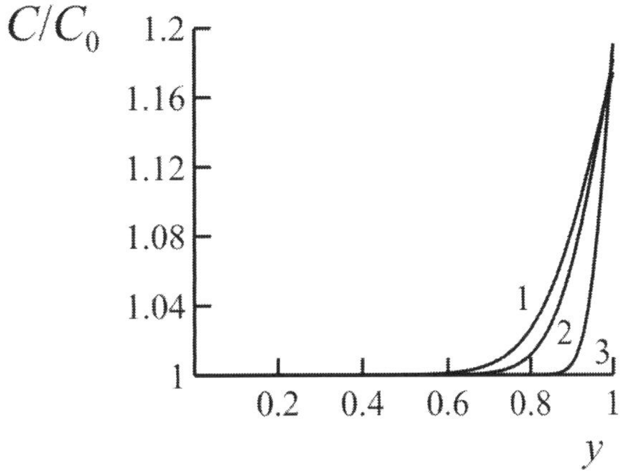

Fig. 3. Profiles of dimensionless concentration in different cross-sections of the film, $\bar{q}_w$ = 0.01. 1 – ξ =0.1; 2 – ξ =0.5; 3 – ξ = 0.8.

3 Conclusions

Mathematical model of absorption on a horizontal film of lithium bromide water solution flowing on a cooled wall under the action of shear stress for the given constant heat flux is numerically investigated in this paper. The shear stress on the film surface is set by the motion of co-current or counter-current surrounding saturated water vapor flow. It is shown that at low values of heat flux the film temperature and vapor concentration in the solution downstream increases due to absorption.

The study was funded by FASO Russia.

References

1. J. Yin, L. Shi, M.-S. Zhu, L.-Z. Han, Appl. Energy **67** (2000)
2. V.E. Nakoryakov, N.I. Grigorieva, *Non-Isothermal Absorption in Thermal Transformers* (Nauka, Novosibirsk, 2010)
3. V.E. Nakoryakov, N.I. Grigoryeva, M.V. Bartashevich, Int. J. Heat Mass Transfer **54** (21-22) (2011)
4. V.E. Nakoryakov, V.V. Ostapenko, M.V.Bartashevich, Int. J. Heat Mass Transfer **55** (23-24) (2012)
5. V.E. Nakoryakov, V.V. Ostapenko, M.V.Bartashevich, Int. J. Heat Mass Transfer **89** (2015)
6. M. Mittermaier , F. Ziegler, Int. J. Refrig. **59** (1) (2015)
7. M. Mortazavi, S. Moghaddam, Int. J. Refrig. **66** (2016)
8. E. Garcia-Rivera, J. Castro, J. Farnos, A. Oliva, Int. J. Therm. Sci. **109** (2016)
9. A. N. Pavlenko, V. P. Koverda, V. N. Skokov, A. V. Reshetnikov, A. V. Vinogradov, A. S. Surtaev, J. Eng. Thermophys. **18** (1) (2009)
10. S.Ya. Misyura, Sci. Rep. **7**, 14759 (2017)
11. S.Ya. Misyura, Int. J. Heat Mass Transfer **116** (2017)

MATEC Web of Conferences **194**, 01009 (2018) https://doi.org/10.1051/matecconf/201819401009
HMTTSC-2018

Experimental study of a drop “evolution” under conditions of its free fall on a heated surface

Kseniya Batishcseva[1,*], and *Konstantin* Ponomarev[1]

[1]National Research Tomsk Polytechnic University, 634050 Tomsk, Russia

Abstract. Time dependences of distilled water drop diameter were obtained experimentally after its falling on a heated copper substrate from a height of 0.09 m. The temperature of the solid surface (T_w) varied from 333 to 413 K in steps of 20 K. The effect on the drop dynamics after falling and the maximum ratio of the spreading diameter to the drop diameter in flight (β_{max}) were determined.

1 Introduction

Drop falling is implemented in the technological processes of inkjet printing, spraying, plasmodynamic synthesis, fuel injection in the internal combustion engine, gas turbine operation, drop-cooling systems [1-6]. It is necessary to control their spatial distribution and spreading when they impact the surface in order to prevent the merging of adjacent drops. A drop spreading over a solid surface is found to depend on the gravity forces, inertia, viscosity, friction and surface tension [7-10]. The weight of the drop affects the gravity force; liquid composition affects the viscous friction force and surface tension; mass of liquid and surface roughness on mechanical friction.

The dynamics of a drop on a solid surface is characterized by $\beta = f(\tau)$ [11], where β is the ratio of the spreading diameter to the drop diameter in flight dimensionless contact diameter, τ is a dimensionless time.

The behavior of a liquid on a heated surface was determined by the dimensionless temperature T^* [11]:

$$T^* = \frac{T_w}{Ts}, \quad (1)$$

where T_w is the surface temperature, K; Ts is liquid saturation temperature, K.

The aim of the paper is to determine the dynamics of a drop (change of the geometric shape) of distilled water when it falls freely onto a heated copper surface.

* Corresponding author: bka1801@mail.ru

2 Experimental setup

Experimental studies were carried out on a polished copper substrate M1. Latter was a solid cylinder with a diameter of 50 mm and a thickness of 4 mm. The surface of the substrate was polished with diamond pastes and grinding wheels. The roughness was characterized by the mean surface average roughness (Ra = 0.361 μm). The substrate was purified with isopropanol followed by rinsing with distilled water.

The scheme of the experimental setup is shown in Fig. 1.

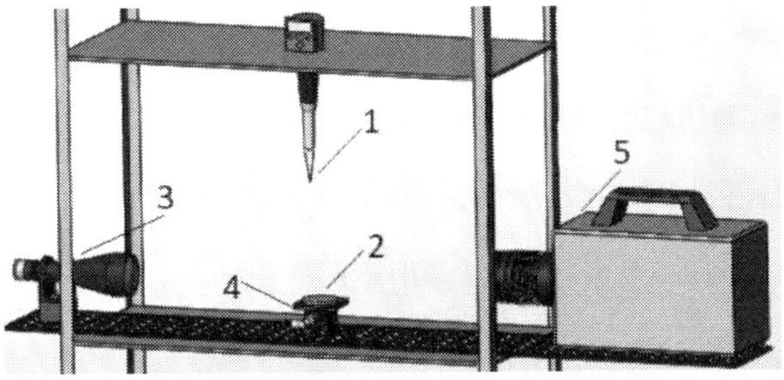

Fig. 1. Schematic diagram of the experimental setup: 1 – syringe dispenser; 2 – substrate; 3 – source of plane-parallel light; 4 – electric heating element; 5 – high-speed video camera.

The distilled water drop with a volume of 10 ± 0.25 μl was formed on the tip of the dispenser (1), a place at a distance of 0.09 m from the surface of the substrate (2). The source of the plane-parallel light (3) illuminating the drop at the moment of incidence onto the substrate mounted on the electric heating element (4). Shadow images of the drop were obtained with the FastCAMSA1.1 model 675K-M1 high-speed video camera (5) with a shooting speed of 5000 frames per second.

The temperature was recorded by a thermocouple of type K (OMEGA 5TC-TT-K-40-36) standing under the substrate. The thermocouple error under the conditions of the investigated temperature range did not exceed 0.1 K. The substrate temperature varied from 333 to 413 K (T^*=0.6 ÷ 1.4) in 20 K increments.

The drop diameter (D_0) during the fall was determined using expression (3) [11]. While falling, the liquid took the form of a sphere:

$$D_0 = \sqrt[3]{\frac{6 \cdot \sigma \cdot d_c}{\rho \cdot g}} \tag{2}$$

According to (2), the drop diameter was D_0=2.80 mm formed by the dispenser with the tip d_c=0.5 mm, under atmospheric pressure and ambient air temperature T_{air}=298 K.

Speed (U_0) of the falling drop U_0=1.32 m/s was determined by the shadow images analysis results. The results agree well (error <1%) with the free fall speed (3) U_0^*=1.324 m/s and the empirical equation of K. Range [12] (4) U_0^{**}=1.324 m/s:

$$U_0^* = \sqrt{2 \cdot H \cdot g}, \tag{3}$$

$$U_0^{**} = \sqrt{\frac{g}{A} \cdot (1 - e^{-2 \cdot A \cdot H})} \text{ with } A = \frac{3}{4} \cdot \frac{c_f \cdot \rho_{air}}{\rho_{drop} \cdot D_0}, \tag{4}$$

where H is the fall height of the drop, m; c_f is the ball drag coefficient; ρ_{air} is the density of the environment (air), kg/m^3; ρ_{drop} is drop density, kg/m^3.

The dimensionless quantity (c_f), called the ball drag coefficient [13] was determined by the formula:

$$c_f = \frac{24}{\mathrm{Re}} + \frac{4}{\sqrt[3]{\mathrm{Re}}} \tag{5}$$

3 Results and discussions

According to the results of experimental studies, a change in the geometric shape of the drop under the conditions of its impact on a heated (T_w=333 ÷ 413 K) copper polished surface was established (Fig. 2-4).

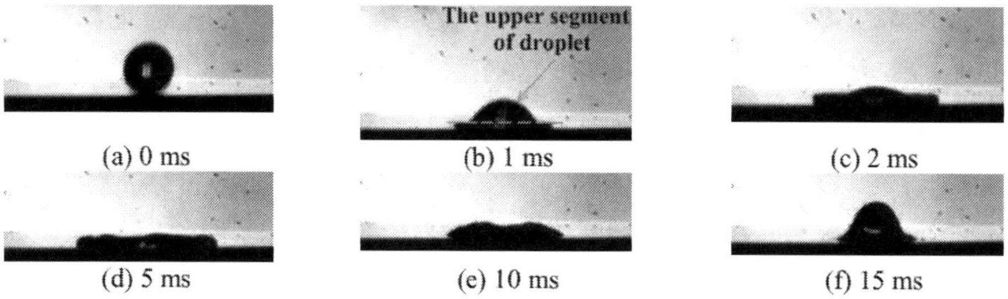

Fig. 2. Change the geometric shape of the drop, at T_w=333 K. Time after the drop touches the surface, ms: (a) 0 ms; (b) 1 ms; (c) 2 ms; (d) 5 ms; (e) 10 ms; (f) 15 ms.

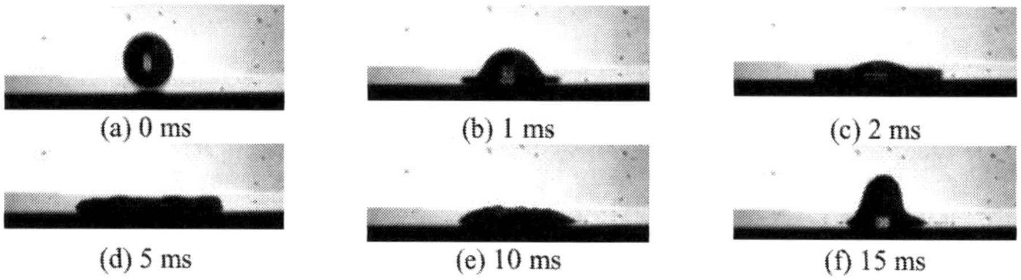

Fig. 3. Change the geometric shape of the drop, at T_w=373 K. Time after the drop touches the surface, ms: (a) 0 ms; (b) 1 ms; (c) 2 ms; (d) 5 ms; (e) 10 ms; (f) 15 ms.

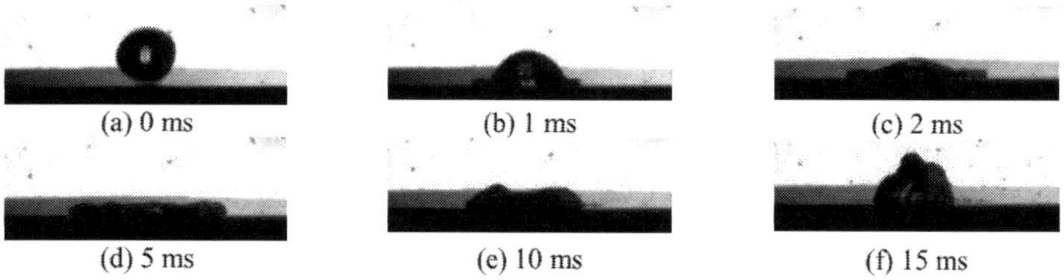

Fig. 4. Change the geometric shape of the drop, at T_w=413 K. Time after the drop touches the surface, ms: (a) 0 ms; (b) 1 ms; (c) 2 ms; (d) 5 ms; (e) 10 ms; (f) 15 ms.

The moment of the drop contact with the surface (Fig. 2 (a)) was taken as the time reference (time=0 ms). From this moment, the drop begins spreading and its upper segment (Fig. 2 (b) without shape changing continues to approach the solid surface at a speed U_0

[14]. The growth rate of the contact diameter decreases due to the viscous friction (the lower layers of the drop are retarded, dragging adjacent faster layers) (Fig. 2 (c)). The liquid molecules on the drop surface attract the liquid from the side in comparison with the ambient air. The surface layer (about 10 A thick), where the molecules approach to move into the depth, exerts pressure on the liquid. The free energy of the surface layer was directed at all points perpendicular to the surface and tends to reduce it to a minimum.

The maximum contact diameter (D_{max}) was reached when the drop spreads into a thin film (Fig. 2 (d)). The latter begins to compress (Fig. 2 (e)) due to capillary forces. At the center of the film, the liquid moves upward, giving it the shape of a drop again (Fig. 2 (f)). Then the contact diameter was reduced to achieve a final equilibrium diameter D_c that was smaller that D_{max}.

We note that at substrate temperatures ($T_w \geq 373$ K) after reaching D_{max} the liquid film was compressed into a drop, but its surface becomes wave-like (Fig. 3 (e), Fig. 4 (e)). The latter can be explained as follows. The liquid on the drop periphery overheats. When the liquid was heated to a temperature at which the film begins to "bounce" off the surface, the liquid at the edges of the drop moves upward, entraining the liquid behind it. The drop surface prevents the rupture of the layer. At the maximum surface temperature ($T_w = 413\ K$) small drops were released into the ambient air from the drop (Fig. 4 (f)).

To quantitatively estimate the drop dynamics after falling on the heated surface in Fig. 5, the dependences of the dimensionless diameter of the drop spread $\beta = D / D_0$, dimensionless time $\tau = t \cdot U_0 / D_0$, and dimensionless temperature $T^* = T_w / T_s$ =0.6 ÷ 1.4 are used.

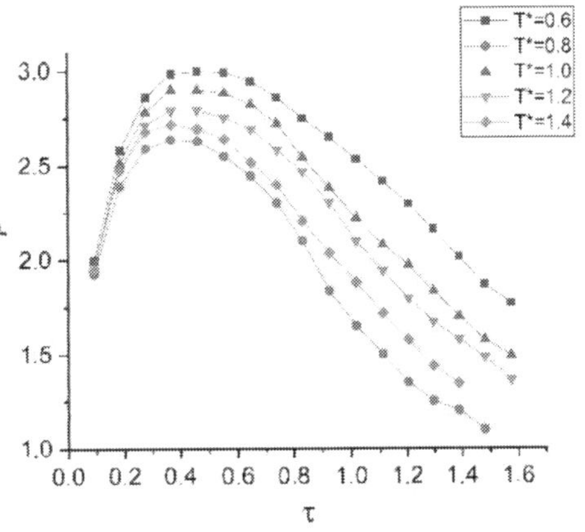

Fig. 5 Dimensionless drop spreading diameter from a dimensionless time at T^* =0.6 ÷ 1.4.

Dependencies (Fig. 5) show maximums of the dimensionless diameter (β_{max}). With a small kinetic energy of drops and low substrate surface temperatures (T^* 0.6 ÷ 0.8) β_{max} of spreading drop decreases from 3.00 to 2.63 (Fig. 5). The increase in the wall temperature increases the part of the heat accumulated by the expanding drop due to the high heat capacity. The evaporation rate from the drop-free surface increases, cooling the latter. Under such conditions, the drop surface tension and its viscosity increase, not allowing the latter to greatly spread.

At the substrate temperature T_w=373 K the dimensionless drop spreading diameter increased to 2.90. This was due to the fact that when the drop spreads its periphery with the lowest height warms up faster than the liquid in the center. The drop warmed up quickly in contact with the substrate. The increase in temperature reduced the effect of capillary forces and viscous friction forces, which led to rising. The drop did not break up into small parts.

At surface temperatures above the boiling point of the liquid (T^*=1.2÷1.4) β_{max} decreases up to 2.78 at T_w=413 K. Under these conditions, steam bubbles of micron size are created on the drop periphery heated to a temperature higher 373 K which bursting, separating the microdrops from the liquid. The higher the substrate temperature the more microbubbles are generated, and the secondary drops are more separated. The maximum drop contact spreading diameter decreases.

4 Conclusion

Based on the results of the experimental studies the dependences of the dimensionless diameter (β) of the distilled water drop spreading on time (τ) were established. The maximum dimensionless diameter (β_{max}) under these conditions was in the range β_{max} =2.63÷3.00.

A qualitative and quantitative analysis of the drop dynamics after its fall on a heated polished (Ra=0.361 µm) copper substrate was carried out. It was found that at a dimensionless temperature $T^* < 1$ upon reaching β_{max} a liquid film was compressed in a drop with a regular rounded shape. When $T^* = 1$ the surface of the compressed drop was a waveform. The latter was due to overheating and the motion of the liquid in the drop. When $T^* > 1$ in addition to the drop waveform it was possible to observe the formation of steam microbubbles in it, which, when detached, separate microdrops from the liquid bulk. The latter was the cause of the decrease in β_{max} with growth of T_w.

The reported study was supported by RFBR, research project No. 18-38-00315 mol_a.

References

1. V. E. Nakoryakov, S. Ya. Misyura, S. L. Elistratov, R. A. Dekhtyar, J. Eng. Thermophys. **23**, 257 (2014), DOI: 10.1134/S1810232814040018
2. V. E. Nakoryakov, S. Y. Misyura, J. Eng. Thermophys. **25**, 24 (2016), DOI: 10.1134/S1810232816010033
3. M. Popov, O. Maniv, MATEC Web Conf. **141**, 01060 (2017), DOI: 10.1051/matecconf/201714101060
4. A. Sivkov, A. Pak, I. Shanenkov, Y. Kolganova, I. Prosvirin, Adv. Mater. Res. **880**, 36 (2014), DOI: 10.4028/www.scientific.net/AMR.880.36
5. M. V. Piskunov, P. A. Strizhak, Exp. Therm. Fluid Sci. **91**, 103 (2018), DOI: 10.1016/j.expthermflusci.2017.10.018
6. D. O. Glushkov, D. P. Shabardin, P. A. Strizhak, K. Yu. Vershinina, Fuel Process. Technol. **143**, 60 (2016), DOI: 10.1016/j.fuproc.2015.11.014
7. E. G. Orlova, D. V. Feoktistov, G. V. Kuznetsov, K. O. Ponomarev, Eur. J. Mech. B Fluids **68**, 118 (2018), DOI: 10.1016/j.euromechflu.2017.12.002
8. D. V. Feoktistov, E. G. Orlova, A. G. Islamova, MATEC Web Conf. **23**, 01054 (2015), DOI: 10.1051/matecconf/20152301054

9. K. A.Batischeva, E. G. Orlova, D. V. Feoktistov, EPJ Web Conf. **82**, 01055 (2015), DOI: 10.1051/epjconf/20158201055
10. D. V. Zaitsev, D. P. Kirichenko, V. S. Ajaev, O. A. Kabov, Phys. Rev. Lett. **119**, 094503 (2017), DOI: 10.1103/PhysRevLett.119.094503
11. M. Bornhorst, O. Deutschmann, Int. J. Heat Fluid Flow **69**, 55 (2018), DOI: 10.1016/j.ijheatfluidflow.2017.10.007
12. K. Range, F. Feuillebois, J. Colloid Interface Sci. **203**, 16 (1998), DOI: 10.1006/jcis.1998.5518
13. N. A. Fuchs, *The Mechanics of Aerosols* (Ed. Pergamon Press, London, 1964)
14. M. B. Lesser, Proc. R. Soc. Lond. A **377**, 289 (1981), DOI: 10.1098/rspa.1981.0125

Identification of kinetic triplets by results of derivatographic analysis

Ivan S. Bondarchuk[1,4], *Sergei* S. Bondarchuk[2,3], and *Boris* V. Borisov[4,*]

[1]National Research Tomsk State University, 634050 Tomsk, Russia
[2]Tomsk State Pedagogical University, 634061 Tomsk, Russia
[3]Institute for Problems of Chemical and Energetic Technologies SB RAS, 659322 Biisk, Russia
[4]National Research Tomsk Polytechnic University, 634050 Tomsk, Russia

Abstract. The method for identification of the triplet of kinetic parameters of a heterogeneous reaction using the data of the derivatographic analysis is proposed. This method is characterized by high accuracy and relative simplicity and it can be effectively realized using MS Excel software.

Metal nanopowder is one of the most prospective high-energy materials that is widely used in many spheres [1, 2]. This explains long and sustained interest in studying the features of metal nanopowder, in particular the kinetics of oxidation [3]. The proceeding heterogeneous reaction can be in full measure characterized by the so-called "kinetic triplet" – the frequency factor A, the activation energy E_a and the mathematical description of the reaction mechanism $f(\alpha)$ [4, 5, 6].

To identify these three parameters, the data of the derivatographic experiment conducted in the non-isothermal mode are used, when the change in the degree of transformation of the material α over time t is described by the differential equation

$$\frac{d\alpha}{dt} = A \cdot \exp\left(-\frac{E_a}{RT}\right) \cdot f(\alpha), \quad R = 8.3144621 \ \frac{\mathrm{J}}{\mathrm{mol} \cdot \mathrm{K}}. \tag{1}$$

Under the temperature-linear heating where the rate of temperature increase T with a given parameter $\beta = dT/dt$ and a specific mechanism $f(\alpha) = (1-\alpha)^n$, the kinetic equation (1) can be represented as

$$\frac{d\alpha}{dt} = \frac{A}{\beta} \cdot \exp\left(-\frac{E_a}{RT}\right) \cdot (1-\alpha)^n, \tag{2}$$

where n – order (index) of reaction.

Equation (2) has no analytical solution and numerical methods or approximate analytical ones are used to determine the kinetic parameters.

The aim of this paper is to represent a quantitative result analysis of the derivatographic study of oxidation on the basis of a new approach to calculating the kinetic parameters of a

---

* Corresponding author: bvborisov@tpu.ru

heterogeneous reaction, as well as to obtain relationships for the estimation of these parameters.

Identification of the kinetic parameters is performed with measured values $\{T_i, \alpha_i\}$ ($i = 0,1,\ldots,N$ and is determined by minimizing the coefficient of variation calculated in accordance with (2) the values of the frequency factor A_i

$$A_i = \beta \frac{\int_{\alpha_0}^{\alpha_i} \frac{d\alpha}{(1-\alpha)^n}}{\int_{T_0}^{T_i} \exp\left(-\frac{E_a}{RT}\right) dT}, \quad i = 1,\ldots,N. \tag{3}$$

The integral in the numerator of equation (3) has a following quadrature

$$\int_{\alpha_0}^{\alpha} \frac{d\alpha}{(1-\alpha)^n} = -\begin{cases} \ln[(1-\alpha_0)/(1-\alpha)], & \text{if } n = 1, \\ [(1-\alpha_0)^{1-n} - (1-\alpha)^{1-n}]/(1-n), & \text{if } n \neq 1. \end{cases}$$

The denominator integral of the Arrhenius shape is calculated by approximation [7]

$$\int_0^T \exp\left(-\frac{E_a}{RT}\right) dT \approx \frac{RT^2}{E_a} \cdot \frac{\frac{E_a}{RT} + 0.25403 \ln\left(\frac{E_a}{RT}\right) + 0.36665}{\frac{E_a}{RT} + 0.24598 \ln\left(\frac{E_a}{RT}\right) + 2.41457} \exp\left(-\frac{E_a}{RT}\right),$$

that provides calculation accuracy with a relative error approximately estimated to 10-4%.

Fig. 1 shows a solution of the example from the monograph [8] by using the method described above; discrepancy between parameters n and E_a is less than 0.01%.

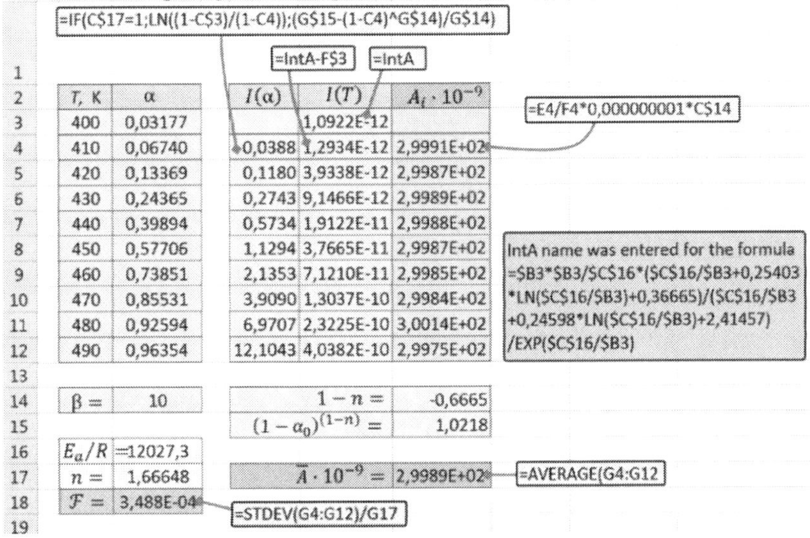

1						
2	T, K	α		$I(\alpha)$	$I(T)$	$A_i \cdot 10^{-9}$
3	400	0,03177			1,0922E-12	
4	410	0,06740		0,0388	1,2934E-12	2,9991E+02
5	420	0,13369		0,1180	3,9338E-12	2,9987E+02
6	430	0,24365		0,2743	9,1466E-12	2,9989E+02
7	440	0,39894		0,5734	1,9122E-11	2,9988E+02
8	450	0,57706		1,1294	3,7665E-11	2,9987E+02
9	460	0,73851		2,1353	7,1210E-11	2,9985E+02
10	470	0,85531		3,9090	1,3037E-10	2,9984E+02
11	480	0,92594		6,9707	2,3225E-10	3,0014E+02
12	490	0,96354		12,1043	4,0382E-10	2,9975E+02
13						
14	$\beta =$	10			$1-n =$	-0,6665
15					$(1-\alpha_0)^{(1-n)} =$	1,0218
16	E_a/R	=12027,3				
17	$n =$	1,66648			$\bar{A} \cdot 10^{-9} =$	2,9989E+02
18	$\mathcal{F} =$	3,488E-04				
19						

Fig. 1. Screenshot showing the solution of identification of kinetic parameters.

The minimization of the spread of the calculated values (on fig.1 this functional – the objective function – is represented by F) was conducted using Solver tool in the mode (for the given case) of the nonlinear generalized decreasing gradient method [9] for the scalable values of E_a/R and n.

To estimate the kinetic parameters of the reactions A, E_a and n the degree of conversion α as a function of temperature in some approximation can be described (approximated) by equation.

$$\alpha = \frac{1}{1 + \exp\left[\frac{-2\pi(T - T_0)}{\Delta T}\right]}, \quad T \in \left[T_0 - \frac{\Delta T}{2};\, T_0 + \frac{\Delta T}{2}\right], \tag{4}$$

that depends on two parameters: distinctive temperature T_0 of the highest reaction rate, that defines half reaction of a matter $\alpha = 0.55$ and temperature range ΔT of an area $(1 + e^{\pi})^{-1} \le \alpha \le (1 + e^{-\pi})^{-1}$ of aprox. 90% of material transformation (fig. 2).

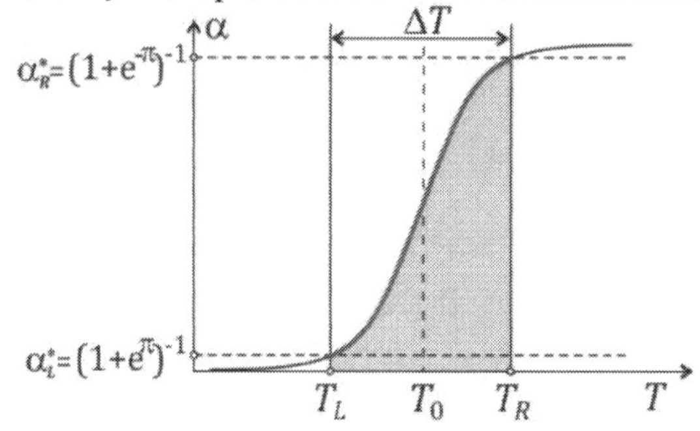

Fig. 2. Qualitative character of the derivatogramm

Using the approximation differential (4) and the kinetic equation of the process (2) for the representative values of the temperature $T_L = T_0 - \Delta T/2, T = T_0$ and $T_R = T_0 + \Delta T/2$ we can determine the main parameters of the reaction:

$$n = \frac{2 \cdot \ln\left(\mathrm{ch}\frac{\pi}{2}\right)}{\ln\left(\mathrm{ch}\frac{\pi}{2}\right) + \frac{\pi}{4}\frac{\Delta T}{T_0}} \approx \frac{2}{1 + 0.8537\frac{\Delta T}{T_0}},$$

$$E_a = \pi n R \frac{T_0^2 - (\Delta T/2)^2}{\Delta T}, \tag{5}$$

$$A = 2^{n-1}\frac{\pi\beta}{\Delta T} \cdot \exp\left(\frac{E_a}{RT_0}\right).$$

Output parameters T_0 and ΔT for the ratios (5) are defined by temperatures T_L and T_R using for example a method of proportional parts of initial relation $\{T_i; \alpha_i\}$.

The analysis of the parameters of the same example [4] according to the ratios (5) results in the following deviations from the precise solution: 3% deviation from the reaction order n and 2.9% deviation from the activation energy E_a.

Thus, in this paper the method for identification of the triplet of kinetic parameters of a heterogeneous reaction using the data of the derivatographic analysis is proposed. This method is characterized by high ac-curacy and relative simplicity and it can be effectively realized using MS Excel software, which is available on almost every personal computer whether at a laboratory, school or at home. In addition, the method of approximate estimation of kinetic parameters on the basis of dependence approximation between the degree of material transformation and the temperature of the logistic curve is proposed.

This work was supported in part by the Russian Foundation for Basic Research (Project No. 16-38-60031 mol_a_dk).

References

1. S. Vorozhtsov, I. Zhukov, A. Vorozhtsov, A. Zhukov, D. Eskin, A. Kvetinskaya, Adv. Mater. Sci. Eng. **2015**, 718207 (2015)
2. I. Zhukov, S. Vorozhtsov, V. Promakhov, I. Bondarchuk, A. Zhukov, A. Vorozhtsov, JPCS **652** (2015)
3. A. B. Vorozhtsov, M. Lerner, N. Rodkevich, H. Nie, A. Abraham, M. Schoenitz, E. L. Dreizin, Thermochim. Acta. **636**, 48 (2016)
4. I. S. Bondarchuk, I. A. Zhukov, V. V. Promakhov, S. A. Vorozhtsov, *Multifunctional Chemical Materials and Technologies: Abstracts International Scientific Conference* (TSU, Tomsk, 2015)
5. J. Zaripov, B. Borisov, and S. Bondarchuk, EPJ Web Conf. **76**, 01032 (2014)
6. I. S. Bondarchuk, I. A. Zhukov, V. V. Promakhov, *PFSD-2018* (TSU, Tomsk, 2015)
7. C. Junmeng, H. Fang, W. Yi, Y. Fusheng, Chem Eng J. **124**, 18 (2006)
8. J. E. House, *Principles of Chemical Kinetics* (Academic Press, Cambridge, 2007)
9. I. S. Bondarchuk, I. A. Kurzina, S. S. Bondarchuk, Higher Education Today **9**, 22 (2014)

Numerical research of metal tungsten fluoride process

Roman Brendakov[1], *Alexander* Shvab[1], and *Vladimir* Brendakov[1,*]

[1]National Research Tomsk State University, 634030 Tomsk, Russia

Abstract. In this paper we consider the mathematical modelling of the process of obtaining tungsten hexafluoride gas, based on the reaction between tungsten powder and fluorine in a chemical reactor, which has a rectangle cross section channel as a working area. The modelling is conducted on the basis of solving three-dimensional equations of impulse, heat and substance transfer in the binary compound, taking into account the heterogeneous reaction undergoing on the bottom of the channel, between the tungsten powder layer and the fluorine gas flow. As a result of numerical solving the problem, the speed, concentration and temperature vector fields' distributions were revealed, as well as integral characteristic of the completeness of fluorine use. The credibility of the results obtained is confirmed by test researches and the comparing of calculation results with the empirical evidence.

1 Introduction

The obtaining of wear-resistant coatings, products made of refractory complex form materials or other complex parts of machines, working at high temperatures, pressures and shearing stresses is a pressing issue nowadays. As it is known, in practice there are lots of problems in the mechanical processing of refractory materials. However, the most promising way of getting wear-resistant coatings or products made of refractory materials has recently been a method, based on their precipitation from steam-gas phase with the help of the recovery of their fluorides or chlorides [1]. To carry out this method, it is necessary to initially undertake the process of fluoridation of metal tungsten. That is why the first phase of this method is the process of getting gaseous tungsten hexafluoride, which is formed as a result of the reaction between gaseous fluorine and tungsten powder. The process of tungsten fluoridation has the following reaction,

$$W + 3F_2 = WF_6 \tag{1}$$

which starts taking place at about 300 °C.

In the work the numerical calculation of tungsten powder fluoridation in the chemical reactor was made, which has an extended rectangle cross section channel as a working area.

*Corresponding author: vnbrendakov@mephi.ru

2 Mathematical model of the process

This process of tungsten powder fluoridation is made due to undergoing of the gaseous fluorine above the layer of tungsten powder, which is at the bottom of the channel. Going through the active zone of the reactor, fluorine interacts with tungsten powder, forming tungsten hexafluoride. The aim of the work was the research of hydrodynamics and thermal mass transfer influence on the process of getting gaseous metal tungsten hexafluoride. To reveal this influence, the numerical solving of three-dimensional equation system of Navier-Stokes of heat and substance transfer with a heterogeneous reaction on the lower surface of the reactors working area was made. For more full use of fluorine in a chemical reaction, a very low speed of its movement in the reactor is set. Therefore, the components of the binary compound can be considered as an incompressible environment. However, the density of the compound changes sufficiently due to the change of concentrations of the compound components due to the chemical reaction (1). That is why we consider the density a variable, though dependant only on the compound components concentrations and using the concept of additivity for gaseous compounds, we assume the compound density to be:

$$\rho_{sm} = \rho_f m_f + \rho_g m_g,$$

where m_f and m_g are voluminous concentrations of fluorine and tungsten hexafluoride accordingly. If we take fluorine density as a scale and consider for the binary compound that $m_f + m_g = 1$, then we will have a dimensionless compound density ρas a result, which is

$$\rho = k_\rho + \left(1 - k_\rho\right) m_f$$

We impose relative mass fluorine concentrations $\varepsilon_f$and tungsten hexafluoride ε_g. Expressing the compound density through the mass concentration ε_f [2], we get:

$$m_f = \frac{\varepsilon_f}{\varepsilon_f + \dfrac{M_f}{M_g}\left(1 - \varepsilon_f\right)}$$

where M_f and M_g are molecular masses of fluorine and tungsten hexafluoride accordingly.

One of the efficient ways of solving equations of environment movement in the derivatives "speed-pressure" is a method of physical fission along the time of speed and pressure fields.

3 Numerical results

The credibility of the numerical solving of the liquid dynamics for the set course in the square section channel was defined by the matching with the analytical solution, which is presented in the Fig. 1. The analytical solution has the following form [3]:

$$u_x = \sum_{i=1,3,5...}^{\infty} \sum_{j=1,3,5...}^{\infty} \frac{\sin\left(\dfrac{\pi y i}{b}\right)\sin\left(\dfrac{\pi z j}{h}\right)}{i\,j\left(\dfrac{i^2}{b^2} + \dfrac{j^2}{h^2}\right)}. \quad (2)$$

Here, u_x is a ratio of dimensional speed to mid-dimensional and the dimensionless square sides are $b=B/H=1$ and $h=H/H=1$.

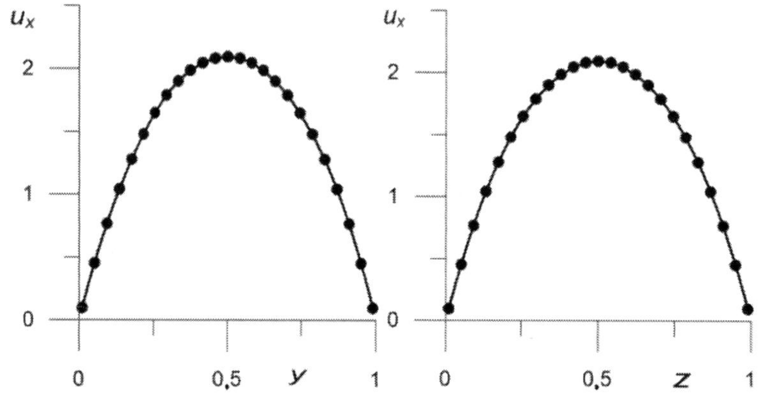

Fig.1. Comparison of the calculation (solid line) with the analytical dependence (2), shown by points, for steady flow in a channel with a square cross-section, a - on the central line in z, b - on the center line along y.

The credibility of the physics-mathematics model of the process of tungsten fluoridation offered was checked by comparing with experimental data [4] in distribution of the cross-section channel middle value of the mass concentration of tungsten hexafluoride, depending on the dimensionless channel length $x=X/L$. The comparison of the calculated and experimental data is shown in Fig.2.

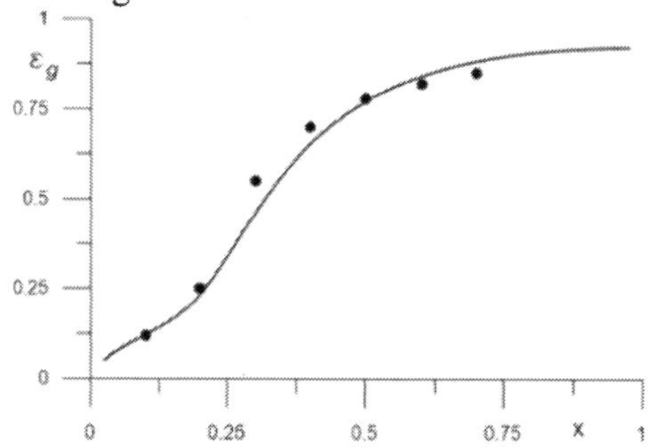

Fig. 2. The distribution of the average concentration of the concentration ε_g of tungsten hexafluoride as a function of the dimensionless length of the reaction zone x for parameters Re =11; k_ρ =7.6; m = 2; Pr = 1; Pr_d =1; k_μ = 0.75; U = 32951 J/mole; k_0= 438; k_c = 0.45.

The distribution of the isolines of the relative mass concentration ε_f, temperature isolines ϑ and compound densityρin the OXZ plane for the middle section of the channel, so$y$ =0.5 is shown accordingly in Fig.3-Fig.4.

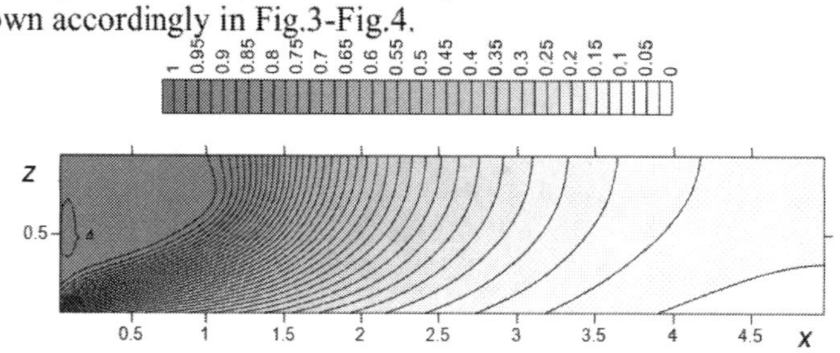

Fig. 3. The distribution of the isolines of the relative mass concentration of fluorine ε_f in the middle section of the channel y = 0.5.

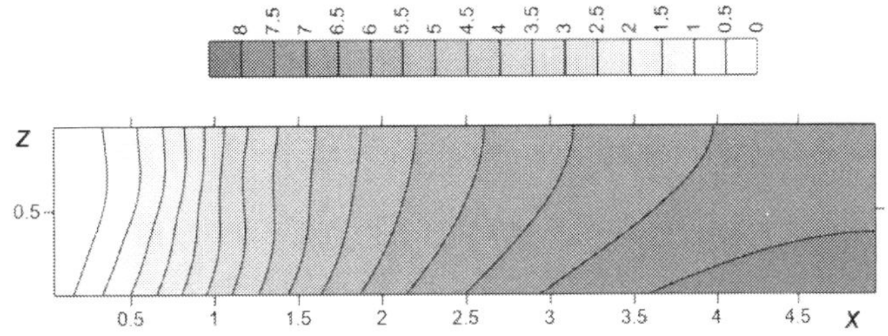

Fig. 4. The distribution of temperature isolines ϑ in the middle section of the channel $y = 0.5$.

The adequacy of the model created is confirmed by the results, shown in the Fig.1-2. In the Fig. 3-4 the typical distribution of the mass fluorine and temperature concentration in the reactor is depicted. As it is seen from the figures, the process of stationary layer of tungsten powder fluoridation by breezing it with gas fluorine is quite mixed. This is primarily connected with the non-linear temperature field nature along the working area of the chemical reactor due to the additional thermal flow output , connected with endothermic reaction, as well as the significant increase of the compound density and as a result, blocking the access of the fluorine component to the reaction area, An important factor, influencing on the efficiency of fluoridation process is the formation of the diffusion border layer, which is the reason for the hexafluoride formation speed decrease. A complicated non-linear nature of the fluoridation process implementation doesn't allow to effectively apply methods of the planning of the experiments for getting regressive dependencies, connecting the main indicators of the process with regulated factors (the fluorine expense, the process temperature, etc.) As a conclusion, mathematical modeling is the only way to get reliable qualitative information about the process of tungsten powder fluoridation by the elementary fluorine [5].

4 Conclusion

The developed mathematical model can be used for the optimization of the fluoridation process in chemical reactors, as well as it can become a source of getting new detailed information about the mass, impulse and energy transferring processes, which opens the perspectives for creating new original technological decisions, meanwhile providing a higher level of environment protection [6].

References

1. J.-H. Park, J. Chem. Eng. **19**, 391 (2002)
2. K. F. Pavlov, P. G. Romankov, A. A. Noskov, *Examples and tasks in the course of processes and machines of chemical technology* (Chemistry, Moscow, 1981) [in Russian]
3. B. S. Petuhov, *Heat exchange and resistence in laminar course of liquid in tubes* (Energy, Moscow, 1967) [in Russian]
4. Yu. M. Korolev, Russian journal of non-ferrous metal Eng. **4**, 23 (2016)
5. V. N. Brendakov, A. V. Shvab, A. A. Demidenko, N. S. Evseev, R. V. Brendakov Patent No. 2641596 Way of getting tungsten product by applying tungsten layer by layer and device for its implementation (2016)
6. Yu. M. Korolev, Russian journal of non-ferrous metal Eng. **6**, 29 (2016)

Mathematical modeling of processes of heat and mass transfer during drying of wood biomass

Natalya Ivanova[1,*], and *Elena* Bulba[1]

[1]National Research Tomsk Polytechnic University, 634050 Tomsk, Russia

Abstract. For the first time, a mathematical model for the drying of woody biomass during conductive heating with localization of the evaporation front has been formulated. The processes of moisture removal during the filtration of steam through the porous structure of the material at an ambient temperature of Te = 373 K were considered. Humidity was varied (in the range from 6% to 40%) and dimensions of wood blanks (Rd = 0.0035 - 0.035 m). Based on the results of numerical simulation, the conditions and characteristics (evaporation rate Wisp, drying time τ dry) of the process of moisture removal from wood biomass are determined. The mathematical model allows to calculate the drying time, as well as the mass evaporation rate for different sizes of wood sample, humidity and temperature conditions.

1 Introduction

The use of wood waste in the energy sector is ecologically and energy-efficient [1]. The latter is because that the wood biomass is carbon-neutral, the sulfur and nitrogen content in the wood is lower than the traditional solid (coal) [2], liquid (oil, fuel oil) and gaseous (natural gas) energy carriers [3, 4].

Progress in wood drying technologies is possible by studying the regularities of heat and mass transfer processes [5]. Many of physical and structural factors that find the result of intense thermal effects on wood make it impossible to study the mechanism of drying experimentally. For this reason, the most acceptable at this stage in the development of the theory and practice of the drying of moist wood biomass is the use of the apparatus of mathematical modeling.

But it is worth noting that the developed mathematical models of drying are based, as a rule, on substantially simplifying the procedure for modeling the assumptions [6]. For example, in [7], a one-dimensional model of heat and mass transfer of the processes occurring when drying wet wood biomass is considered. Application of the statement [7] is possible only when studying the processes of moisture removal in bodies possessing axial symmetry (a cylinder, a ball). In this case, it is impossible to consider the real configuration of the body under conditions of uneven surface heating.

In [8] the problem of heat and mass transfer is considered taking into account the deepening of the evaporation zone during drying. However, in the formulation of the

* Corresponding author: ivanova_tatka@mail.ru

problem, the authors of Ref. 8 neglected the temperature gradients in the moist zone. It was also believed that the heat supplied by convection is completely absorbed by the evaporation of water. In [9], a system of differential equations is proposed that describes the processes of heat and moisture transfer in capillary-porous materials. But the statement [9] did not take into account the influence of structural factors of wood on the characteristics of its drying.

The temperature non-equilibrium at the boundary of the front of intensive evaporation of water in a porous medium is taken into account in [10]. To solve the problem of motion of the evaporation front, the method of accumulating the time step was used. This approach allows us to take into account a sufficiently large group of factors that significantly influence the dynamics of drying, however, this method is possible only in the case of drying objects having the properties of axial symmetry (for example, a ball).

In this paper, we consider the processes of heat and mass transfer in the convective drying of a moist porous medium in a two-dimensional formulation. The purpose of this work is mathematical modeling of drying processes of wood biomass and analysis of energy costs for carrying out the process of moisture removal.

2 Formulation of the problem and method of solution

The wet wood biomass is considered as an object of research. At the initial time, a sample of wood (cylindrical shape) is introduced into the high-temperature medium and heated by convection. The latter leads to the initiation of a process of evaporation of moisture. The evaporation front moves from the surface into the interior of the wood billet. As a result, a porous skeleton with high thermal resistance is formed. The solution area can be conditionally divided into two zones: dry and moist wood. Water vapor, formed during evaporation, moves from the deep layers to the surface in the radial direction. The drying of the wood ends when the moisture evaporates completely.

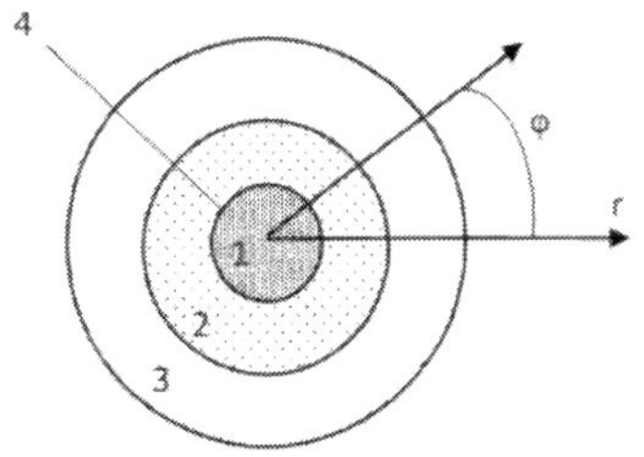

Fig. 1. Scheme of the problem solution area: 1) wet wood; 2) dry wood; 3) the region of saturation of the gaseous medium by water vapor; 4) the interface between the "wet - dry wood" system.

The drying process is described by the following system of non-stationary partial differential equations:

- the energy equation for woody biomass, taking into account the complex of heat and mass transfer processes occurring together, and the formation of a substantially heterogeneous structure, including the wet and dry parts [11]:

$$\rho\frac{\partial h}{\partial \tau}+\rho_s C_s\left(U_s\frac{\partial T}{\partial r}+\frac{V_s}{r}\frac{\partial T}{\partial \varphi}\right)\Psi\left(r_{eva}-r,\varphi_{eva}-\varphi\right)= \\ =\frac{1}{r}\frac{\partial}{\partial r}\left(\lambda r\frac{\partial T}{\partial r}\right)+\frac{1}{r^2}\frac{\partial}{\partial \varphi}\left(\lambda\frac{\partial T}{\partial \varphi}\right)-\frac{W_{eva}Q_{eva}\delta\left(r_{eva},\varphi_{eva}\right)}{\Delta}, \quad (1)$$

where ρ – density of wood, kg/m^3; h – enthalpy, J /kg; τ – time, s; ρ_s – density of water vapor, kg/m^3; C_s - heat capacity of steam, J/(kg·K); U_s – speed of steam in the radial direction, m/s; T – is the temperature at the boundary of evaporation, K; r – radial coordinate, m; V_s - speed of steam in the tangential direction, m/s; φ – is the azimuthal coordinate; ψ – is the Heaviside function; r_{eva} –is the coordinate of the evaporation front, m; λ – thermal conductivity of wood, W/(m·K); W_{eva} – mass evaporation rate, kg/(m^2·s); Q_{eva} – is the heat of vaporization, Дж/кг; δ – is the Dirac function; Δ – is the transformation parameter of the evaporation front;

- energy equation for the gas layer around the wood [12]]:

$$\rho\frac{\partial h}{\partial \tau}+\rho\left[U\frac{\partial h}{\partial r}+\frac{V}{r}\frac{\partial h}{\partial \varphi}\right]=\frac{\lambda}{r}\frac{\partial}{\partial r}\left(r\frac{\partial T}{\partial r}\right)+\frac{\lambda}{r^2}\frac{\partial^2 T}{\partial \varphi^2}; \quad (2)$$

- the equation of piezoelectric conductivity [13,14]:

$$\frac{\partial \Pi P}{\partial \tau}=\frac{1}{r}\frac{\partial}{\partial r}\left(r\frac{K_p}{\beta\gamma}\frac{\partial P}{\partial r}\right)+\frac{1}{r^2}\frac{\partial}{\partial \varphi}\left(\frac{K_p}{\beta\cdot\gamma}\frac{\partial P}{\partial \varphi}\right)+\sum J_i\Psi\left(r_{eva}-r,\varphi_{eva}-\varphi\right), \quad (3)$$

where Π - porosity; K_p - is the coefficient of the proportionality of the medium; P – is the pressure, Pa; γ - kinematic viscosity; β - compressibility coefficient of water vapor; J_i- is the function determining the inflow or withdrawal of the mass.

- the mass evaporation rate was calculated similarly [15]:

$$W_{eva}=V_0\cdot exp\cdot\left(\frac{(Q_{eva}\cdot\mu\cdot(T-T_f))}{(R\cdot T_f\cdot T)}\right), \quad (4)$$

where W_0=2.55·10^{-5} – evaporation rate at freezing point (kg/m^2 s); μ=0.018 – – molar mass of water (kg/mol); T_f=273– freezing point of water (K); R=8.314 – is the universal gas constant (J/mol·K).

The coordinate of the evaporation boundary was found from the expression:

$$r_{eva}=r_0-\int_0^{\tau}Vd\tau, \quad (5)$$

where $V=\frac{W}{\rho w}$, linear velocity of evaporation front advance, m/s.

Boundary conditions:

$$t=0,\ 0<r<r_w,\ T(r,\varphi,0)=T_0,\ r_w<r<r_g,\ T(r,\varphi,0)=T_g; \quad (6)$$

$$t=0,\ 0<r<r_w,\ P(r,\varphi,0)=P_0=101325\ Pa$$

Border conditions:

$$0<\varphi<\pi,\quad \mathrm{r}=0,\quad \frac{\partial T}{\partial r}=0,\ \frac{\partial P}{\partial r}=0; \quad (7)$$

$$0<\varphi<\pi,\quad \mathrm{r}=r_d,\quad P(\varphi,r)=P_0;$$

$$\lambda_d\frac{\partial T_1}{\partial \varphi}=\lambda_w\frac{\partial T_2}{\partial r},\quad r=r_d; \quad (8)$$

$$T_1(r_d,\tau)=T_2(r_d,\tau);$$

$$r = r_g, \quad \frac{\partial^2 T}{\partial r^2} = 0, \quad 0 < \varphi < \pi; \tag{8}$$

$$\varphi = 0, \quad \varphi = \pi, \ 0 < r < r_g, \quad \frac{\partial T}{\partial \varphi} = 0.$$

Initial data:

The ambient temperature is $T_e = 373$ K; Specific heat of vaporization $Q_{eva} = 2.5 \cdot 10^6$ J / kg; a sample of wood with radius Rd = 0.0035-0.035 m; The humidity of the sample was considered in the range φ = 6-40%.

The thermophysical characteristics of the interacting substances were taken as follows [16]: $\lambda_w = 0.598$ W / m · K; $\rho_w = 998.2$ kg / m^3; $C_w = 4190$ J / kg · K; $\lambda_{air} = 0.026$ W / m · K; $\rho_{air} = 1.205$ kg / m^3, $C_{air} = 1005$ J / kg · K; $\lambda_d = 0.4$ W / m · K; $\rho_d = 510$ kg / m^3, $C_d = 2400$ J / kg · K.

The system of equations (1-6) is solved by the method of finite differences using the algorithm [17]. Equations (1-3) are solved using quasilinear evaporation frontification using an implicit four-point difference scheme [18]. The process of heat and mass transfer in wood biomass under drying conditions is non-linear, which is due to the joint effect of a complex of physical processes occurring under conditions of intense phase transformations. As a result of evaporation, a very small, almost infinitely thin zone with an intense heat sink is formed in the porous structure of the wood [19]. Therefore, the drying process is one of the costliest technological stages in the preparation of fuelwood.

It should be noted that the results of mathematical modeling correlate well with known (published earlier) experimental data [20]. The drying time established in experiments [20] (under conditions of natural convection) is 9600 seconds. The moisture content of the wood preform was $\varphi \approx 40\%$, the temperatures in the drying chamber varied [20] in the range from 333 to 413 K.

Numerical studies were carried out in a fairly wide range of changes in the main parameters and process characteristics at atmospheric pressure of the environment.

3 Results and discussion

Based on the results of numerical simulation, the values of the total drying time for samples of woody biomass of cylindrical shape from 0.0035 m to 0.035 m were obtained.

Fig. 2 shows the characteristic drying times of the samples, determined from the results of a numerical solution of Eqs. (1-6).

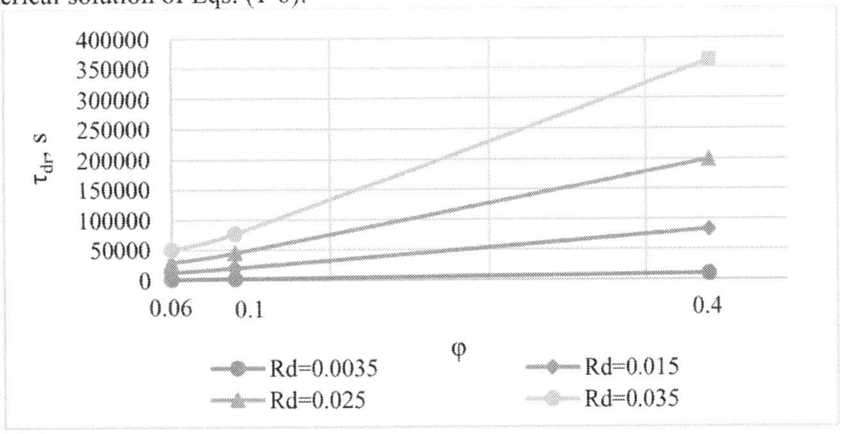

Fig. 2. Dependence of drying time of wood biomass from moisture at typical characteristic sizes of samples and ambient temperature $Te = 373$ K.

Analysis of the results showed that the wood blank of a cylindrical shape with a radius of Rd = 0.035 m with a moisture content of 6% is completely dried in 14 hours, and at a humidity of 40%, the drying time is 100 hours. Accordingly, it can be said that increasing the humidity by almost 7 times (from 6% to 40%) results in a seven-fold lengthening of the drying period.

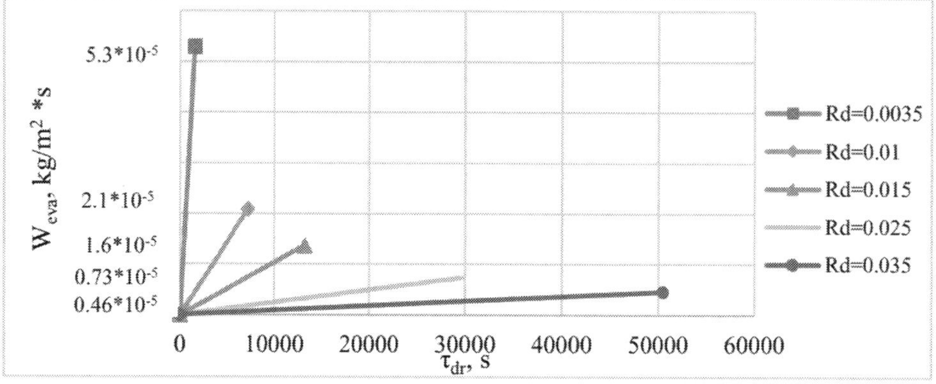

Fig. 3. Dependence of the mass evaporation rate on the drying time at ambient temperature $Te = 373$ K and moisture content $\varphi = 6\%$ for the characteristic size of wood biomass.

The curves in Fig. 3 results allow us to conclude that from a sample of wood of cylindrical shape with a radius of Rd = 0.0035 m with an initial humidity of 6 % and an evaporation rate of $W_{eva} = 5.3 \cdot 10^{-5}$ kg / m^2 · s, all moisture is removed in 25 minutes, from the sample by radius Rd = 0.035 m at $W_{eva} = 0.46 \cdot 10^{-5}$ kg / m^2 · s after 14 hours. Consequently, an increase in the size of the test sample by a factor of 10 leads to a substantial increase in the drying time (by more than 30 times).

4 Conclusion

A new mathematical model of the processes of heat and mass transfer, proceeding together when drying wet wood, taking into account nonequilibrium parameters at the boundary of phase transitions, is developed. Based on the results of mathematical modeling, the main integral characteristics of the processes of moisture removal from wood biomass are determined. Dependences of the drying time on the initial moisture content of the wood are obtained. It is shown that an increase in humidity from 6 % to 40 % of a cylindrical tree billet with a radius of 0.0035 m results in an extension of the total drying period from 25 minutes to 2.7 hours of drying (more than 6 times).

It is established that the dimensions of the wood biomass preform have a significant effect on drying characteristics. An increase in the radius of the wood billet by a factor of 10 (from 0.0035 to 0.035 m) at a humidity of 40% leads to a significant extension of the drying period (from 2.7 hours to 100 hours). Most likely, this is due to the fact that as a result of moisture removal a porous layer of dry material with high thermal resistance is formed. The latter leads to a significant delay in the heat transfer process. Also, as a result of numerical simulation, the values of the mass velocities of moisture removal from the wood billet are established for different sizes and humidity.

References

1. L.V. Zysin, N.L. Koshkin, Teploenergetika **4**, 23 (1997) [in Russian]
2. L.J. R Nunes, J.C. O Matias, P. S. Catalo, Renew Sustain. Energy Rev. **40**, 153 (2014)
3. M.V. Vasilevsky, E.G. Zykov, A.S. Razva, Theor. Found. Chem. Eng. **45**, 304 (2011)

4. A.S. Razva, M.V. Vasilevsky, V. Rykov, MATEC Web Conf. **23**, 01035 (2015)
5. O.R. Dorniak, Forest Journal **5**, 100 (2012) [in Russian]
6. R.G. Safin, R.R. Safin, R.R. Khasanshin, Forest Journal **4**, 64 (2006) [in Russian]
7. N.F. Timerbaev, A.N. Grachev, R.G. Safin, Chem. Chem. Technol. **49**, 103 (2006) [in Russian]
8. P.V. Akulich, N.N. Grinchik, Engineering Physics Journal **71**, 225 (1998)
9. M.M. Razin, Engineering and Physics Journal **74**, 290 (2001)
10. G.V. Kuznetsov, V.V. Salomatov, S.V. Syrodoy, The physics of burning and explosion **51**, 409 (2015)
11. S.V. Syrodoy, G.V. Kuznetsov, N.Y. Gutareva, K.A. Bugaeva, Combust. Sci. Technol. **190**, 4 (2017)
12. A.M. Grishin, S.P. Sinitsyn, I.V. Akimova, The physics of burning and explosion **6,** 17 (1991)
13. Y. Chen, K. Aanjaneya, A. Atreya, Safety Journal **91**, 820 (2017)
14. A. Atreyaa, P. Olszewskib, Y. Chena, H.R. Baumc, Int. J. Heat Mass Transfer **107**, 319 (2017)
15. I.I. Markov, A.A. Hashchenko, O.V. Vecher, Proceedings - North Caucasian Branch of Technological Sciences of the Russian Federation **6,** 48 (2002) [in Russian]
16. I.S. Grigoryev, E.Z. Meilikhov, *Physical quantities* (Energoatomizdat, Moscow, 1991) [in Russian]
17. S.V. Syrodoy, G.V. Kuznetsov, V.V. Salomatov, Heat power engineering **10**, 1 (2015)
18. V.M. Paskonov, V.I. Polezhaev, L.A. Chudov, *Numerical Modeling of Heat and Mass Transfer Processes* (Science, Moscow, 1984)
19. G.V. Kuznetsov, P.A. Strizhak, Engineering and Physics Journal **87**, 103 (2014)
20. E.E. Bulba, N.A. Ivanova, MATEC Web Conf. **141**, 01014 (2017) DOI:10.1051/matecconf/201714101014

Determination of heat-mass transfer coefficients within the apparatuses with jet-film contact devices

Ilnur N. Madyshev[1], *Oksana* S. Dmitrieva[2], and *Andrey* V. Dmitriev[2,*]

[1]Kazan National Research Technological University, 420016 Kazan, Russia
[2]Kazan State Power Engineering University, 420066 Kazan, Russia

Abstract. One of the ways of intensifying the heat and mass transfer processes in gas-fluid systems is designing jet-film contact devices with a developed surface of phase contact at high flow velocities. A contact jet-film device has been developed. The results of numerical investigation of the operation of this device are presented. There were determined the coefficients of heat transfer to the air from the surface of liquid inside of the drain cup of contact device. The criterion equations of convective heat transfer for engineering calculations of jet-film contact devices is corrected.

1 Introduction

Various heat-mass transfer processes and apparatuses are widely used in chemical, petrochemical, energy and other industries [1-3]. Efficiency of these processes is mainly determined by improvement of the used contact devices. Moreover, material and energy costs significantly depend upon the design of contact device [4-6].

Most of the currently used tray and nozzle contact devices have the following disadvantages. Separation areas shall be provided within the tray column apparatuses, which significantly increase the volume of apparatus, without being the main zone of interaction of gas-liquid flows. Nozzle columns have an uneven distribution of liquid and gas phases in the cross section of working area, which significantly reduces the efficiency of heat-mass transfer processes. In addition, at high gas flow rates, there is a problem of liquid droplets entrainment that largely restricts the increase in efficiency of the current heat-mass transfer devices and can lead to the "choking" of the column.

2 Description of jet-film contact device

In order to eliminate the specified disadvantages, the authors developed the jet-film contact device for the heat-mass transfer apparatuses [7], shown in Fig. 1. The proposed device consists of parallel square drain cups *1* with vertical walls, required for maintaining the

* Corresponding author: ieremiada@gmail.com

liquid level inside of them. Vertical perforated baffles *2* are the supports for drain cups *1* and have slotted holes for the installation of appropriate drain cups *1*. The drain cups *1* are open from the top end, but recurved leaves *3* in the form of circular segments are executed in the bottom of cups. These segments are required for the liquid distribution over the surface of vertical perforated baffles *2*.

Fig. 1. Jet-film contact device: *1* – drain cup; *2* – baffles; *3* – recurved leaves.

The developed jet-film contact device operates as follows. Liquid, through the plurality of recurved leaves *3*, executed in the bottom of drain cups *1*, is dispersed in the form of jets to the below located vertical perforated baffles *2*. When liquid jets move along the surface of vertical baffles *2*, the liquid is distributed with formation of a stable film flow. Thus, the coming down liquid film interacts with an ascending gas flow. After that the formed film, colliding with liquid surface inside of drain cups *1*, collapses. When distances between the drain cups, arranged at the same level, are equal to width of a drain cup, it provides stable, uniform gas flow, which leads to decrease in hydraulic resistance of the proposed jet-film contact device.

3 Analysis of results of numerical studies

An important step in the development of engineering methods for calculating the apparatuses with jet-film contact devices is to determine the heat-mass transfer coefficients from the flowing liquid film to the gas flow. In this regards, numerical studies were carried out by means of the software complex ANSYS Fluent, in which the interaction of the gas and liquid flows was simulated as exemplified by air-water system at the temperature of 20°C. The device consisted of 3 contact stages and in course of studies, geometrical dimensions of device were changed in the scale, proportional to its characteristic dimensions, namely to the width of the drain cup, taken equal to 100 mm and height of the wall equal to 10 mm. Moreover, actual gas rate was changed from 1 up to 20 m/s, which corresponded to the values of Reynolds number from $6.6 \cdot 10^3$ up to $6.0 \cdot 10^5$. The liquid level in the drain cup was taken as maximum.

In order to simulate the turbulent flows, modification of the two-parametric turbulence model k-ω SST, showing satisfactory conformance with experimental data, obtained in earlier studies, was used [8-10].

The research results showed that the well-known criterion equation for the case of liquid flow over the flat plate does not correctly describe the heat-mass transfer processes within the contact stage. The classical equation for determining the average heat transfer coefficient at turbulent air flow regime is the following [11, 12]:

$$\mathrm{Nu}_G = 0.032\mathrm{Re}_G^{0.8} \tag{1}$$

where $Nu_G = \alpha l/\lambda$ – Nusselt number; α – heat transfer coefficient, W/(m²·K); l – characteristic linear dimension (length of the plate), m; λ – heat conduction coefficient, W/(m·K); $Re_G = Wl/\nu_G$ – Reynolds number; W – actual gas rate, m/s; ν_G – coefficient of kinematic gas viscosity, m²/s.

The diagram in Fig. 2 shows that an increase in actual gas rate and geometrical dimensions of contact device leads to increase in values of Nusselt number. However, regarding the heat transfer coefficient values, their highest values correspond to the low heat transfer surfaces. Thus, for example, decrease in the surface of vertical baffles by 4 times leads to an increase in the values of heat transfer coefficients by 10.8-37.5%, depending on the actual gas rate.

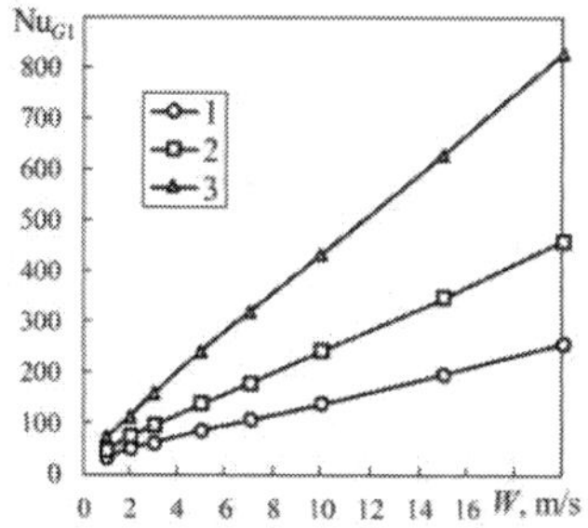

Fig. 2. Dependency of Nusselt number on actual gas rate at different geometrical constants of contact device M_b: 1 – 0.5; 2 – 1; 3 – 2.

Processing of the numerical simulation results allowed to correct the multiplier value and index of Reynolds number in the convective heat transfer equation in case of liquid flow over the flat plate for the jet-film contact device as follows:

$$Nu_{G1} = 0.0184 Re_G^{0.8038}, \tag{2}$$

approximation accuracy coefficient of which was 0.9989 in the range of Reynolds numbers from $6.6 \cdot 10^3$ up to $6.0 \cdot 10^5$ (Fig. 3a). It shall be noted, that the width of drain cup was taken as characteristic linear dimension, equal to the length of plate.

In the course of numerical studies, there were also determined the coefficients of heat transfer to the air from the surface of liquid inside of the drain cup. Processing of the numerical simulation results allowed to obtain the following criterion equation:

$$Nu_{G2} = 0.0052 Re_G^{0.8685}, \tag{3}$$

approximation accuracy coefficient of which was 0.997 in the range of Reynolds numbers from $6.6 \cdot 10^3$ up to $6.0 \cdot 10^5$ (Fig. 3b).

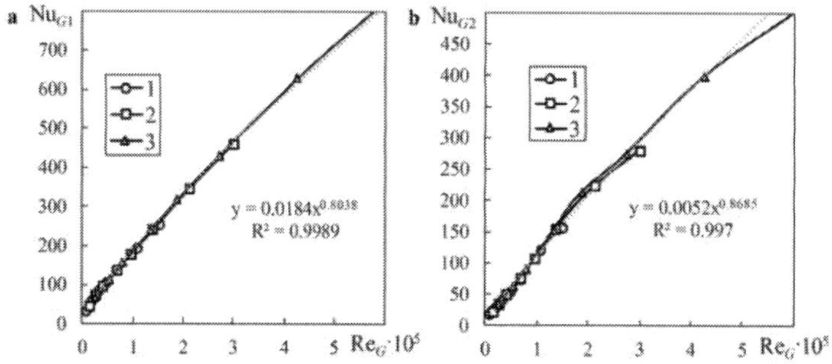

Fig. 3. Dependency of Nusselt number on Reynolds number at the heat transfer from the surface of vertical baffles (a) and from the surface of liquid inside of the drain cup (b) at different geometrical constants of contact device M_b: 1 – 0.5; 2 – 1; 3 – 2.

Comparing the heat transfer coefficients from the surface of vertical baffles and from the surface of liquid inside of the drain cup, it can be seen that, upon other equal conditions, the heat transfer from the surface of the baffles is the most intensive. Thus, for example, when the width of drain cup is equal to 100 mm, the heat transfer coefficients from the surface of vertical baffles by 1.54-2.08 times greater than the similar coefficients from the liquid surface depending on the Reynolds numbers taken.

The mass transfer coefficient in the gas phase can be determined on the basis of a well-known Lewis analogy:

$$\beta_y = \frac{\alpha}{\rho_G c_p}\left(\frac{\text{Pr}}{\text{Sc}}\right)^{2/3}, \tag{4}$$

where ρ_G – gas density, kg/m^3; c_p – heat capacity at constant pressure, J/(kg·K); Pr = ν_G/a_G – Prandtl number; Sc = ν_G/D_G – Schmidt number; a_G – temperature conductivity coefficient, m^2/s; D_G – diffusion coefficient, m^2/s.

4 Conclusion

Thus, it can be stated that the use of corrected criterion equations of convective heat transfer will improve the accuracy of engineering calculations for design of heat-mass transfer apparatuses with the proposed jet-film contact devices.

The reported study was funded by RFBR, according to the research project No. 16-38-60081 mol_a_dk.

References

1. Z. J. Wei, Z. L. You, S. Q. Gui, Ind. Eng. Chem. Res. **49**, 3732 (2010)
2. K. Ponomarev, E. Orlova, D. Feoktistov, EPJ Web Conf. **110**, 01060 (2016)
3. V. Maksimov, T. Sharifulin, L. Zhidkova, K. Eloyan, EPJ Web Conf. **159**, 00033 (2017)
4. J. Mackowiak, Chem. Eng. Res. and Des. **89**, 1308 (2011)
5. A. E. Orlando, L. C. Medina, M. F. Mendes, E. M. A. Nicolaiewsky, Braz. J. Chem. Eng. **26**, 619 (2009)
6. G. Q. Wang, X. G. Yuan, K. Yu, Ind. Eng. Chem. Res. **44**, 8715 (2005)
7. A. V. Dmitriev, O. S. Dmitrieva, I. N. Madyshev, A. N. Nikolaev, L. V. Kruglov, Patent RU No. 171022 (2017)
8. E. Maslov, I. Zharova, B. Borisov, N. Fedotova, EPJ Web Conf. **76**, 01027 (2014)
9. B. V. Borisov, EPJ Web Conf. **110**, 01012 (2016)
10. O. S. Dmitrieva, A. V. Dmitriev, I. N. Madyshev, A. N. Nikolaev, Chem. Pet. Eng. **53**, 130 (2017)
11. G. V. Kuznetsov, J. Eng. Phys. Thermophys **89**, 1241 (2016)
12. O. S. Dmitrieva, A. V. Dmitriev, L. V. Kruglov, Procedia Eng. **150**, 753 (2016)

Combustion of mixtures containing the activated aluminium powder

Yana Dubkova[1,*], *Andrei* Mostovchikov[2], and *Valery* Kuznetsov[1]

[1]National Research Tomsk State University, 634050 Tomsk, Russia
[2]National Research Tomsk Polytechnic University, 634050 Tomsk, Russia

Abstract. The ballistic characteristics of the mixture compositions containing aluminum powder ASD-6M have been studied. It is shown that the activation of aluminum powder by microwave radiation leads to an increase in the combustion heat of the mixture composition by 11%.

1 Introduction

Nanoscale and micron powders of metals are increasingly used in various fields of engineering, chemical and metallurgical industries, medicine, metal-working manufacturing, etc. Also, metal powders are widely used in power engineering: they are used as an additive that increases the energy of combustion of high-energy materials. Thus, an increase in the stored energy in the added metal powder will allow the intensification of the combustion of high-energy material. It has now been established that the effect of accelerated electron flows [1], as well as microwave radiation [2], makes it possible to increase the specific thermal effect of oxidation of micro- and nanopowder metals. Thus, it is relevant to study the effect of the irradiation of a metal powder on the energy of combustion of a high-energy material that has such a modified powder in its composition.

The purpose of this work is to study the combustion rate and c heat of combustion of high-energy materials containing in their composition the micron powder of aluminum type ASD-6M, both untreated and processed by microwave radiation.

2 Materials and methods of the experimental research

To conduct experiments on the activation of aluminum powder, microwave electromagnetic radiation with a carrier frequency of 2.8 GHz, a pulse repetition frequency of 25 Hz, and a pulse duration of 25 ns was used. Scheme experimental stand for irradiation experiments powders previously described in [2].

Pre-measured the initial dispersion and the activated alumina powder microwaves ASD-6M laser diffraction instrument “Mastersizer 2000”.

The efficiency of the microwave activation of ASD-6M powder was determined using the differential thermal analysis method on the STD Q600 instrument to change the value of the exothermal effect of aluminum oxidation when a sample weighing 5 mg was heated

* Corresponding author: y.a.dubkova@niipmm.tsu.ru

in air at a rate of 10 °C/min. The error in determining the values of the thermal effects was ± 1.8%.

To conduct studies on the calorific value and the burning rate, mixtures were prepared consists the basis of ammonium perchlorate and divinyl rubber containing 15 wt. % of the investigated aluminum powder.

After curing the samples, tests were conducted to determine the burning rate in air at atmospheric pressure and tests to determine the calorific value of combustion in a calorimeter bomb. All studies to determine the calorific value were carried out in accordance with the procedure described in [3, 4].

3 Experimental results

Results of powder analysis of ASD-6M and ASD-6M SHF (super-high frequencies) powders are presented in the form of histograms of the particle size distribution in Fig. 1 and in the form of a table (Table 1). It can be noted that the average volume diameter D (4.3) of the particles of the activated aluminum powder is 22% less than the original one. The ASD-6M microwave powder differs from the ASD-6M powder with a more developed surface and a narrower distribution width. Table 1 shows that the difference between the ASD-6M and ASD-6M microwave powders in D (0.50) is 19.5%, in D (3.2) about 18.3%, D (4.3) is 21.7%.

Table 1. Key data of ASD-6M and ASD-6M SHF powders dispersion analysis.

Characteristic, μm	**Type**	
	ASD-6M	ASD-6M (SHF)
D(0.00)	0.42	0.42
D(0.10)	1.95	1.58
D(0.50)	4.19	3.37
D(0.90)	7.82	6.31
D(0.100)	24.50	11.53
D_{10}, mean diameter	1.14	1110
D_{32}, average surface diameter	3.32	2.71
D_{43}, average volume diameter	4.74	3.71

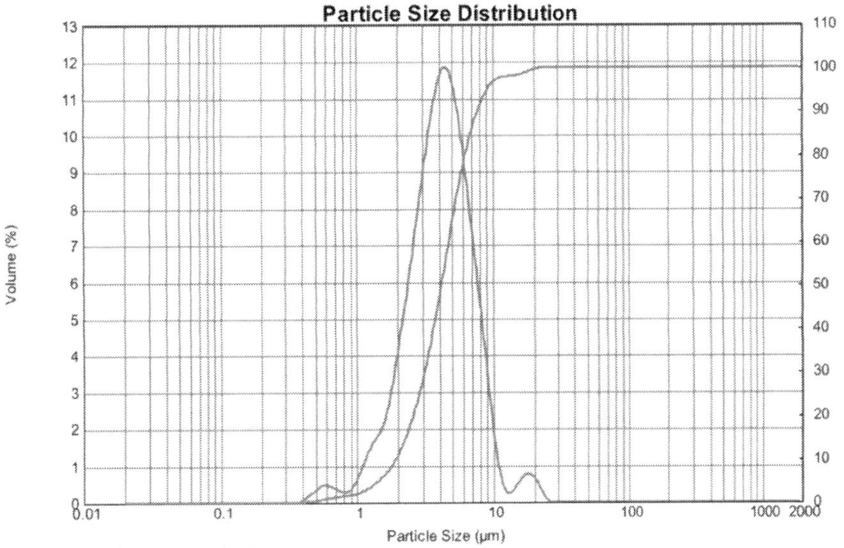

(a)

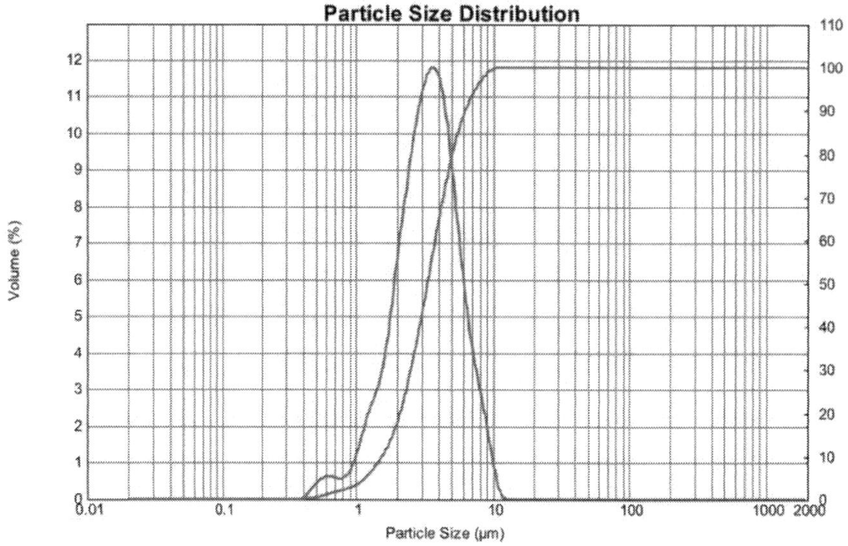

(b)

Fig. 1. Histogram of the distribution of aluminum powder particles: (a) ASD-6M, (b) ASD-6M(SHF).

Fig. 2 shows thermograms of thermal decomposition of aluminum nanopowder before (a) and after irradiation (b). When comparing these thermograms, it is seen that in the irradiated powder, the mass gain is less by 3.1% than in the irradiated powder. At the same time, in the irradiated aluminum powder, the specific thermal oxidation effect (total for two stages) increased by 780 J / g (~ 10.3%). In addition, the specific thermal effect of aluminum melting in micron powder decreased by ~ 9.9%. Taking into account that the degree of oxidation slightly decreased after irradiation, the increase in the heat of oxidation and the decrease in the heat of fusion can be explained by the additional allocation of the energy of radiation stored by the powder. Thus, this additional stored energy of radiation can be used to increase the energy of oxidation of mixed fuel compositions.

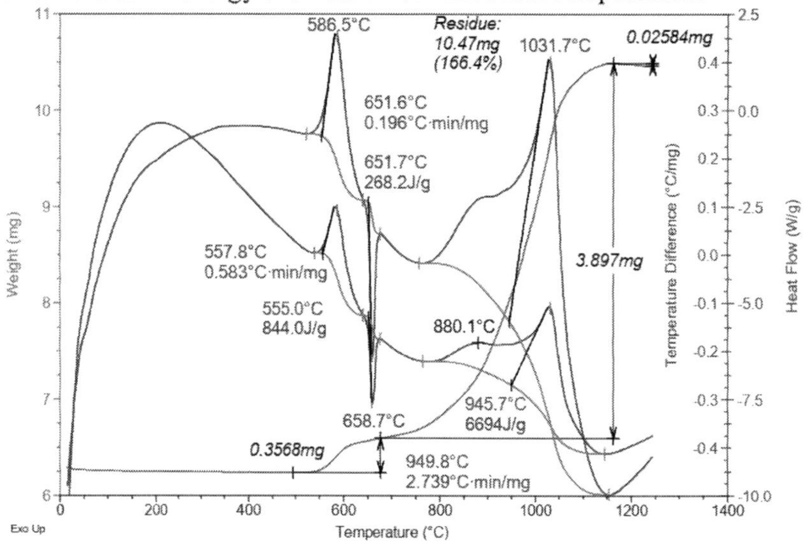

(a)

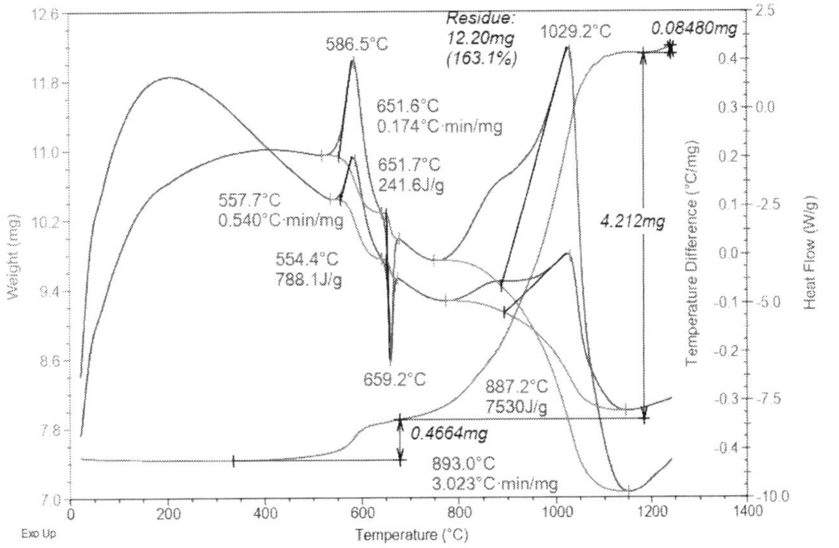

(b)

Fig. 2. Aluminum powders thermograms: (a) ASD-6M, (b) ASD-6M (SHF).

During the studies of heat and combustion rate, it was experimentally established that the linear combustion rate of samples containing 15% aluminum powder is practically independent of the type of powder (initial and activated by microwave radiation). However, the heat of combustion of the blended composition containing the activated aluminum powder is 11% higher than the heat of combustion of the composition with the original aluminum powder.

A summary of the study results o is presented in Table 2.

Table 2. The results of the investigation of aluminum powder and compositions based on it.

Aluminum powder characteristics				**Mixtures characteristics**	
Powder type	Particle size, *D* (4.3)	Oxidation onset temperature, $T_{o.t.}$, °C	Oxidation thermicity, ΔH, J/g	Burning rate, mm/s	Heat of combustion, MJ/kg
ASD-6M	4.74	~ 540	7270	1.57±0.04	5.60±0.1
ASD-6M SHF	3.71	~ 435	8076	1.52±0.03	6.22±0.1

4 Conclusion

Thus, the conducted studies showed that the activation of aluminum powder ASD-6M by exposure to microwave radiation, leads to a change in its dispersion and an increase in the heat of combustion of the mixture compositions containing activated aluminum. It was found that activation of aluminum powder by microwave radiation increases its dispersion, and the average volume diameter of the particles is reduced by 22%. The oxidation temperature of the activated powder decreases by more than 100 °C, and the total thermal oxidation effect increases by 11%. Replacing the batch of industrial aluminum powder with the activated microwave radiation practically did not affect the value of the linear combustion rate of the mixture composition. When combusted in a calorimeter bomb, samples of fuel with activated aluminum are separated by 11% more heat compared to samples containing non-activated aluminum powder. Thus, the obtained results indicate that activation of aluminum powder ASD-6M with the help of microwave radiation increases heat release during oxidation of aluminum under non-isothermal conditions and during combustion in the composition of the mixture composition.

This work was financially supported by the Ministry of Education and Science of the Russian Federation, Project No. 11.1928.2017/4.6 and by the Ministry of Education and Science of the Russian Federation in terms of state order, project code 9.9036.2017.

References

1. A. P. Il'in, L. O. Root, A. V. Mostovshchikov, Tech. Phys. **57**, 1178 (2012)
2. A. V. Mostovshchikov, A. P. Il'in, P. Yu. Chumerin, Yu. G. Yushkov, V. A. Vaulin, B. A. Alekseev, Tech. Phys. Letters, **42**, 344 (2016)
3. Ya. Dubkova, A. Zhukov, MATEC Web Conf. **92**, 01068 (2017)
4. S. Basalaev, A. Usanina, Ya. Dubkova, MATEC Web Conf. **110**, 01022 (2017)

Experimental investigation of the process of hydrate-formation by the method of explosive boiling of liquefied freon 134a in the water volume during decompression

Anton V. Meleshkin[1,*], and *Dmitriy* S. Elistratov[1]

[1]Kutateladze Institute of Thermophysics, 630090 Novosibirsk, Russia

Abstract. The experiments on obtaining gas hydrate of refrigerant 134a were carried out by the method, based on explosive boiling-up of a layer of liquefied gas in a bulk of water at decompression. It is shown that this method combines several factors, leading to intensification of hydrate formation process, resulting in the fast gas hydrate growth. The effect of the decompression rate on the volume of produced hydrate was studied experimentally.

1 Introduction

As far back as in the beginning of the 20th century, the technological offers on the storage and transport of natural gas in the hydrate state appeared. This method attracts the experts attention for quite long time due to the property of gas hydrates to concentrate large volumes of gas at relatively low pressures. Without a pipeline a promising way of gas transportation is its conversion into the gas hydrate (solid) state and its transportation at the atmospheric pressure and low temperature (-10 ... -20°C) [1]. In [2-4] and some other paper, the properties of gas hydrates, basic conditions and features of their formation are described; the types of their crystallization, mechanisms of gas hydrate formation and decomposition are presented [5-11].

A shock-wave method for obtaining gas hydrates was proposed, the intensification of the hydrate formation process resulted from the imposition of several factors, such as an increase in the contact surface of the hydrate-forming gas-water due to the shattering of the bubbles by the shock wave, the transfer of the system far into the region of hydrate formation, and the acceleration of the bubbles with respect to water [12-13]. We continue to search for effective ways to intensify hydrate formation with the help of mechanical and thermal effects on gas-liquid media [14-17].

* Corresponding author: mav@itp.nsc.ru

2 Experimental setup and methods

The studies were carried out using the setup in the form of a high-pressure reactor of up to 25 MPa. The working section was made as a stainless steel vessel (diameter of 100 mm, height of 300 mm) with a water jacket for thermal stating (Fig.1).

The experiment has been performed in the following manner. Water in the amount of 200 ml was into the reaction vessel and cooled by the cryostat through the outer walls to the temperature of hydrate formation. Further, 200 g of Freon 134a from the tank was fed into the water-filled vessel (with the temperature of 2-8°C). The pressure in the tank is much higher than the pressure in the autoclave, and the temperature of gas in the tank is higher than the water temperature. Entering the autoclave, gas started pressurization. Gas was cooled to the temperature of water, causing it liquation and accumulation in the form of a layer on the autoclave bottom. Then the pressure was decreased, liquefied gas under the water layer started boiling over the entire volume, so that there was an increase in interfacial surface. On the bubble walls, the growth of the gas hydrate layer started. Due to active boiling, the bubbles with the hydrate film, formed on the interface, collided constantly and deformed, causing the hydrate film peeling, and the rate of hydrate formation did not decreased. The heat released during hydrate formation was compensated by heat absorption during boiling. Due to superposition of these processes, there was a fast increase in the mass of gas hydrate.

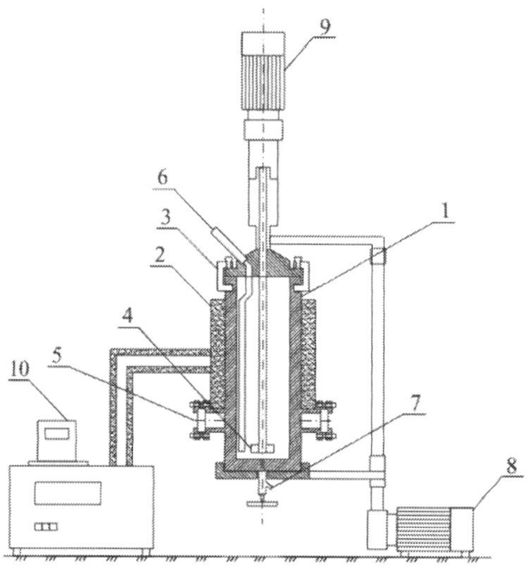

Fig. 1. Scheme of experimental setup: 1 – vessel; 2 – water jacket; 3 – lid; 4 – mixer; 5 – viewing windows; 6 – bottom tube; 7 – bottom valve; 8 – drive for working section lowering/lifting; 9 – mixer drive; 10 – cryostat.

In the course of experimental studies, gas-hydrates of Freon R134a, carbon dioxide and propane were obtained. In Fig. 2,3 shows the dependence of pressure and average temperature on the working site on time, when obtaining Freon gas hydrate 134a. The ideal formula for the unit cell of freon gas hydrate 134a is known to be 8CH2FCF3*136H2O, the crystal structure of hydrate II. Thus, the mass fraction of gas k in pure hydrate is 25%. As can be seen in Fig. 3, as a consequence of the decompression, a rapid pressure drop occurred in the system, which led to an explosive boiling of the liquefied gas. Intensive boiling resulted in a slight increase in pressure, after which the boiling intensity decreased and the pressure dropped to the ambient pressure. In addition, the average temperature decreased (due to phase transition and external cooling) and reached minus levels.

In Fig. 4 shows the same process on the P-T diagram. The process begins at a place where the coolant passes into the region where it exists in a gas-different state. It is worth noting that formation of gas hydrate is possible in both regions. But the process of explosive boiling initiated by decompression significantly improves the formation of the gas hydrate (the rate of hydrate formation increases by several orders of magnitude).

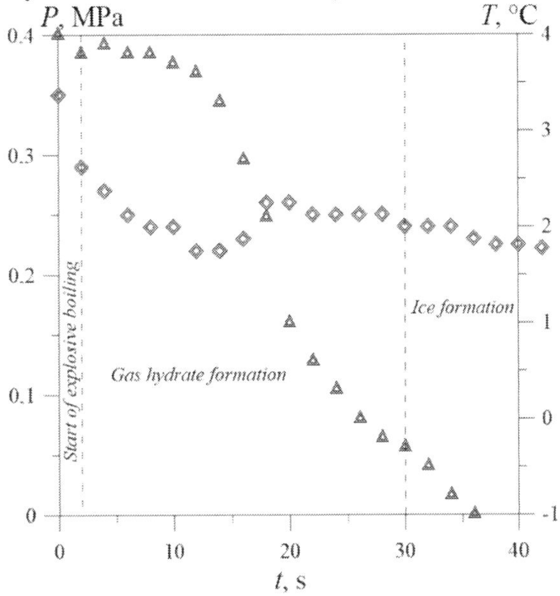

Fig. 2. Dependence of pressure P (rhombus) and average temperature (triangles) on the working site on time.

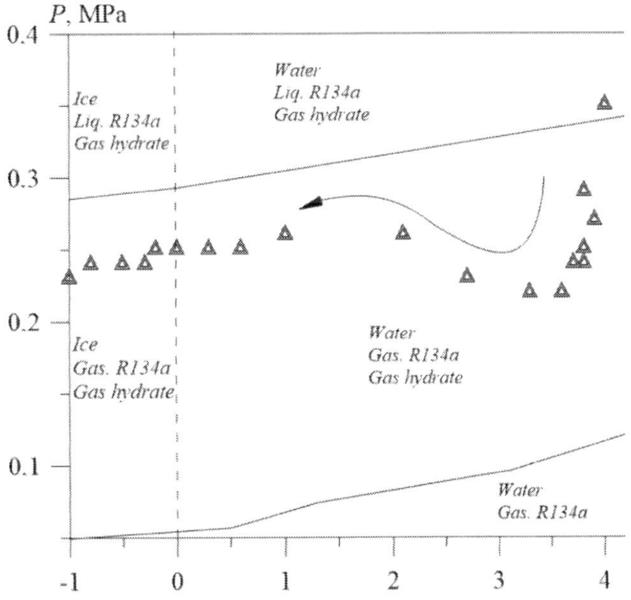

Fig. 3. P-T diagram of the process (triangles); the upper solid line is the Freon 134a boiling curve; the lower solid line is the equilibrium line of freon gas hydrate 134a.

The decomposition of the obtained sample showed that the initial mass concentration of the gas was about 20%. Thus, about 70% of the sample obtained in the experiment is a gas hydrate.

Conclusion

In this work, a new method of obtaining gas hydrates based on the boiling of liquefied gas under the surface of water as a result of decompression was experimentally investigated. The process of obtaining Freon 134a gas hydrate has been described, the advantages of this method over others have been revealed. The study of decomposition of the obtained gas hydrate samples showed that the gas mass in it was about 20%, which implies that the sample was 70% gas hydrate.

The study was funded by FASO Russia.

References

1. J. Gudmundsson, M. Mork, O. Graff, *Proc. Of the 4th Intern. Conf. on gas hydrates* (2002)
2. Y. F. Makogon, *Hydrates of Hydrocarbons* (PennWell Books, Tulsa, 1997)
3. V. S. Yakushev, V. A. Istomin, *Physics and Chemistry of Ice* (Hokkaido University Press, Hokkaido, 1993)
4. E. D. Sloan, C. A. Koh, 3rd ed Chemical industries series **721** (2008)
5. V. Sh. Shagapov, N. G. Musakaev, M. K. Khasanov, Int. J. Heat Mass Transfer **84**, 1030 (2015)
6. V. Sh. Shagapov, M. K. Khasanov, N. G. Musakaev, Int. J. Heat Mass Transfer **107**, 347 (2017)
7. N. G. Musakaev, M. K. Khasanov, S. L. Borodin, Int. J. Heat Mass Transfer **118**, 455 (2018)
8. M. Z. Faizullin, A. V. Vinogradov, V. P. Koverda, Chem. Eng. Sci. **130**, 135 (2015)
9. S. Y. Misyura, Sci. Rep. **6**, 30324 (2016)
10. V. E. Nakoryakov, S. Y. Misura, S. L. Elistratov, A. Yu. Manakov, A. A. Sizikov, J. Eng.Thermophys. **22**, 169 (2013)
11. V. E. Nakoryakov, S. Y. Misura, S. L. J. Eng. Thermophys. **25**, 24 (2016)
12. V. E. Dontsov, A. A. Chernov Int. J. Heat Mass Transfer **52**, 4919 (2009)
13. A. A. Chernov, V. E. Dontsov, Int. J. Heat Mass Transfer **54**, 4307 (2011)
14. A. A. Chernov, A. A. Pil'nik, D. S. Elistratov, I. V. Mezentsev, A. V. Meleshkin, M. V. Bartashevich, M. G. Vlasenko, Sci. Rep. **7**, 40809 (2017)
15. A. A. Chernov, D. S. Elistratov, I. V. Mezentsev, A. V. Meleshkin, A. A. Pil'nik, Int. J. Heat Mass Transfer **108**, 1320 (2017)
16. V. E. Nakoryakov, A. N. Tsoi, I. V. Mezentsev, A. V. Meleshkin, J. Eng. Thermophys. **23**, 1 (2014)
17. V. E. Nakoryakov, A. N. Tsoi, I. V. Mezentsev, A. V. Meleshkin, T and A **21**, 279 (2014)

Finite element analysis of a thermally insulated infrared radiant emitter

Anton N. Ermolaev[1,*], *Olga* V. Khaustova[2], and *Anastasia* P. Yakovets[2]

[1]Tyumen industrial University, 625000 Tyumen, Russia
[2]National Research Tomsk Polytechnic University, 634050 Tomsk, Russia

Abstract. The purpose of this work is to analyze the efficiency of the technical solution that is based on thermal insulation of the infrared radiant burner outer surface and to identify of the characteristics of combustion and heat and mass exchange processes. Calculation of the heat balance is performed for thermal capacities of 5, 10, 15, 20, 30, 40 kW and proceeding from the main goal that was to generate directed local heating of workplaces and production areas. We used Ansys Multiphysics software and Fluent CFD solver to implement finite element analysis. Calculation of the thermal insulation layer thickness for a given external surface temperature was additionally performed. The technical solutions offered provides an optimal thermal regime in the whole building and enable us to increase the efficiency of the high temperature infrared radiant emitter to 2-17% and a consequence of the system as a whole.

1 Formulation of the aim

Radiant heating systems based on gas-fired infrared heat emitters are used to effectively heat large-scale premises [1-5]. Temperature of emitter's ceramic surfaces can exceed 1000 °C [3, 4, 6]. This issue creates a list of problems that require a number of technical solutions to solve them [3-5, 7-10].

The purpose of this work is to analyze the efficiency of the technical solution that is based on thermal insulation of the infrared radiant burner outer surface (Fig. 1) and to identify of the characteristics of combustion and heat and mass exchange processes. Series of numerical and experimental studies was carried out using licensed software and a certified attorney base to solve the tasks [1]. The results of the development of a finite element parametric model of a gas infrared heating system containing a radiation source and irradiation surfaces are presented. Standard model of infrared radiant gas burner and its modification with a low thermal conductivity coefficient of the reflector were examined (Fig. 1, b).

Calculation of the heat balance is performed for thermal capacities of 5, 10, 15, 20, 30, 40 kW and proceeding from the main goal that was to generate directed local heating of workplaces and production areas. Calculation of the thermal insulation layer thickness for a

* Corresponding author: ermolaevanton03@gmail.com

given external surface temperature was additionally performed. The thermal insulation thickness is equal to 50 mm according to the calculation results.

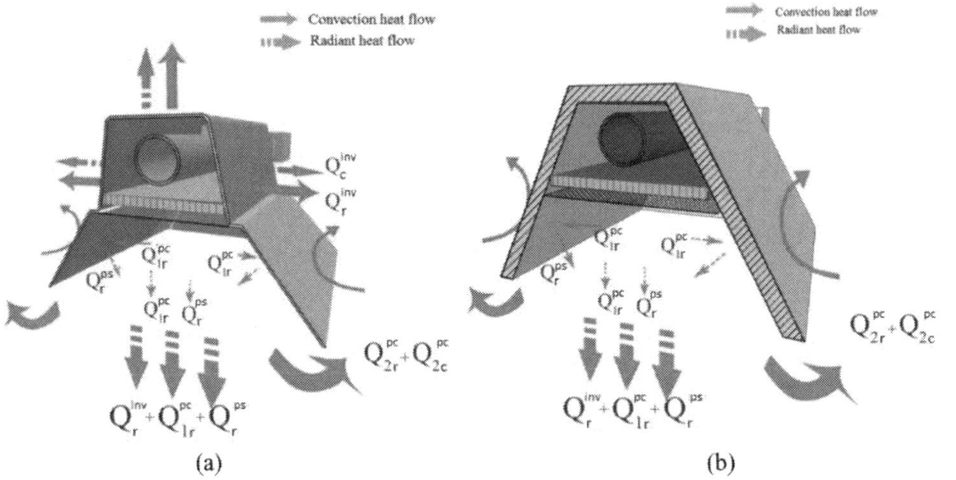

Fig.1. Diagram of the movement of heat flows during the operation of the standard (a) and isolated (b) model of high-temperature radiator.

2 Method of the research

We used Ansys Multiphysics software and Fluent CFD solver to implement finite element analysis. A number of heat transfer problems [3-6, 11-15] in ventilated rooms with radiation sources of heating have been solved by simulating the diffusion combustion processes in infrared radiant gas burner and turbulent gas flow using the Discrete Ordinates DO model for calculating radiant heat transfer:

$$\frac{dI(\vec{r},\vec{s})}{ds}+(a+\sigma_s)I(\vec{r},\vec{s})=an^2\frac{\sigma T^4}{\pi}+\frac{\sigma_s}{4\pi}\int_0^{4\pi}I(\vec{r},\vec{s}')\Phi(\vec{s},\vec{s}')d\Omega' \qquad (1)$$

where T – temperature; σ – The Stefan's constant; I – intensity of radiation; Φ – phase function; Ω – solid angle; r – position vector; s – direction vector of radiation; s' – direction vector of scattering; a – coefficient of absorption; n – refractive index; σ_s – coefficient of scattering.

To solve the differential equations of heat and mass exchange it was considered that the gas-fired infrared heat emitter is made of steel 12Cr18Ni10Ti of 2 mm, the gas pressure before the inlet nozzle was 1270 Pa, the air-fuel ratio (AFR) is 9.996, fuel is 100% methane and oxidizer is 21% oxygen and 79% nitrogen. Gas-fired infrared heat emitters with heat output of 5, 10, 15, 20, 30, 40 kW were investigated.

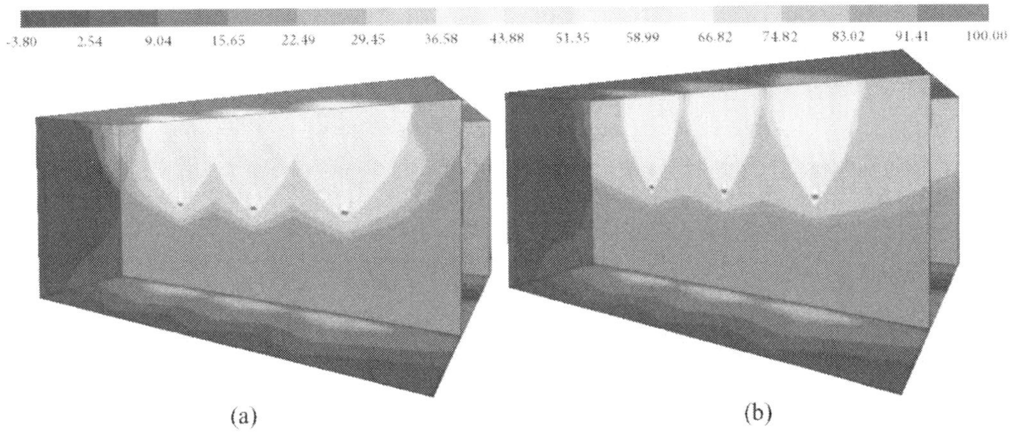

Fig. 2. Temperature fields in the volume and on the surfaces of the production room (°C) when the emitters of typical (a) and insulated (b) model with a capacity of 10 kW are placed at a height of 6 m.

3 Results and Conclusion

Analysis of the obtained data indicates that efficiency of radiator is increased with increasing of its thermal power. This is due to the fact that the heat exchange area (the area of the burner construction) increases with increasing of thermal power disproportionally. As a result, the heat losses of the burner outer surface are reduced to a unit of power. In turn, heat losses with outgoing gases increase because the number of ceramic plates increases with increasing of heat power with an average unit of power of 1.5 kW. It leads to increasing of the combustion products volume and to their less intensive dilution in the air masses.

The use of insulation has made it possible to achieve the following efficiency indicators:

• the volume of heat transferred in the direction of the working area was 3089-24715 W or 63% of the available heat;

• the heat losses with outgoing gases were 1841-14730 W or 37%;

• efficiency increase of 2-5% depends on the thermal output.

The increase of heat losses with outgoing gases in the insulated model is explained by a decrease of the heat transfer intensity from the reflector internal surface to the outer one. As a result, there is a decrease of the convective heat transfer from the combustion products of the region bounded by the reflector to the wall of the reflector inner surface.

The technical solutions offered provides an optimal thermal regime in the whole building and enable us to increase the efficiency of the high temperature infrared radiant emitter to 2-17% and a consequence of the system as a whole. In addition, a part of heat losses by the burner with outgoing gases will decrease, that will solve the problem of overheating of the upper room zone.

References

1. Y. M. Luo, L. Chevalier, F. Utheza, X. Nicolas, Int. Polym. Process. **30**, 554 (2015)
2. M. Ostafin, M. Balkenhol, J. Erlemeyer, M. Köhler. Mater. Werks. **41**, 11 (2010)
3. A. N. Ermolaev, O. V. Khaustova, MATEC Web Conf. **141**, 01019 (2017)
4. A. N. Ermolaev, S. A. Khaustov, MATEC Web Conf. **110**, 01024 (2017)

5. A. N. Ermolaev, S. A. Khaustov, MATEC Web Conf. **110**, 01038 (2017)
6. S. Dolgov, E. Savchenko, S. Khaustov, R. Tabakaev, A. Zavorin, EPJ Web Conf. **110**, 01074 (2016)
7. S. A., Khaustov Belousova Y. A., Buvakov K. V., Dolgih A. Y., Kulesh R. N. IFOST 2016 Proceedings **2**, 548 (2016)
8. J. A. Oraz, I. E. Korzilova, A. N. Ermolaev, S. A. Khaustov, A. Y. Dolgih, MATEC Web Conf. **92**, 01001 (2016)
9. K. A. Al-attab, J. C. Ho, Z. A. Zainal, Exp. Therm Fluid Sci. **62**, (2015).
10. D. Patino, B. Crespo, J. Porteiro, J. L. Miguez, Appl Therm Eng. **100**, 849 (2016)
11. P. S. Gergelizhiu, S. A. Khaustov, R. B. Tabakaev, P. Y. Novoseltsev, A. V. Kazakov, A. S. Zavorin, *Proc. of 2014 Int. Conf. on Mechanical Engineering, Automation and Control Systems* (Tomsk, 2014)
12. S. A. Khaustov, A. S. Zavorin, K. V. Buvakov, N. A. Zakharushkin, MATEC Web Conf. **19**, 01020 (2014)
13. S. A. Khaustov, A. S. Zavorin, K. V. Buvakov, L. D. Kudryashova, A. V. Tshelkunova, EPJ Web Conf. **82**, 01041 (2015)
14. S. A. Khaustov, A. S. Zavorin, K. V. Buvakov, V. A. Sheikin, EPJ Web Conf. **82**, 01039 (2015)
15. S. A. Khaustov, A. V. Kazakov, G. A. Cherkashina, L. A. Sobinova, EPJ Web Conf. **110**, 01028 (2016)

Improvement air condensers evaluation model

Svyatoslav Tsibulskiy[1,*], *Nikolay* Galashov[1], *Denis* Mel'nikov[1], *Alexandr* Kiselev[1], and *Al'bina* Bannova[1]

[1]National Research Tomsk Polytechnic University, 634050 Tomsk, Russia

Abstract. The results of analysis of the literature on the calculation of the heat transfer coefficient of an air condenser in the flow past a bundle of finned tubes by an air flow. The methods of calculation are disassembled, marked advantages and disadvantages of each. Calculations of the heat transfer coefficient for each method are given; the results compared with the experimental data.

1 Introduction

Air condensers (AC) are now widely used in a number of foreign TPPs with power units up to 1000 MW and at CCGTs. The main disadvantage of air condensers is a low heat transfer coefficient and large dimensions [1]. Therefore, it is advantageous to use them at TPPs, where turbines are installed with low steam flow rates to the condenser. This is primarily the turbines for utilization of combined-cycle gas turbines (CCGT), the steam consumption in the condenser is three times less than for conventional steam-turbine power units of the same unit capacity [2]. At the same time, TPP based on CCGT is becoming profitable to establish in the area of gas production and low temperatures of outside air.

One of the most serious problems of AC operation in winter during the condensation of water vapor is the condensate freezing in the tubes, which can lead to the destruction of tubes and the need to repair the condenser and stop the turbine. The solution to this problem is to use the AC in a CCGT with three cycles, where the low-boiling substances (LBS) [3] operates in the lower cycle, and its condensation can reliably pass at temperatures below -40 °C [4].

The problem of design and application of AC is the lack of reliable methods for their calculation. The most problematic issue is the cooling package finned tubes with air [5].

To create algorithm and program for calculating AC operating on the LBS, are made the analysis and a choice of the currently most acceptable methods for calculating a bundle of finned tubes when they are cooled by air.

2 Methods for calculating the heat transfer coefficient for a bundle of finned tubes

* Corresponding author: stzibulsky@tpu.ru

2.1 Method for calculating the heat transfer coefficient for a package of finned tubes of circular cross-section with a chess arrangement, A.I. Kirillov [6].

Method obtained based on experimental studies of heat treatment of the finned tubes.

The average convective heat transfer coefficient over the surface:

$$\alpha_k = Nu \cdot \lambda / d_H,\ W/m^2 \cdot K, \tag{1}$$

where λ – coefficient of thermal conductivity of air; the hydraulic beam diameter is defined as:

$$d_H = 2[t(s_1-\delta)-2\delta h)/2(h+t)],\ m, \tag{2}$$

where t – step between fins on a tube, m; s_1, s_2 – The transverse and longitudinal pitch of the tubes in the tube bundle, m; h, δ – height and average thickness of the fin, m.

The Nusselt number is determined by the criterial equation:

$$Nu=0.36 \cdot Re_n \cdot Pr^{0.33} C_Z \cdot C_S \cdot \psi^{-0.5}, \tag{3}$$

where:

$$n = 0.6 \cdot \psi^{0.07}; \tag{3a}$$

coefficient of finning:

$$\psi = [1+2 \cdot h(d_0+h+\delta)]/(d_0 t); \tag{3b}$$

C_Z – correction factor for the number of rows of tubes in the bundle along the air flow; coefficient of the tube bundle form:

$$C_S = ((\sigma_1-1)/(\sigma'_2-1))^{0.1}; \tag{3c}$$

relative transverse and longitudinal steps of the tube bundle:

$$\sigma_1 = s_1/d_0,\ \sigma_2 = s_2/d_0; \tag{3d}$$

relative diagonal step:

$$\sigma'_2 = \sqrt{\sigma_1^2/4+\sigma_2^2}; \tag{3f}$$

d_0 – outer diametr of the tube, m.

The Nusselt ratio is applicable for the range of the Renolds number $Re=5 \cdot 10^3 \div 37 \cdot 10^4$ and the shape of the tube bundle $C_S=0.46 \div 2.2$.

The heat transfer coefficient is defined as:

$$\alpha=\alpha_\kappa \eta, \tag{4}$$

where:

$$\eta=1-0{,}058 \cdot \beta \cdot h; \tag{4a}$$

$$\beta = \sqrt{2\alpha_\kappa/(\lambda \cdot \delta)}. \tag{4b}$$

Range of change $\beta h = 0.1 \div 3.7$.

2.2 Method for calculating the heat transfer coefficient for a package of finned tubes of circular cross-section with a chess arrangement, A.N. Bessonny [7].

Reduced heat transfer coefficient:

$$\alpha = \alpha_k[(\psi - 1)E + 1]/\psi, \text{ W/m}^2\cdot\text{K}, \quad (5)$$

where E – fin efficiency factor; $\alpha_к$ is determined by the formula (1),
where:

$$Nu = St\cdot Pr^{2/3}\cdot Re\cdot Pr^{1/3}\cdot C_Z, \quad (6)$$

where:

$$St\cdot Pr^{2/3} = -7.5\cdot 10^{-7}Re + 0.0135; \quad (6a)$$

C_Z – coefficient taking into account the number of rows in the direction of air movement.

2.3 Method for calculating the heat transfer coefficient for a package of finned tubes of circular cross-section with a chess arrangement, E.N. Pis'menny [8].

The method presents a unified system of generalizing equations that allows one to find in the interval $Re=5\cdot 10^3\div 2\cdot 10^5$ the values of heat transfer coefficients for chess beams of cross-finned tubes with a finning coefficient $\psi=1.2\div 39.0$ and relative transverse and longitudinal pitch $\sigma_1=1.7\div 6.5$, $\sigma_2=1.3\div 9.5$, $(\sigma_1/\sigma_2 = 0.3\div 5.2)$.

The average convective heat transfer coefficient over the surface is determined by the formula (1). The hydraulic diameter of the bundle of tubes is determined by the formula (2).

The Nusselt number is determined by:

$$Nu=1.13Re^m Pr^{0.33} C_q C_Z, \quad (7)$$

where for the exponent:

$$m=0.7+0.08th(X)+0.005\psi; \quad (7a)$$

$$X=\sigma_1/\sigma_2-1.26/\psi-2; \quad (7b)$$

$$C_q=(1.36-thX)(1.1/(\psi+8)-0.014). \quad (7c)$$

Coefficient C_Z takes into account the effect on the heat transfer of the number of transverse rows of tubes in the tube bundle z_2.

For chess packages with $\sigma_1/\sigma_2 < 2$ and $z_2>8$ $C_Z=1$, while $z_2<8$:

$$C_Z=3.15z_2^{0.05}-2.5. \quad (8)$$

The reduced heat transfer coefficient is defined as:

$$\alpha=\alpha_k(E\psi_E F_f/F+F_t/F), \quad (9)$$

where surface area of tube without fins:

$$F_t=\pi\cdot d_0\cdot\psi\cdot L; \quad (9a)$$

surface area of the fins:

$$F_f=z_2F_tB/s_1; \quad (9b)$$

total area of finned tube:

$$F=F_t+F_f; \tag{9c}$$

L – tube length; B – the width of the tube bundle; coefficient of theoretical efficiency of the tube fin:

$$E=th(\beta\cdot h'')/(\beta\cdot h''); \tag{9d}$$

where:

$$\beta=\sqrt{2\alpha_k/(\lambda\cdot\delta)}; \tag{9f}$$

$$h''=\left(h+\frac{\delta}{2}\right)\left[1+\left(0.191+0.054\frac{D}{d_0}\right)\ln\frac{D}{d_0}\right]; \tag{9g}$$

D – the outer diameter of the fins; correction for non-uniformity of heat transfer across the surface of the tube:

$$\Psi_E=1-0.016(D/d_0-1)[1+th(2\cdot\beta\cdot h-1)]\cdot \tag{9h}$$

The limits of applicability of equation $D/d_0=1.1\div4.0$; $\beta\cdot h=0.1\div4.0$.

2.4 Method for calculating the heat transfer coefficient for a package of finned tubes of circular cross-section with a chess arrangement, J. Moore [9].

Coefficient of heat transfer:

$$\alpha = 1000St\cdot Pr^{2/3}\cdot\rho\cdot c\cdot\omega/Pr^{2/3},\ W/(m^2K), \tag{10}$$

where ρ, c, ω – air density, heat capacity and velocity; Pr – Prandtl number;

$$St\cdot Pr^{2/3} = 0.134Re^{-0.319}(t/h)^{0.2}(t/\delta)^{0.1134}. \tag{10a}$$

Reynolds number:

$$Re = G\cdot d_H/(\mu\cdot A_{min}), \tag{11}$$

where G, μ – air flow rate and viscosity; A_{min} – minimum sectional area for the passage of air through the tube bundle; h, t, δ – height, pitch and thickness of the fin;
the hydraulic diameter and limits of application of equation:

$$d_H = 4h\cdot t/(h+t). \tag{12}$$

$$Re =1100\div18000. \tag{13}$$

3 Experimental results

According to these methods, for the same initial data, heat transfer coefficient calculations for the AC section.

For the calculation, the following design parameters of the AC section were adopted:

1. Outer diameter of the support tube $d_0 = 0.027$ m.
2. The fin height h = 0.015 m.

3. The average thickness of the fins $\delta = 0.000735$ m.
4. Fin step t = 0.0025 m.
5. Transverse step $s_1 = 0.084$ m.
6. Longitudinal step $s_2 = 0.074$ m.
7. Number of longitudinal rows of tubes $z_1 = 4$ pcs.
8. Number of transverse rows of tubes $z_2 = 6$ pcs.
9. The width of the tube bundle B = 0.945 m.
10. Tube length L = 12 m.

The results of the calculation are given in the Table 1.

Table 1. The results of comparison of the methods for calculating the heat transfer coefficient for the air condenser section.

Autor	d_G, m	$St \cdot Pr^{2/3}$	Re	Nu	α, W/(m^{2}K)
Kirillov [6]	0.0115	-	4929.7	38.9	80
Bessonny [7]	0.0115	0.0098	4929.7	40.9	64.2
Pis'menny [8]	0.0115	-	4929.7	27.8	51.9
Moore [9]	0.0115	0.0072	4805.3	-	67.6

The table shows that the highest heat transfer coefficient was obtained by calculating the formulas Kirillov, Moore and Bessonny, but the error in calculating them with respect to the experimental data is estimated to be above 20%. In this case, even when calculating using the Sleepless method, where the heat transfer coefficient is lower than that of Kirillov and Moore, the heat transfer coefficient is multiplied by 0.8.

When calculating the heat transfer coefficient using the Pis'menny formulas, the greatest number of geometrical and thermophysical parameters characterizing the heat exchange of the tube bundle with air is used. Taken into account the uneven heat transfer of the fin surface. The boundaries of the applicability of the formulas are clearly defined. The technique provides an error in the calculation of the heat transfer of finned tube packs below 20%.

As a result of the analysis performed in the construction of the VC model, the Pis'menny method was chosen for calculating the heat transfer coefficient, since they most accurately reflect the actual process of heat exchange of the finned tube bundle with cooling air.

References

1. K. Savina, A. Abramovskih, V. Haymin, S. Lavrinenko, MATEC Web Conf. **72**, 01042 (2016)
2. K. Larionov, D. Gvozdjakov, A. Zenkov, V. Zaytsev, EPJ Web Conf. **110**, 01034 (2016)
3. N. Galashov, S. Tsibulskiy, T. Serova, EPJ Web Conf. **110**, 01068 (2016)
4. G. Nikolay, S. A. Tsibulskii, G. Vladimir, S. Tatiana, MATEC Web Conf. **37**, 01021 (2015)
5. N. N. Galashov, S. A. Tsibul'ski, Power Technology and Engineering **49**, 365 (2016)
6. A.I. Kirillov, *Thermal and hydraulic calculation of heat exchangers for compressor units* (LPI, Leningrad, 1982) [in Russian]
7. A. N. Bessonny, G. L. Dreitser, V. B. Kuntysh, *Basics of calculating and designing heat exchangers for air cooling* (Nedra, SPb, 1996) [in Russian]
8. E. N. Pis'menny, *Heat exchange and aerodynamics of packages of cross-finned tubes* (Alterpress, Kiev, 2004) [in Russian]
9. J. Moore, R. Grimes, E. Walsh, A. O'Donovan, Energy **13** (2014)

Research of the opportunity of using bran as a building material

Maria Gaydabrus[1,*], *Dias* Mussafirov[1], and *Roman* Tabakaev[1]

[1]National Research Tomsk Polytechnic University, 634050 Tomsk, Russia

Abstract. As a result of the functioning of the agro-industrial complex, several million tons of bran that require recycling are annually produced in Russia. At present, there is no resource-efficient way of utilizing bran, which ensures its ecological compatibility and economic efficiency. In this regard, the purpose of the paper is to investigate the possibility of using wheat bran as a building material. The article presents the prerequisites for the construction of low-rise buildings using wheat bran as a building material: an experiment was carried out for the manufacture of bran blocks, and their thermophysical characteristics, maximum crushing strength, and moisture resistance were measured. Building blocks made from bran have a thermal conductivity comparable to the thermal conductivity of wood. The high strength of the bran blocks to squeeze indicates the possibility of their use for the construction of low-rise buildings without any strengthening of their design. However, this blocks exhibit hydrophilic properties, which requires the application of a protective layer. For example, a conventional wood varnish can be use as such a coating.

1 Introduction

The Russian Federation is one of the leading countries in terms of grain production and export [1]. In 2017, the grain harvest in Russia amounted to a record 136 million tons [2]. However, the technical equipment of Russian grain processing enterprises has a rather low efficiency, not allowing more than 45-65% of the finished product to be produce from the feedstock [3]. This leads to a huge accumulation of agricultural waste – bran.

In practice, bran is used as a feed additive to animals or poultry, amounting to 15-60% of the total weight of mixed fodder [4, 5]. However, this direction is limited by poor palatability and low digestibility by animals [6].

Options are considered for burning bran for energy needs [7, 8]. However, the slagging ability, characteristic for vegetable fuels, leads to low combustion efficiency and quick failure of boiler equipment.

Disposal of bran by burial can lead to severe environmental consequences: a strong and stable crust forms on the surface of the earth, which adversely affects the further use of soil [5].

* Corresponding author: 779030742352@mail.com

It is necessary to continue searching for ways of rational utilization of bran. As such a direction, the manufacture of building materials from bran can be considered. Bran is an environmentally friendly raw material, which makes it possible to obtain environmentally friendly natural materials on their basis.

Today, environmental construction is an actual world trend, which refers to the erection of all types of buildings. It arose in the 60s of the XVIII century on the example of construction of individual housing and remained imperceptible until the end of the XX century. Sharp interest in the issues of ecological housing construction has emerged in the last decade: the world community realized that human civilization plays a huge role in the destruction of our planet, introducing it into its activity in an unstable state.

Strengthens the ecological construction not only the growth of prices for building materials, but also the high toxicity of many of them. In addition, eco-materials are a renewable resource that exists in many parts of the earth. Nature provides at our disposal many wonderful building materials, such as clay, stones, wood, straw, bark, leaves, reeds, etc.

For example, a straw house [9] is one of the types of eco-building objects, for the erection of which blocks from pressed straw are used. The modern construction of thatched buildings began in the late 19th century in the United States, which is considered the world leader in the construction of houses from biomass materials (over 100,000 houses). These houses have low construction costs, excellent microclimate, strong enough; due to low weight do not require a heavy foundation and lifting gear.

Moreover, such houses have a great ability to heat insulation and carbon dioxide absorption, which increases their interest in large metropolitan areas and contaminated areas.

In this regard, the purpose of the study is to investigate the possibility of using wheat bran as a building material. The article provides the prerequisites for the construction of low-rise buildings using wheat bran as a building material.

2 Material and methods

As a research object, we chose wheat bran.

The values of the thermal characteristics of bran (thermal conductivity, heat capacity and thermal diffusivity) are determined on the Discovery Laser Flash DLF-1200 thermal diffusivity analyzer (TA Instruments, USA) in the temperature range 298-423K. For this, the bran was pre-compressed (P≈200 MPa) using a hydraulic press.

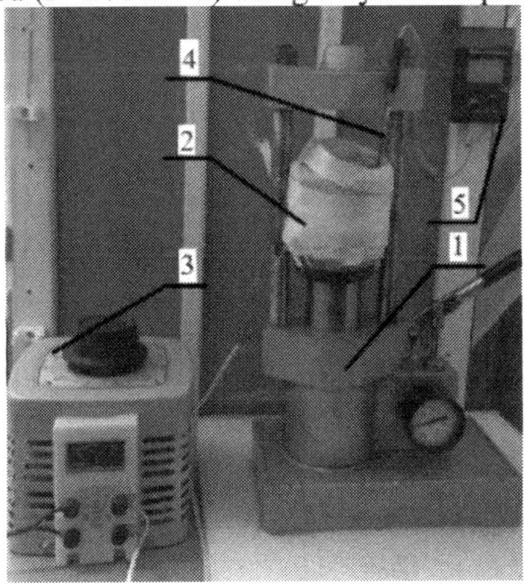

Fig. 1. Laboratory installation for the production of bran blocks: 1 – hydraulic hand press, 2 – indirect heated mold, 3 – auto transformer, 4 – thermal couple (type K), 5 – thermometer.

For manufacturing building blocks from bran, a manual hydraulic press with a heated mold was used (Fig. 1). A sample of bran (weight is 150 g) was poured into a mold and pressed according to recommendations [7] at a pressure of 10-15 MPa and a temperature of 140°C for 10-20 minutes.

The strength of the blocks produced was determined by loading the blocks with pressure until it destroyed. To carry out these tests, was used a hydraulic press. The maximum pressure at which the block was destroyed was taken as the strength of the block for crushing.

Some of the blocks were previously covered with a protective layer – yacht wood varnish. Then the mass of blocks covered with a protective layer and without it was measured. The blocks were wetted with distilled water for 20 minutes, after which they were wiped dry and weighed.

A change in the shape of the water drop on the surface of the block was also investigated. For this purpose, we used a Levenhuk DTX 500 microscope with a magnification of 30^x.

3 Results and discussion

On the Fig. 2 shows the results of determining the thermal characteristics of the studied bran. It is seen that as the temperature increases, the thermal conductivity and heat capacity of bran increases, and the thermal diffusivity decreases. The thermal conductivity of the bran block is comparable to the thermal conductivity of most types of wood (λ = 0.1-0.2 W/(m×K) [10-12]), which indicates the possibility of using blocks in places with low temperatures (down to –40°C) in winter. In this case, the thickness of the walls of the blocks should have the same value as the wood wall recommended for a particular region.

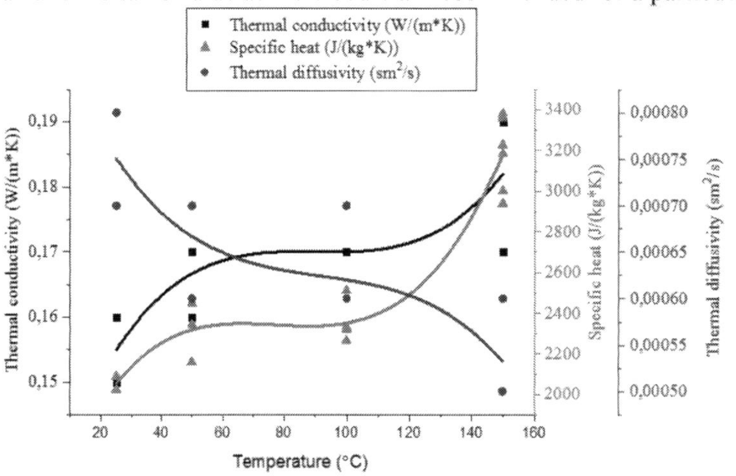

Fig. 2. Measured values of specific heat, thermal conductivity and thermal diffusivity of bran.

The strength of the blocks produced was determined by loading the blocks with pressure until it destroyed. To carry out these tests, was used a hydraulic press. The maximum pressure at which the block was destroyed was taken as the strength of the block for crushing.

According to the recommendations [7] blocks of bran are made (Fig. 3). The strength tests showed that the block was destroyed at a pressure of 35.5 MPa. The strength of the block was about 1112.6 kg/m^3. This indicates that blocks of bran are suitable for the construction of low-rise buildings without any strengthening of their construction.

Fig. 3. Measured values of specific heat, thermal conductivity and thermal diffusivity of bran.

Water resistance tests showed that when exposed to water, blocks that are not protected by the water repellent layer gradually swell, lose their shape and eventually break down. The application of a protective layer, for example, of yacht lacquer, intended for wood processing, allows ensuring water resistance of the blocks.

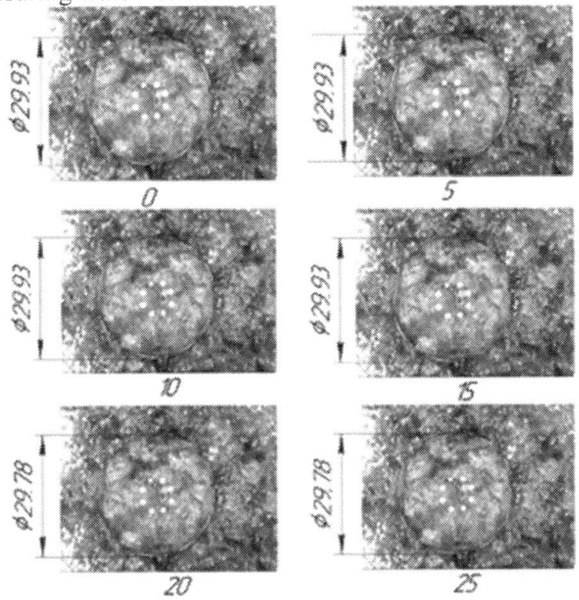

Fig. 4. The shape of a drop of water on the surface of the block of bran, pre-treated with varnish, over a period of time (an increase of 30^x).

As can be seen from Fig. 4, the shape of the drop does not change with time, which indicates the hydrophobic properties of the block. After 20 minutes, the size of the drop decreases, which can be explained by the gradual evaporation of water from the surface of the block.

4 Conclusions

Building blocks made from bran have a thermal conductivity comparable to the thermal conductivity of wood. Given the widespread prevalence of wood as a material for the construction of low-rise buildings, it can be assumed that bran due to similar thermal characteristics will be able to compete with it in this segment.

The high strength of the bran blocks to squeeze indicates the possibility of their use for the construction of low-rise buildings without any strengthening of their design.

Blocks from bran exhibit hydrophilic properties, which require the application of a protective coating. For example, a conventional wood varnish can be used as such a coating.

The research is funded from Grant Council of the President of the Russian Federation. The experimental calculations are carried out at Tomsk Polytechnic University within the framework of Tomsk Polytechnic University Competitiveness Enhancement Program grant.

References

1. I. G. Generalov, S. A. Suslov, Bulletin NGII **40** (9), 3 (2014)
2. M. A. Nikolaeva, L. V. Kartashova, Food Industry **3**, 77 (2018)
3. A. I. Altukhov, Vestnik of Kursk State Agricultural Academy **5**, 2 (2015) [in Russian]
4. E. B. Etuk, A. V. Ifeduba, U. E. Okata, I. Chiaka, C. Ifeanyi, N. J. Okeudo, B. O. Esonu, A. B. I. Udedibie, J. C. Moreki, J Anim Sci Adv. **2**, 510 (2012)
5. S. V. Antimonov, R. F. Sagitov, S. Yu. Solovykh, Izvestia vuzov.Pishevaya tekhnologia **2-3**, 61 (2008) [in Russian]
6. V. D. Kaminsky, A. Y. Karunsky, M. B. Babich, Hranenie i pererabotka zerna [Storage and processing of grain] **5**, 42 (2000) [in Russian]
7. M. Gaydabrus, I. Razov, R. Tabakaev, V. Lebedev, MATEC Web Conf. **110** (2017)
8. M. Yu. Chernetskiy, A. A. Dekterev, A. Burdukov, Journal of Siberian Federal University.Engineering & Technologies **6**, 625 (2013)
9. M. Myhrman, S. O. MacDonald, *Build it with Bales: A Step-by-step Guide to Straw-bale Construction* (Chelsea Green Publishing Company, London, 1997)
10. A. U. Franchuk, *Tables of heat engineering indicators of building materials* (Research Institute of Building Physics, Moscow, 1969) [in Russian]
11. A. P. Babichev, N. A. Babushkina, A. M. Bratkovsky, *Physical quantities.Reference book* (Energoatomizdat, Moscow, 1991) [in Russian]
12. V. Blazy, *Handbook of the designer.Building physics* (Technosphere, Moscow, 2004) [in Russian]

Numerical Investigation of Tungsten Reduction Process

Natalia Gicheva[1], *Alexander* Shvab[1], and *Vladimir* Brendakov[2,*]

[1]National Research Tomsk State University, 634050 Tomsk, Russia
[2]Seversk Technological Institute, 636036 Seversk, Russia

Abstract. The subject of this research was a chemical reactor for the producing tungsten coverings by means of chemical vapor deposition. The physical and mathematical model of viscous incompressible fluid motion and heat and mass transfer in a vortex chamber has been described. The velocity field, the temperature and the concentration distributions in a vortex chamber were obtained. Calculations reliability was verified by different methods. Parametric studies on effects of the Reynolds, Prandtl and Rossbi criteria on the flow characteristics were performed.

1 General mathematical model

1.1 System of equations

In this paper the process of tungsten deposition from the tungsten hexafluoride and hydrogen mixture is considered. The stoichiometry of that is:

$$WF_6 + 3H_2 = W + 6HF.$$

The numerical simulation of hydrodynamics and heat and mass transport for viscous compressible gas in a cylindrical reactor is carried out in this work (Fig. 1) [1]. The gas having axial velocity U_0, temperature T_0 and concentration C_0 enters from the pipe in the center of chamber, and then exits through the annual channel on the periphery of that. The top and the bottom of a vortex chamber are rotating with a constant angular velocity ω. A susceptor with a temperature T_I fills a part of the bottom, the tungsten reduced deposits there.

In this model a binary mixture is considered, the first component of one is tungsten hexafluoride with hydrogen and the second component is hydrogen fluoride. By virtue of low velocities and temperature differences the fluid considered is incompressible, but density depends on concentration because component ratio changes during the reaction. The density of a mixture is determined by the additive principle by means of the mass concentrations and the true densities of the corresponding components.

* Corresponding author: vnbrendakov@mephi.ru

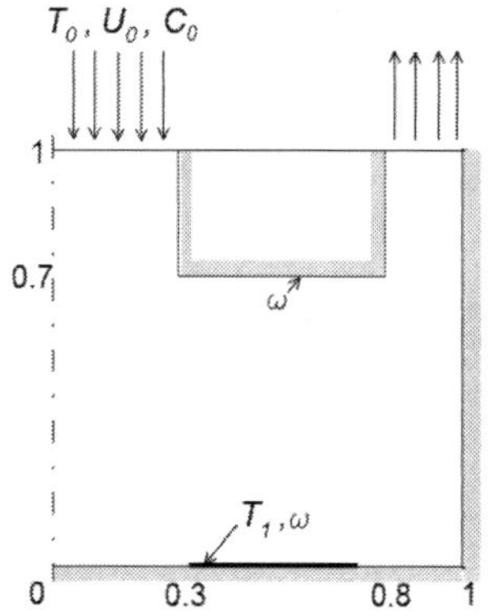

Fig. 1. A schematic image of a vortex chamber of reactor.

The system of equations describing processes in a vortex chamber contains viscous incompressible fluid dynamics (1-3), mass conservation (4), energy (5) and mass transport (6) equations. It can be represented in dimensionless form in a cylindrical coordinates:

$$\frac{\partial(\rho u_r)}{\partial t}+\frac{\partial(\rho u_r)^2}{\partial r}+\frac{\partial(\rho u_z u_r)}{\partial z}=-\frac{\partial p}{\partial r}-\frac{\rho u_r^{\,2}}{r}+\frac{\rho u_\varphi^{\,2}}{r}+$$

$$+\frac{2}{\mathrm{Re}}\frac{\partial}{\partial r}\left(\mu\frac{\partial u_r}{\partial r}\right)+\frac{1}{\mathrm{Re}}\frac{\partial}{\partial z}\left[\mu\left(\frac{\partial u_r}{\partial z}+\frac{\partial u_z}{\partial r}\right)\right]+\frac{1}{r}\frac{1}{\mathrm{Re}}\left(2\mu\frac{\partial u_r}{\partial r}-2\mu\frac{u_r}{r}\right); \quad (1)$$

$$\frac{\partial(\rho u_z)}{\partial t}+\frac{\partial(\rho u_r u_z)}{\partial r}+\frac{\partial(\rho u_z)^2}{\partial z}=-\frac{\partial p}{\partial z}-\frac{\rho u_r u_z}{r}+$$

$$+\frac{1}{\mathrm{Re}}\frac{\partial}{\partial r}\left[\mu\left(\frac{\partial u_z}{\partial r}+\frac{\partial u_r}{\partial z}\right)\right]+\frac{2}{\mathrm{Re}}\frac{\partial}{\partial z}\left(\mu\frac{\partial u_z}{\partial z}\right)+\frac{1}{\mathrm{Re}}\frac{\mu}{r}\left(\frac{\partial u_z}{\partial r}+\frac{\partial u_r}{\partial z}\right)+\frac{Gr}{\mathrm{Re}^2}\theta; \quad (2)$$

$$\frac{\partial(\rho u_\varphi)}{\partial t}+\frac{\partial(\rho u_r u_\varphi)}{\partial r}+\frac{\partial(\rho u_z u_\varphi)}{\partial z}=-\frac{2\rho u_r u_\varphi}{r}+$$

$$+\frac{1}{\mathrm{Re}}\frac{\partial}{\partial r}\left[\mu\left(\frac{\partial u_\varphi}{\partial r}-\frac{u_\varphi}{r}\right)\right]+\frac{1}{\mathrm{Re}}\frac{\partial}{\partial z}\left(\mu\frac{\partial u_\varphi}{\partial z}\right)+\frac{1}{\mathrm{Re}}\frac{2\mu}{r}\left(\frac{\partial u_\varphi}{\partial r}-\frac{u_\varphi}{r}\right); \quad (3)$$

$$\frac{\partial\rho}{\partial t}+\frac{\partial(\rho u_r)}{\partial r}+\frac{\partial(\rho u_z)}{\partial z}+\frac{\rho u_r}{r}=0; \quad (4)$$

$$\frac{\partial\rho\theta}{\partial t}+\frac{\partial(\rho u_r\theta)}{\partial r}+\frac{(\partial\rho u_z\theta)}{\partial z}=\frac{1}{\mathrm{Re}\cdot\mathrm{Pr}}\left(\frac{\partial^2\theta}{\partial r^2}+\frac{1}{r}\frac{\partial\theta}{\partial r}+\frac{\partial^2\theta}{\partial z^2}\right)+$$

$$+\frac{(1-k_c)}{\mathrm{Re}\,\mathrm{Pr}_D}\left[\frac{1}{r}\frac{\partial}{\partial r}\left(\theta\rho r\frac{\partial C}{\partial r}\right)+\frac{\partial}{\partial z}\left(\theta\rho\frac{\partial C}{\partial z}\right)\right]-\frac{\rho u_r\theta}{r}; \quad (5)$$

$$\frac{\partial(\rho C)}{\partial t}+\frac{\partial(\rho u_r C)}{\partial r}+\frac{(\partial\rho u_z C)}{\partial z}=\frac{1}{\mathrm{Re}\cdot\mathrm{Pr}_D}\left[\frac{\partial}{\partial r}\left(\rho\frac{\partial C}{\partial r}\right)+\frac{\rho}{r}\frac{\partial C}{\partial r}+\frac{\partial}{\partial z}\left(\rho\frac{\partial C}{\partial z}\right)\right]-\frac{\rho u_r C}{r}. \quad (6)$$

For non-dimensional system (1-6) to obtain the following parameters were used: radius of the vortex chamber, linear velocity at the input, density and viscosity of a gas mixture at the input and maximal temperature difference. For a numerical solution of the momentum transport equations the time-splitting method and the method of pressure and velocity correction were used. Transport equations are solved by dint of the alternating direction implicit scheme. The solving is realized on a staggered grid.

1.2 Boundary conditions

For a unique solution to obtain, the boundary conditions in dimensionless form are set. At the input of chamber we have:

$$u_r = 0, \quad u_z = -1, \quad u_\varphi = 0, \quad \theta = 0, \quad C = 1;$$

at the symmetry axis by $r = 0$ we have:

$$u_r = 0, \quad \frac{\partial u_z}{\partial r} = 0, \quad u_\varphi = 0, \quad \frac{\partial \theta}{\partial r} = 0, \quad \frac{\partial C}{\partial r} = 0;$$

at the bottom on a susceptor:

$$u_r = 0, \quad u_z = 0, \quad u_\varphi = R\omega \cdot r, \quad \theta = 1, \quad \frac{\partial C}{\partial z} = kC^n.$$

The reaction rate constant k basically depends on a susceptor temperature and size of particles deposited. Outside the susceptor and at the top wall of chamber:

$$u_r = 0, \quad u_z = 0, \quad u_\varphi = R\omega \cdot r, \quad \frac{\partial \theta}{\partial z} = 0, \quad \frac{\partial C}{\partial z} = 0;$$

at the right wall and at the walls of input and output channels (n is normal direction):

$$u_r = 0, \quad u_z = 0, \quad u_\varphi = 0, \quad \frac{\partial \theta}{\partial n} = 0, \quad \frac{\partial C}{\partial n} = 0;$$

at the chamber output, the Neumann conditions are used for all functions: $\partial/\partial z = 0$.

2 Results and discussion

As a result of a numerical solution the distributions of velocity, temperature and concentrations were obtained. Influence of flow velocity, natural convection and rotation degree is studied.

An analysis performed let conclude that the parameters of a gas flow significant effect on its behavior and therefore on the matter layer depositing on a susceptor. This let regulate the reactor operating conditions for the optimal result to obtain. So, a rotation leads to a more uniform distribution of a temperature field and a matter near the susceptor. This occurs due to a uniform thickness boundary layer generation. At the figure 2 the streamlines (a), the temperature field (b) and the concentration contour lines (c) in a reactor chamber are represented. The flow parameters: Re = 50, Rω = 5, Gr = 10^4, Pr = 1, Pr_D = 1.

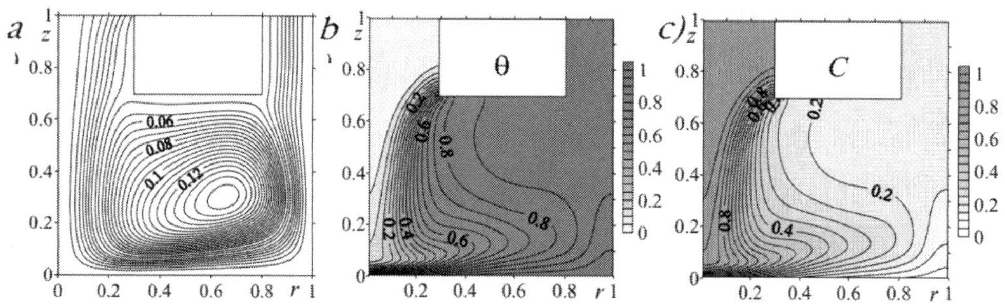

Fig. 2. The streamlines, temperature and concentration distributions in a vortex chamber depending on rotation availability.

3 Reliability proof

Verification of model created was performed by different methods: comparison of two solving methods, comparison with analytical solution for annual channel by low Reynolds number, test problem solution and qualitatively comparison with results known in literature.

The flow due to a rotating disk was considered. The model developed was applied to this problem solving. Fig. 3 shows a comparison between the calculation results and Cochran's exact solution. Dots denote exact values; solid lines represent the model results.

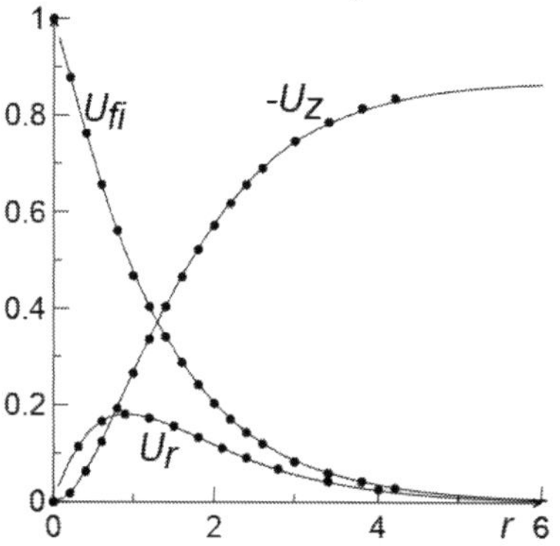

Fig. 3. Velocities distribution near a rotating disk.

4 Conclusions

The physical and mathematical model of viscous incompressible fluid motion and heat and mass transfer in a vortex chamber has been described. Parametric studies were performed. A significant influence of a natural convection and flow rotation was found. The study performed allows selecting reactor operating conditions for the optimal result and uniform tungsten covering to obtain.

References

1. V. N. Brendakov, A. A. Demidenko, A. V. Shvab, N. S. Evseev, The method for obtaining a tungsten ware by layer-by-layer deposition of tungsten and a device for its implementation, Pat. No. 2641596 (2018) [in Russian]

MATEC Web of Conferences **194**, 01020 (2018) https://doi.org/10.1051/matecconf/201819401020
HMTTSC-2018

Research of parameters of the steam boiler BKZ-220-100 at joint burning of natural gas and low-grade fuel

Elena Vorontsova[1], *Andrey* Gil[1,*], and *Alexander* Romanenko[1]

[1]National Research Tomsk Polytechnic University, 634050 Tomsk, Russia

Abstract. In this article the research of prospects of use of low-grade brown coal of the Talovsky Deposit of the Tomsk region as fuel for local power is carried out. The study is carried out by checking calculations of the steam boiler BKZ–220–100. The result of the study is to obtain data on the parameters of the boiler during the combustion of brown talovsky coal as the main fuel, as well as in a mixture with natural gas or Kuznetsk coal.

1 Introduction

Brown coal of the talovsky Deposit of the Tomsk region has the following characterized by high humidity in a wide range (20 % – 60 %), low heat of combustion, large yield of volatile. Such parameters of brown coal indicate that this fuel is low-grade and its use is economically feasible only for the needs of local energy [1].

The object of the study is a steam boiler with natural circulation BKZ-220-100. The boiler unit is operated at Tomsk SDPP-2.

The aim of the work is to study the parameters of the steam boiler in the combined combustion of brown talovsky coal and natural gas.

2 Initial data and research methods

Due to the wide range of moisture content of brown coal of the talovsky Deposit, using the methods of numerical simulation of combustion processes, the optimal for combustion in the boiler furnace BKZ-220-100 values of fuel moisture. Fig. 1 shows the temperature change in the height of the combustion chamber of the boiler under study when burning tall brown coal with a moisture content of 20-50% and Kuznetsk coal.

* Corresponding author: angil@tpu.ru

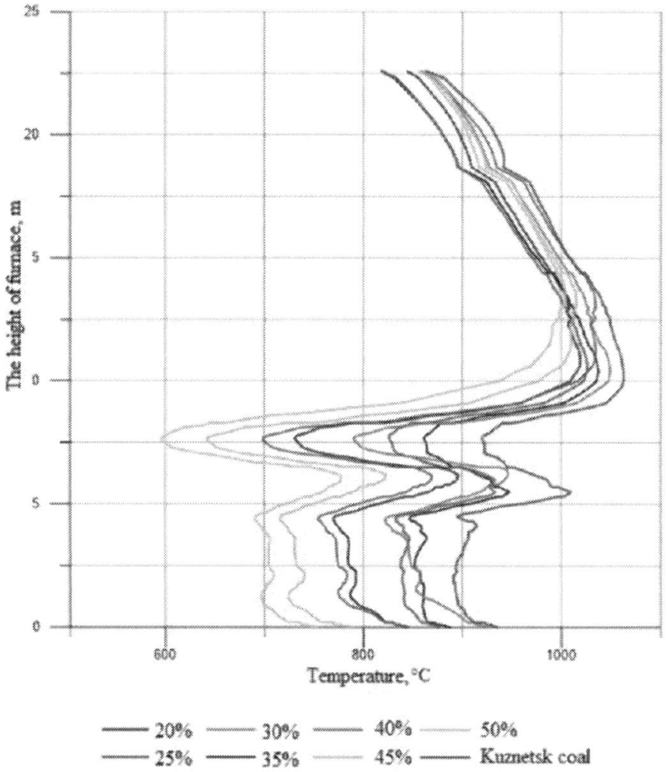

Fig. 1. The temperature Distribution along the height of the boiler furnace.

Fig. 1 shows that the temperature distribution in the boiler furnace height depends on the humidity of the talovsky coal, i.e. as the humidity of the burned talovsky coal increases, the temperature in the area of the burner devices decreases. In addition, there is a similar pattern of temperature distribution in the combustion of talovsky coal with a moisture content in the range of 25-30% and Kuznetsk coal [2].

As a result of numerical simulation of the combustion process of tall brown coal in the furnace of the steam boiler BKZ–220–100, the optimal range of fuel moisture was determined, which was 25% – 30%. The conclusion of numerical modeling is the conclusion that the mathematical model of the boiler furnace

BKZ-220-100 is suitable for further research. The obtained concentration fields of particles, water vapor and temperature regime in the combustion volume correspond to the accepted combustion scheme [2]. For further research, the selected coal with a moisture content of 25%.

The initial characteristics of the talovsky brown coal with a moisture content of 25% are presented in table 1.

Table 1. Features Talovsky brown coal with a moisture content of 25% in working condition.

Moisture content W^r, %	Ash content A^r, %	Volatile yield V^{daf}, %	Low heat of combustion Q_i^r, MJ/kg	Ultimate analysis (dry ash-free basis, %)				
				C^r	H^r	N^r	S^r	O^r
25.00	27.69	63.1	14.05	30.55	3.19	12.93	0.09	0.55

After determining the optimal humidity, the calibration calculation of the boiler BKZ–220–100 was performed for the burning of brown coal from the Talovsky Deposit, as well as mixtures of brown coal with natural gas and brown coal with Kuznetsk coal, which is designed for the boiler unit under study.

Calculations for the combustion of a mixture of coal is made as an alternative, as natural gas is more expensive fuel compared to Kuznetsk coal and its combustion is less cost-effective.

Table 2 presents the characteristics of natural gas, table 3 – characteristics of Kuznetsk coal.

Table 2. Characteristics of natural gas.

Methane CH_4 %	Dioxide CO_2 %	Ethane C_2H_6 %	Propane C_3H_8 %	Bhutan C_4H_{10} %	Nitrogen N %	Low heat of combustion, Q_d^r MJ/m^3
98.72	0.14	0.12	0.01	0.01	1.00	34.86

Table 3. Characteristics of Kuznetsk coal.

Moisture content W^r, %	Ash content A^r, %	Volatile yield V^{daf}, %	Low heat of combustion Q_i^r, MJ/kg	Ultimate analysis (dry ash-free basis, %)				
				C^r	H^r	N^r	S^r	O^r
13.8	14.4	24.0	20.92	55.7	4.0	9.9	0.3	1.9

3 Results

The parameters of the boiler operation at the brown angle with a moisture content of 25 % are presented in table 4.

Table 4. Parameters of the boiler BKZ-220-100 in the combustion of brown coal.

Name	Dimension	Value
Low heat of fuel combustion	MJ/kg	14.052
Excess air ratio at the exit of the furnace	–	1.2
Boiler efficiency	%	92.237
Exit-gas temperature	°C	136
Fuel consumption	kg /c	10.97
Flue gas temperature at the exit of the furnace	°C	900
Temperature of superheated steam	°C	454

Check calculation of the boiler during combustion in its combustion chamber of talovsky coal with a moisture content of 25 % showed that without the reconstruction of the heating surfaces the required temperature of superheated steam is not achieved.

The results of verification calculations for the combustion of talovsky coal with natural gas and Kuznetsk coal are presented in tables 5 – 6.

Table 5. Parameters of the boiler BKZ-220-100 when burning a mixture of tall brown coal and natural gas.

Name	Dimension	Brown coals content, %			
		30	40	50	60
Low heat of fuel mixture combustion	MJ/kg	46.84	35.13	28.10	23.42
Boiler efficiency	%	93.11	93.22	93.32	93.19
Exit-gas temperature	°C	135	134	133	135
Coal consumption	kg /c	3.39	4.50	5.60	6.68
Gas consumption per 1 kg of coal	m^3/ kg	0.94	0.61	0.40	0.27
Gas consumption	m^3/c	3.19	2.72	2.26	1.80
Flue gas temperature at the exit of the furnace	°C	1132	1081	1043	1006
The temperature of cold air	°C	52	51	50	55
The temperature of hot air	°C	348	338	337	320
The temperature of hot water	°C	303	298	296	284
Temperature of superheated steam	°C	510	510	510	497

The test calculations of the boiler under study when using a mixture of talovsky brown coal and natural gas as a fuel showed that the required parameters of superheated steam are achieved at a fraction of brown coal in the range from 30 % to 50 %.

With a decrease in the proportion of brown coal below 30%. there is a danger of exceeding the temperature of the flue gases at the exit of the furnace over the temperature of the beginning of deformation of the ash of the brown coal. which is equal to 1180 ° C. The maximum share of brown coal in the fuel mixture at which the required temperature of superheated steam is reached is 50%.

The increase in the share of talovsky brown coal leads to a decrease in temperature parameters. however. the boiler efficiency increases. This is due to the decrease in the volume of flue gases and their temperature.

Table 6. Parameters of the boiler BKZ-220-100 when burning a mixture of tall brown coal and Kuznetsk.

Name	Dimension	Brown coals content. %			
		10	20	30	40
Low heat of combustion of fuel mixture	MJ/kg	20.23	19.55	18.86	18.17
Boiler efficiency	%	92.2	92.25	92.36	92.57
Exit-gas temperature	°C	134	135	135	136
The flow rate of the mixture	kg /c	7.94	8.18	8.45	8.72
Flue gas temperature at the exit of the furnace	°C	1098	1080	1061	1038
The temperature of cold air	°C	47	50	50	54
The temperature of hot air	°C	352	349	347	335
The temperature of hot water	°C	304	301	297	292
Temperature of superheated steam	°C	510	510	510	503

Results of verification calculations of the steam boiler BKZ-220-100 at burning of a mixture of brown coal with project Kuznetsk coal (table 6) allow to draw a conclusion that the maximum share of brown coal with a moisture content of 25% in the mixture is 30%. With the increase of the share of talovsky brown coal. the temperature parameters are reduced. due to the decrease in the heat of combustion of the mixture. Boiler efficiency increases by reducing the volume of flue gases and their temperature.

4 Conclusion

The study showed that the use of brown coal from the talovsky Deposit of the Tomsk region with a moisture content of 25% as a fuel for local energy is possible. However use it for combustion in the boiler BKZ-220-100 working as a part of the power unit of Tomsk SDPP–2, it is advisable only in a mixture with natural gas or Kuznetsk coal.

References

1. E. V. Chernyaev, V. K. Bernatonis, G. Yu. Boyarko, *Materials of the International conference "100 years in the service of science and production". Regional geology. Geology of mineral deposits* (TPU publ., Tomsk, 2001)
2. T. S. Taylasheva, A. V. Gil, E. S. Vorontsova, Proceedings of Tomsk Polytechnic University. Georesources engineering **327** (1), 128 (2016)

Mathematical simulation of thermal state of digital current and voltage transformer in unfavourable weather conditions

Elena E. Gotovkina[1], *Grigory* I. Parfenov[1], *Nikolay* N. Smirnov[1,*], *Vladimir* D. Lebedev[1], and *Vladimir* V. Tyutikov[1]

[1]Ivanovo State Power Engineering University, 153003 Ivanovo, Russia

Abstract. This study examines issues relating to heat exchange between an innovative 6(10) kV current and voltage transformer (and its components) and its environment in unfavourable weather conditions. Existing designs of digital current and voltage transformers are featured in systems of commercial and technical electric power accounting, and also relay protection and automation. We have presented the results of mathematical simulation of the thermal state of both a single resistor included in our measuring device, and the digital transformer as a whole. Simulation was performed using the finite element method. We have taken into account the impact of such factors as direct and diffused solar radiation, high ambient temperatures and various voltage levels on the thermal state of the functioning equipment. To verify the developed mathematical model we have presented the results of a series of experiments conducted in a high-voltage environmental chamber. Our study conclusively shows that voltage, insolation and ambient temperature significantly affect the thermal state of a functioning current and voltage transformer.

1 Introduction

The digital instrument transformer [1] consists of a resistive voltage divider and current transformer (see Figure 1). Several resistors are contained within the body of the voltage instrument transformer.

The electrical processes occurring in resistive voltage dividers comprising a group of resistors entail heat emission. Overheating of a resistor may result in failure of the measuring equipment under review.

Methods of thermal protection against overheating of power transformers are widely known [2]. Unfortunately, thermal tests during operation of power transformers, and even more so in the case of current and voltage instrument transformers (especially new types) are often performed according to maintenance schedules [3], which negatively impacts the reliability of such electrical equipment [4]. Developing new systems of self-diagnostics checks for the thermal state of digital instrument transformers that determine temperature

* Corresponding author: nsmirnov@bk.ru

according to the most heated up electrical equipment part in real time is therefore an important task, especially in the development of "smart" power grids.

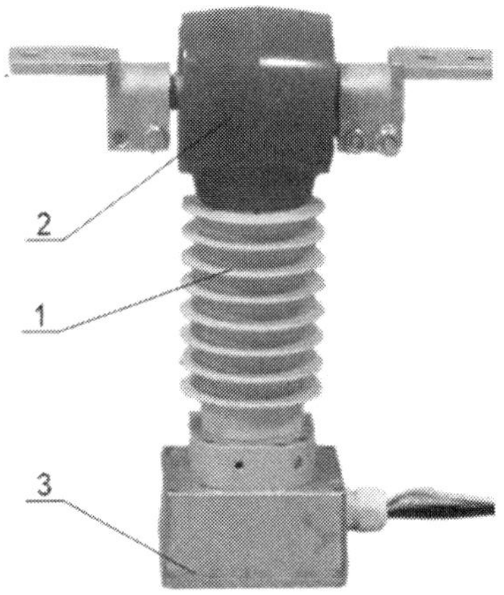

Fig. 1. External view of a 6 (10) kW digital combined transformer: 1 – primary voltage converter; 2 – primary current power converter; 3 – low voltage electronic unit.

Instrument transformers may be located either inside electric power substations or in the open air [1]. If in the former case a mechanical ventilation system to cool the transformer may be provided, it is impossible to control ambient air parameters in the open air. Instrument transformer operation in no-wind conditions (natural convection), high ambient temperatures and insolation, or in emergency mode (for example, with intermittent arcing faults) is the least favourable option from the point of view of heat exchange.

An indispensable stage of research in developing algorithms for transformer self-diagnostics is mathematical simulation of transformer thermal state under the above unfavourable weather conditions. Among the well known methods of solving differential equations that describe heat exchange, including methods of finite differences and algorithms for their implementation [5, 6], we chose the finite elements method [7], which has demonstrated highly accurate results in the simulation of complex heat exchange. The mathematical model was realized using COMSOL Multiphysics.

2 Description of mathematical model of heat exchange between a transformer (and resistor) with its environment

Mathematical simulation of heat exchange was performed both for the entire digital current and voltage transformer and for the cylindrical measuring resistor with R_l resistance, placed in insulation material identical to the one used in the transformer.

A diagram of the mathematical model of heat exchange between a transformer and its environment in conditions of natural convection is shown in Figure 2. The model included description of thermal conductivity inside the transformer, heat transfer from its surface, heat emissions from heating elements, and also diffused and direct solar radiation.

Since the transformer is intended to operate in the open air and exposed to insolation (see Figure 2), with surface temperatures t_{surf} exceeding ambient temperature t_{amb}, the thermal balance for this model may be written as follows:

$$Q_{el} + Q_{ins}^{dir} + Q_{ins}^{diff} = Q_{conv} + Q_{rad}, \quad (1)$$

where Q_{el} – heat, released according to the Joule-Lenz law during passage of electric current through the conducting parts of a combined transformer, W; Q_{ins}^{dir} – heat from direct solar radiation, W; Q_{ins}^{diff} – heat from diffused solar radiation, W; Q_{conv} – convection heat flow from heated transformer surface to ambient air, W; Q_{rad} – radiation heat flow from heated transformer surface to surrounding objects, W.

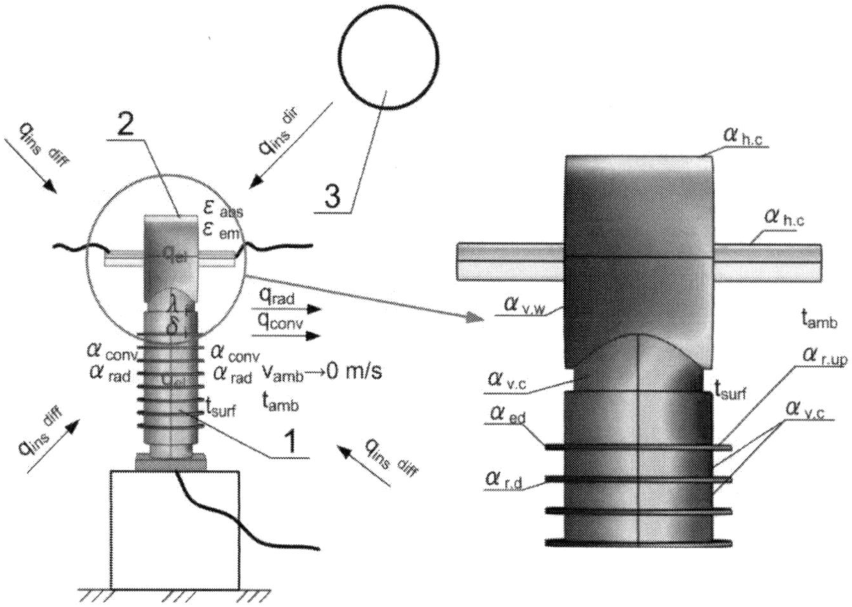

Fig. 2. Diagram of the mathematical model of heat exchange between a combined transformer and its environment: 1 – primary voltage converter; 2 – primary current power converter; 3 – Sun.

When electric current passes through the transformer with voltage supply, heat energy q_{el}, W, is released in the primary current and voltage converters according to the Joule-Lenz law. Solid state insulation, comprising layers of material of thickness δ and heat conductivity coefficient λ, creates additional resistance to removal of heat from the resistors. The physical heat properties of the solid materials were set with the aid of constants [8].

Ambient air is the operating environment, receiving heat emissions form electrical equipment. Heat transfer from the heated surface of a transformer to its environment is due to radiation q_{rad} and convection q_{conv}. Convection heat transfer is characterized by heat flux density q_{conv}, W/m^2, and heat transfer coefficient α_{conv}, W/(m^2·°C).

Functions for determining Nusselt *(Nu)*, Grashoff (*Gr*), Prandtl (*Pr)*, Raleigh (*Ra*) criteria, dependences for calculating convection heat transfer coefficients, volumetric expansion coefficients etc. for the corresponding surfaces and instances of heat exchange were described on the basis of heat exchange theories [8, 9], taking into account the work of Academician M.A. Mikheyev [8] and other researchers [10], and also using data from the authors' physical experiments.

Thus, for example, after converting our values into the formula for determining the heat transfer coefficient in the vertical cylinder, we obtained:

$$\alpha_{v.c} = \frac{0.54 \cdot (\frac{g \cdot R_0^3}{\nu^2} \cdot \frac{1}{T_0} \cdot \left| t_{surf} - t_{amb} \right| \cdot Pr)^{0.25} \cdot \lambda_f}{R_0}, \qquad (2)$$

where g – free fall acceleration, $g=9.81$ m/s^2 ; R_0 - determining size (cylinder height), ν – kinematic air viscosity, m^2/s; $\Delta T=\left|t_{surf}-t_{amb}\right|$ – temperature difference module between wall and ambient air, °C (K); λ_f – air thermal conductivity, W/(m·K); T_0 – determining temperature, K.

Coefficients of radiation heat transfer from the surface of the digital transformer to its environment α_{rad} were set according to the Stefan-Boltzmann formula [8], using the procedure incorporated in our software. Heat supply from direct q_{ins}^{dir} and diffused q_{ins}^{diff} solar radiation was set by means of a function. The quantity of obtained solar energy also depends on coefficients of absorption ε_{abs} and emission ε_{em} of the materials used in making the transformer outer casing.

This study is also interesting because account is taken of heat emissions from working equipment in both voltage and current transformers in our combined measuring device, and their total impact on the thermal state of the combined digital transformer during a certain time. Heat emissions from precision resistors in the voltage transformer were calculated according to the Joule-Lenz law and set with the aid of a function depending on electric grid mode, simulating various values of voltage and current.

Thermo-physical properties of ambient air were described with the aid of a function, based on reference literature [8].

3 Results of mathematical simulation of heat exchange between a resistors placed in electrical insulation material and its environment

The resistor (Figure 3, item 3) was placed in electrical insulation sheathing (Figure 3, item 4) with thermal conductivity λ .

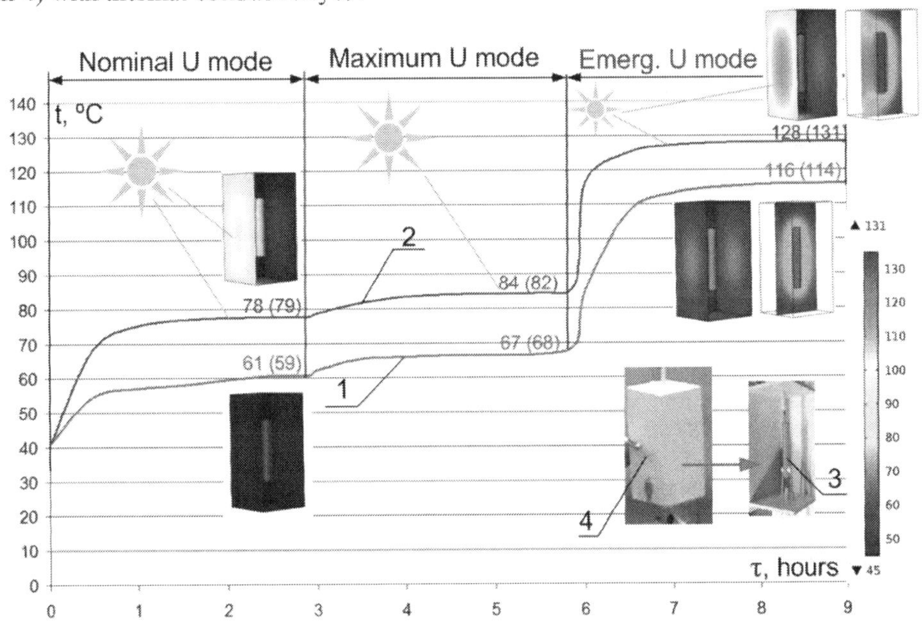

Fig. 3. Dependence of temperature on the lower cap of a resistor (3) placed in insulation material (4) on voltage and insolation: 1, 2 – in the absence and under the impact of insolation on the insulation surface, respectively.

We simulated step changes in the values of phase voltage supplied to the resistor in the following modes: nominal $U_{nom.ph} = 1.92$ kV, maximum $U_{max.ph} = 1.2 \cdot U_{nom.ph} = 2.31$ kV, and emergency $U_{emerg.ph} = 2.1 \cdot U_{nom.ph} = 4.04$ kV. Ambient temperature was set at 40 ^{0}C.

Figure 3 shows the results of our simulation (for resistor temperatures obtained during a physical experiment in a high-voltage environmental chamber being given in brackets). We should note that the difference between the results of physical and computer simulation is 1-3 °C, which reflects the accuracy of data obtained from simulation using COMSOL software.

In the absence of insolation with nominal voltage $U_{nom.ph,}$ resistor temperature rose to t_r=61 °C in 3 hours; with voltage increased to the maximum value, the temperature rose to 67 °C (an increase of 6 °C). Changing over to emergency mode resulted in the temperature rising to 116 °C (an increase of 49 °C). Insolation significantly increased warming of both insulation material and the resistor. Thus, with nominal voltage, resistor temperature increased by a further 17 °C (up to 78 °C), with maximum voltage – by 17 °C (up to 84 °C), with emergency mode – by a further 12 °C (up to 128 °C).

4 Results of mathematical simulation of heat exchange between a digital transformer and its environment

Studies (see Figure 4) were performed for a combined digital 6(10) kV transformer (taking boundary alignment of measuring resistor parts [1]). We should note that simulation was envisaged on two kinds of mesh: large, with around 22 thousand elements, and small 39 thousand elements. The calculation process on the large mesh took about 16 minutes, on the small mesh it took 7 times longer, or around 2 hours. Errors occurred during simulation on the large mesh due to the finite element not matching some of the transformer rib dimensions. The temperature difference was 0.5 - 4 ^{0}C.

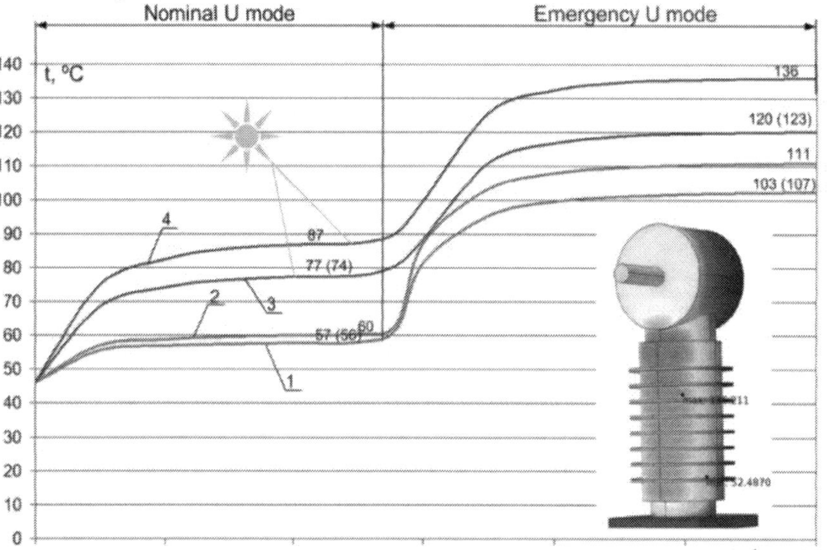

Fig. 4. Charts showing changing resistor temperature over time in nominal and emergency voltage modes: 1, 2 – temperatures at the lower and most heated resistor, respectively, without insolation; 3, 4 – the same, with insolation.

Two modes of phase voltage were set: nominal and emergency. Simulation was performed both with and without insolation on the transformer surface. The results may be seen in Figure 4, temperatures obtained during physical experiment in an environmental chamber being given in brackets. In the interests of electrical safety during physical experiment, temperatures were measured only on the lower caps of the lower resistors, the ambient temperature being 40 ^{0}C.

From the results of our simulation, we may conclude that the difference between temperatures obtained during physical experiment and mathematical simulation did not exceed 4 degrees, which may be related to inaccurate information regarding the physical properties of the materials used. With these allowances, however, the mathematical model under review may be regarded as accurate.

A significant rise in temperature occurred when voltage was increased from nominal to emergency values. Thus, the temperature of the most heated resistor rose from 60 to 111 ^{0}C, while with insolation the temperature of the same resistor increased from 87 to 136 ^{0}C. The difference between the lower and most heated resistors ranged from 3 (nominal mode, without insolation) to 16 ^{0}C (emergency mode, with insolation), which is related to the cumulative capability of the adjacent resistors and their mutual warming.

5 Conclusion

The developed models of heat exchange between both a resistor and a digital transformer with its environment may be regarded as adequate. These models may be used in the further development of systems of self-diagnostics for the thermal condition of a transformer. Voltage, insolation and ambient temperature significantly influence the thermal condition of a voltage and current transformer.

This research was conducted with financial support from the Ministry of Education and Science of the Russian Federation at Ivanovo State Power Engineering University within the framework of a federal target programme, "Research and Development in Priority Areas of Developing the Russian Scientific and Technological Complex in 2014-2020" concerning "A Multifunctional System based on Digital Current and Voltage Transformers for a Digital Substation" (Agreement No. 14.577.21.0276 on Granting Subsidies, dated 26 September 2017, Unique Identifier for Applied Scientific Research (Projects) RFMEFI57717X0276).

References

1. A.A. Yablokov, N.N. Smirnov, V.V. Tyutikov, V.A. Gorbunov, MATEC Web Conf. **141** (2017)
2. R. Hunt, M.L. Giordano, *Thermal overload protection of power transformers – operating theory and practical experience* (Georgia Tech, Atlanta, 2005)
3. Z. Liang-dong, S. Hui, Power System Protection and Control **04** (2010)
4. Ju.P. Aksenov, A.V. Golubev, V.I. Zavidey, Elektro **1** (2006)
5. O.V. Vysokomornaya, G.V. Kuznetsov, P.A. Strizhak, Rus. J. Phys. Ch. B **5** (2011)
6. V.L. Strakhov, A.N. Garashchenko, G.V. Kuznetsov, V.P. Rudzinskii, Comb. Exp. Sh. W. **37** (2001)
7. G.V. Kuznetsov, V.I. Maksimov, M.A. Sheremet, J. Appl. Mech. Tech. Phys. **54**, 4 (2013)
8. V.V. Buhmirov, *Heat and mass transfer* (ISPEU, Ivanovo, 2014)
9. M.A. Ulchiekov, E.E. Bulba, MATEC Web Conf. **92** (2017)
10. O.G. Martynenko, Yu.A. Sokovishin, *Free-convective heat transfer* (Nauka i technika, Minsk, 1982)

Experimental determination of the heat transfer coefficient during evaporation and boiling of thin liquid film

Anastasia Islamova[1,*]

[1]National Research Tomsk Polytechnic University, 634050 Tomsk, Russia

Abstract. Thin films evaporation of distilled water, ethanol and HFE-7100 liquid was experimentally studied. The dependences of heat transfer coefficients in time were determined. It has been established that with a decrease in the layer thickness of distilled water and ethanol, the heat transfer coefficient increases. For the HFE liquid, the nature of the change is different: as the time increases, the heat transfer coefficient decreases.

1 Introduction

At present, development of industries (metallurgy, microelectronics, etc.) is proceeding along the path of increasing the energy saturation of equipment [1-3]. One particular problem of technique is the creation of systems for thermal operating conditions of the equipment [4, 5]. To solve this, it is necessary to create energy efficient systems based on phase transitions [6-8]. Thin films of liquids are used in industrial technologies, for example, for cooling locally heated surfaces [9, 10].

The heat transfer coefficient determines the intensity of heat transfer and characterized the liquid-vapor or liquid-air interface. The layer thickness is one of the significant factors affecting the heat transfer coefficient under evaporation conditions of a thin liquid film on the heat exchange surface.

The objective of the present work was to determine experimentally the change of the heat transfer coefficient in time when thin engineering fluids evaporate.

2 Methods and materials

The research was conducted on the experimental setup simulating the conditions of boiling and evaporation of liquid thin layers [11]. The principle of its operation and the procedure for carrying out the experiments are given in [12].

In the experiments, we used a stainless steel substrate with a diameter of 50 mm and a thickness of 2 mm. Distilled water, ethanol and methoxy-nonafluorobutane (HFE-7100) were used as the test liquids. The material was chosen due to wide usage in heat exchangers, significant differences in thermal properties (the effect of properties on the characteristics of

* Corresponding author: agi2@tpu.ru

heat transfer). The initial thickness of the layer for ethanol was 1 mm, distilled water and liquid HFE-7100 was 1.5 mm, which corresponds to liquid volumes of 2 ml for ethanol and 3 ml for distilled water and HFE-7100 liquid. The choice of layer thickness was due to the ratio of adhesion forces to internal cohesion, which prevented a uniform spreading of the layer over the substrate surface. It is revealed that the stainless steel substrate used in the experiment investigation was not wetted by distilled water (contact angle more than 90°).

The heat flux from the surface of heat exchange to the layer was determined from the temperature gradient in two different sections along the height of the plate:

$$Q = \lambda \frac{\pi d^2}{4l} \cdot \Delta T, \tag{1}$$

where λ is the thermal conductivity of the copper plate, W/(m·K); l is thickness of copper plate (10 mm), m; d is plate diameter (50 mm), m; ΔT is the temperature gradient between the upper and lower sections of the plate, K. The temperature of the plate section is determined as the average value of the thermocouple readings located in one cross section in height.

When analyzing data, it was assumed that the curvature of the interface with a change in layer height occurs only in a very narrow region at the edge of the layer.

The density of the heat flux was determined by the formula:

$$q = \frac{Q}{S}, \frac{\text{W}}{\text{m}^2}, \tag{2}$$

where S is the heat exchange surface area, m^2.

The heat transfer coefficient was calculated by the formula:

$$\alpha = \frac{q}{T_S - T_W}, \tag{3}$$

where q is heat flux density, W/m^2; T_S is vapor temperature above the surface of the liquid, °C; T_W is the temperature of the substrate was assumed to be equal to the temperature of the upper surface of the metal plate in the center, °C.

3 Results and discussion

The dependences of heat transfer coefficient on time for ethanol, distilled water and HFE-7100 liquid are presented in Figures 1-3.

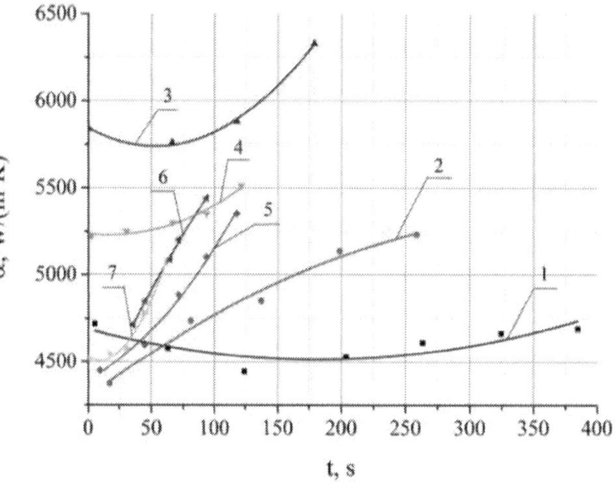

Fig. 1. Dependences of ethanol heat transfer coefficient on the time at the substrate temperature: 1 - 65 °C, 2 - 70 °C, 3 - 75 °C, 4 - 80 °C, 5 - 85 °C, 6 - 90 °C, 7 – 95 °C.

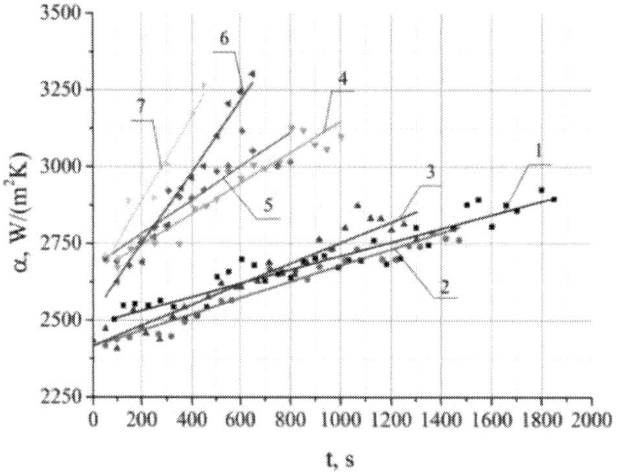

Fig. 2. Dependences of distilled water heat transfer coefficient on the time at the substrate temperature: 1 – 70 °C, 2 – 75 °C, 3 – 80 °C, 4 – 85 °C, 5 – 90 °C, 6 – 95 °C, 7 – 100 °C.

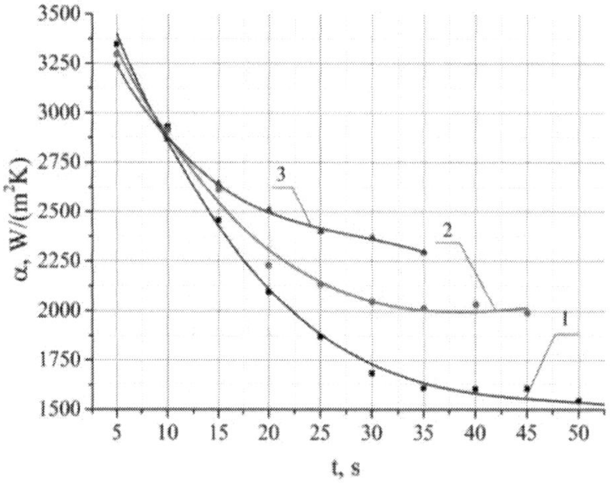

Fig. 3. Dependences of HFE-7100 heat transfer coefficient on the time at the substrate temperature: 1 – 55 °C, 2 – 60 °C, 3 – 65 °C.

According to the analysis of dependences (Fig. 1-3) allows us to conclude that with a decrease in the thickness of the liquid layer, the heat transfer coefficient increases for ethanol and distilled water. The rate of change in the heat transfer coefficient is shown in Fig. 4.

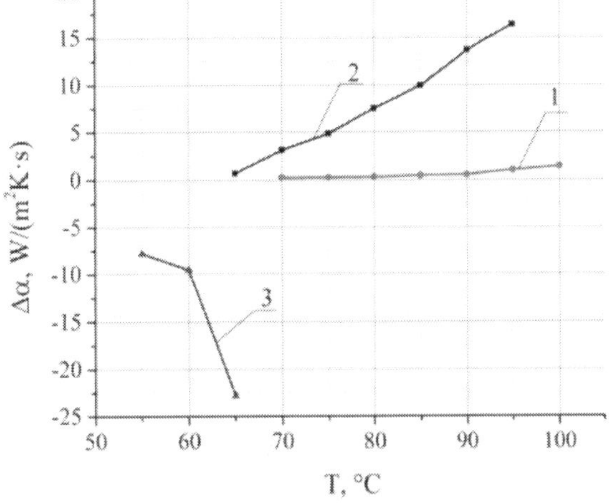

Fig. 4. The rate of change in the heat transfer coefficient from the substrate temperature. Liquids: 1 – distilled water, 2 – ethanol, 3 – HFE 7100.

As seen in Fig. 4, the heat transfer coefficient for ethanol and distilled water increases with increasing temperature. What is more, for ethanol this change is more significant (up to 15 W/(m^2 Ks).

For ethanol, the heat transfer coefficient varies from 4500 to 6250 W/(m^2°C), and at 65 °C it practically does not change during the experiment. At temperatures from 70 to 95 °C with an increase in the time, the coefficient increases by 250-1000 W/(m^2°C) as a function of temperature. The latter is explained by the fact that temperatures over 78 °C are greater than the boiling point. When underheated to boiling (10-15 °C), the heat transfer coefficient varies insignificantly compared with evaporation at boiling point and overheating. With a decrease in underheating to the boiling point, a significant increase in the heat transfer coefficient occurred. At temperature of 75 °C (close to the boiling point), the heat transfer coefficient is the maximum 5750-6250 W/(m^2°C) (Fig. 1). At temperatures above the boiling point, the coefficient decreases, and with a superheat of 10-20 °C, the nature of the latter's change is the same.

Three regions of the change in the heat transfer coefficient for distilled water could be distinguished (Fig. 2): at low temperatures for distilled water (70-80 °C) is the first region of underheating (in 500 seconds the heat transfer coefficient changes by 100 W/(m^2°C)); at temperatures 85, 90 °C – the second region of underheating, at the same time the coefficient changes by 200 W/(m^2°C); and at 95, 100 °C is the third region of underheating, the coefficient changes by 500-600 W/(m^2°C). At boiling (temperature 100 °C) the heat exchange processes took place most intensively; therefore the change in the heat transfer coefficient is maximal.

For the liquid HFE, the nature of the change is different: with increasing time, the coefficient decreases by 1000-1750 W/(m^2°C). The latter is due to the rapidity of the evaporation process, the large molecular mass of the liquid HFE and properties specificity of the refrigerant. In 30-35 seconds, a quasi-stationary regime occurs, the heat transfer coefficient varies insignificantly, as can be seen from Fig. 3.

4 Conclusion

Experimental studies on the thin layers evaporation of ethanol, distilled water, HFE-7100, have been carried out. The changes in the heat transfer coefficient over time were determined. It has been established that with decrease in the thickness of the liquid layer, the heat transfer coefficient increases for ethanol (from 4500 to 6250 W / (m^2°C)) and distilled water (from 2400 to 3400 W / (m^2°C)). For the HFE liquid, the nature of the change is different. As the time increases, the heat transfer coefficient decreases by 1000-1750 W / (m^2°C).

The reported study was supported by RFBR, research project No. 18-38-00315 mol_a.

References

1. D. O. Glushkov, G. V. Kuznetsov, P. A. Strizhak, Combust. Explos. Shock Waves, **50**, 670 (2014), DOI: 10.1134/S0010508214060082
2. D. V. Feoktistov, E. A. Vympin, A. E. Nurpeiis, MATEC Web Conf. **72,** 01081 (2016), DOI: 10.1051/matecconf/20167201081
3. A. A. Sivkov, A. Pak, I. Shanenkov, Y. Kolganova, I. Prosvirin, Adv. Mat. Res. **880**, 36 (2014), DOI:10.4028/www.scientific.net/AMR.880.36
4. E. G. Orlova, D. V. Feoktistov, G. V. Kuznetsov, K. O. Ponomarev, European J. Mech. - B/Fluids, **68,** 118 (2018), DOI: 10.1016/j.euromechflu.2017.12.002
5. I. N. Madyshev, O. S. Dmitrieva, A. V. Dmitriev, A. N. Nikolaev, CAPE, **52**, 299 (2016), DOI 10.1007/s10556-016-0189-2
6. G. V. Kuznetsov, D. V. Feoktistov, E. G. Orlova, S. Y. Misyura, V. S. Morozov, A. G. Islamova, Int. J. Heat Mass Transfer, **68**, 118 (2018), DOI: 10.1016/j.euromechflu.2017.12.002
7. V. E. Nakoryakov, S. Y. Misyura, J. Eng Thermophys. **25**, 24 (2016), DOI: 10.1134/S1810232816010033
8. S. Y. Misyura, Sci. Rep. **6**, 30324 (2016), DOI: 10.1038/srep30324
9. K. Ponomarev, E. Orlova, D. Feoktistov, EPJ Web Conf. **110**, 01060 (2015), DOI: 10.1051/epjconf/201611001060
10. D. V. Zaitsev, M. L. Aviles, H. Auracher, O. A. Kabov, Microgravity Sci. Technol. **19**, 71 (2007), DOI: 10.1007/BF02915756
11. D. Feoktistov, S. Misyura, A. Islamova, K. Batishcheva, V. Velicanov, EPJ Web Conf. **159**, 0011 (2017), DOI: 10.1051/epjconf/201715900011
12. V. D. Velicanov, E. G. Orlova, A. S. Saigash, O. S. Korneva, MATEC Web Conf. **72**, 01028 (2016), DOI: 10.1051/matecconf/2016720102

The comparison of complexities of the Chinese and Russian languages on the example of terminology of chemistry

Zhanna K. Ivanova[1,*], *Juri* V. Kobenko[2], *Mariya* A. Kulkova[3], and *Elena* S. Riabova[4]

[1]National Research Tomsk Polytechnic University, 634050 Tomsk, Russia
[2]National Research Tomsk State University, 634050 Tomsk, Russia
[3]Kazan Federal University, 420008 Kazan, Russia
[4]Samara University, 443086 Samara, Russia

Abstract. The article presents the examples of comparing chemical terms in the Chinese and Russian languages and the results of two experiments conducted with Chinese and Russian students. The first part of the work deals with a comparison of the one–component chemical element names from the periodic table. Then, it continues with dividing different chemical terms in Chinese into several groups: two–component, three–component, four–component, multicomponent group and comparing them with its analogical terms in Russian to analyze the main difficulties in understanding these terms in both languages. The second part of the work presents the main difficulties, revealed during two experiments, conducted with Chinese and Russian students. The students were given texts in their mother tongues with a chemical context, which they wrote and read noting the time. At the end of the work, all the advantages and complexities revealed in the results of the first and second parts of the work were calculated. Similar language comparisons are becoming more relevant in the last few years, for they contribute to development of methods of analyzing languages and discovering new and improved approaches to the study and use of languages.

1 Introduction

Many linguists are wondering whether it is possible to assess the complexity of languages objectively, that is, without attracting native speakers of a particular language. If so, what criteria should be chosen. A. Piperski gives an example of a situation when the Martian has flown to our planet and needs to learn the human language – which language would be easier for him, and which would be more difficult [1]. Linguists studying language complexity are trying to answer this question.

This science field is relatively young, its active development began only in the last 20–25 years and therefore it still does not have a universal theory and methods of verification. Nevertheless, the science of language complexity has already managed to make three main theoretical conclusions, which A. Berdichevsky summarized in his work: 1. The widespread

* Corresponding author: ivanova.happy@yandex.ru

idea that all languages are equally complex is incorrect. 2. You can not only rank languages by complexity, but also try to measure the complexity of a language – or at least a fragment of a language – quantitatively. 3. Quantitative measurements, as well as some qualitative studies show that the complexity of the language depends on social parameters. [2].

In the Chinese and Russian languages most of the terminology consists of borrowing primarily from the English language. Each of these languages has its own ways of borrowing and translating Western terms, which are unique in their own way and have both positive and negative sides.

In the Chinese language most of the borrowings refer to a group of semantic borrowings, while phonetic borrowings are less represented. The main way of forming a term is composition as the most productive way of replenishing the Chinese terminology of material science.

In the Russian language, unlike the Chinese, borrowing from the English language, indirect borrowing, or borrowing in parallel with analogues in the Russian language prevail. These types of borrowings are good in terms of the terminology internationalization and the cumulative function of the English language, but are difficult for understanding by the speakers of the recipient language. Of course, the use of such terms is not a problem when a person knows the Latin and Greek languages, and nowadays even the knowledge of English is enough. However, a foreign language with respect to the Russian language is an external knowledge (meta-knowledge), the volume of which should be minimized.

2 The results of the comparison of the Russian and Chinese languages

As the result of the analysis of chemical elements terms in the Chinese and Russian languages, the main conclusion was made in the context of the two different systems of borrowings of the chemical element terms in the periodic table – this is the advantage of the Chinese terms over the Russian ones, which consists in the translation systematicity of Chinese terms from the original language with the help of the keys. Moreover, while analyzing single–component names of 118 chemical elements of the D.I. Mendeleev's table in the Chinese language the following features were revealed:

– all elements have a key denoting the category of the element that helps to achieve systematicity in translating borrowings of these terms. For example, all the metals have the key 钅 "metal" (钠 – sodium, 钾 – potassium, etc.), nonmetals – keys石 "stone" (硫 – sulfur, 硅 – silicon), 气 "gas" (氯 – chlorine, 氮 – nitrogen), 氵 "water" (溴 – bromine);

– out of 118 terms, 95 characters have a key "metal", 10 – "stone", 12 – "gas", 2 – "water";

– 11 names of chemical elements reflect certain properties or characteristics of these elements, for example, 溴xiù "bromine" from Greek βρῶμος is a "stench"" because the simple substance bromine is a heavy caustic liquid with a strong unpleasant "heavy" smell, and this property is reflected in the meaning of the often used in everyday speech hieroglyph 臭xiù "stinky, rotten", which is used in the right part of the term;

– in the remaining 107 terms, the hieroglyphs used in the right part do not reflect any properties of these elements, performing only the function of phonetics in the terms.

Unlike Chinese terms there is no systematicity in Russian naming units that consists of "hints" or category designations of chemical elements. The only common element in Russian naming units is similar suffixes at the end of the terms. For example, in 65 names of metals out of 96 ones there is a suffix of masculine gender –i at the end of the term. Therefore, almost 68% of the names of metals are systematized to some extent. However, some nonmetals also have the suffix –i kremni (silicon), geli (helium), which confirms the ineffectiveness of suffixes as a possible method of chemical names systematizing. Further, the chemical terms of the Chinese language, which include more than one hieroglyph, are analyzed. The results of the analysis show the transparent semantics of Chinese terms, which clearly reflects the meaning of the term and promotes the understanding of terms by native speakers. Russian terms in comparison with Chinese ones are less clear since most terms are borrowed from the ancient Greek and Latin languages. Let us give some examples.

Two-component term 加 嗅 "odorization": add + smell. In the dictionary, the term "odorization" is defined as giving (to the gas) a characteristic odor. As it can be seen, the translation of the term hieroglyphs ideally reflects the essence of the term.

Three-component term 催化剂 "catalyst" is a chemical that accelerates the reaction. This word comes from the Greek katalysis – destruction. In Chinese the hieroglyphs in this term mean "hurry, induce + chemical + substance, means" – the meaning of the hieroglyphs clearly describes the term.

Four-component term 两性 离子 "amphion" or otherwise "amphoteric ion". In this phrase, the most unclear word is amphoteric, which comes from the Greek word "amphotero" – "dual", "reciprocal" In the Chinese term, hieroglyphs consistently mean: "two + quality, property, suffix –ost + part, leave + particle." We see that the first two hieroglyphs clearly denote the concept of "duality", since amphotericity is the ability of certain chemicals and compounds to have both acidic and basic properties depending on the conditions. The answer to the question, why the term "ion" has the hieroglyph 离 "to part, to leave" is also quite simple – since an ion is produced in the process of separation or addition of an electron to an atom or group of atoms.

Multicomponent term 非压缩氢气 "uncompressed hydrogen". Let us analyze the components of this term: no + pressure + compress + hydrogen + gas. In the Russian version of this term we come across a word of Latin origin "compression", which means "limit the amount of gas under the influence of external forces in order to reduce its volume, increase of pressure and temperature.

2.1 The results of the first experiment

During the first experiment with the writing, the following Chinese language difficulties that Chinese students experienced were revealed.

1. Comparatively long average time of a small text writing – a little more than half a page in 19 min. 6 sec.
2. The problem with punctuation and paragraph indention that makes the text look unstructured.
3. The big difference between classical written and spoken Chinese.
4. The difficulty of hieroglyphs spelling – native speakers forget how to write some hieroglyphs or certain parts of hieroglyphs. For example, students forgot how to write 醛 quán (aldehyde) and 粪 fèn (manure).
5. Homonymy of the Chinese language – because of the consonance of the words hieroglyphs that were not in the text were written. For example, 摄入 shè rù (absorption) 涉入 shè rù (penetrate).
6. The confusion of words, which partly sound the same and are partly similar in spelling. For example, 非 fēi (not–) and 没 méi (not).
7. Words are not separated from each other by spaces in the written form, as in Russian, thus, it is sometimes visually difficult to determine the boundaries of words.

Difficulties of the Russian language, which Russian students experienced during the first experiment.

1. Complexity of grammatical forms that makes it easy to confuse letters in endings, roots or suffixes of words, even simply by inattention. For example, some students wrote "YestestvennYJ and naturalnYJE flavors" (Natural (singular form) and natural (plural form) flavors).

2. Fast fatigue while writing full words without abbreviations.

3. Voiceless/voiced letters at the end of the word – the need to check the spelling of words by changing the form of the word, or simply memorizing the rules. For example, students wrote CinnamaldegiT, instead of CinnamaldegiD.

4. A large number of words of Latin and Greek origin (aromatizator, cinnamaldegide, vanillin, orchideya, glutamat natria, etc.).

The advantages of the Chinese language, revealed during the first experiment:

5. A comparatively small amount of text that takes up less space on the paper.

6. Borrowed words with transparent semantics, which makes terms more understandable in writing. For example, the Chinese word 香草 "scented grass," means "vanilla", but in Russian this word is borrowed, or, for example, the word 香精 "fragrant extract", which is translated into Russian by the Latin word "aromatizator".

The advantages of the Russian language, found out during the first experiment:

1. Comparatively high average speed of text writing – 20 min. 6 sec., especially with the reduction of words – 15 min. 30 sec.

2. The presence of grammatical forms allows you to get more information, which contributes to more detailed understanding and translation. For example, the verb "pridumala" (invent): first, we immediately see that the subject of the action is a woman (feminine gender); secondly, we know for sure that the subject is one (singular), thirdly, the action was in the past (the past tense), fourthly, the action did not take place for a certain period of time, but was completed at a certain stage (the perfect aspect of the verb). In the Chinese translation "发明" there are not such informative grammatical categories.

3. In Russian it is impossible to forget how the whole word is written, you can only make a mistake in writing a letter or a syllable (in Chinese you can forget how to write the whole hieroglyph or confuse and write another hieroglyph that sound the same).

4. A more sophisticated system of punctuation (in Russian it was easier to indent paragraphs and place commas).

2.2 The results of the second experiment

The difficulties of the Chinese language that were revealed during the second experiment with oral speech.

1. One character can be read in different ways and there are no clear rules for this or that variant of reading, you can check only in the dictionary. For example, in the hieroglyph 差, in addition to reading chā, there are several other readings: chà, chāi, chài, cī, cuō, jiē.

2. Homonymy of words.

3. Tones that involve a large articular work.

4. Great semantic value of the tones – if the tone changes, the writing of the word hieroglyph and its meaning change dramatically. Very often, native speakers themselves during the speech clarify the meaning of a word depending on the tone.

The difficulties of the Russian language that were revealed during the second experiment.

1. Voiceless / voiced consonants at the end of the word: because of the fact that voiced consonants at the end of the word are devoiced, it is easy to make a mistake in writing (the Russian student made a mistake by writing not cinnamaldegiD, but cinnamaldegiT, it means

that in oral speech she would incorrectly remember the term and would wrongly use it later, for example, she would then say "CinnamaldegiTa, cinnamaldegiTu", etc.).

The advantages of the Russian language that were revealed during the second experiment.

2. A higher average text reading speed – 2 minutes 41 seconds, which is twice as large as Chinese, it proves that it is relatively easier to read in Russian.

3. The presence of the alphabet in contrast to the Chinese language, where the hieroglyphs are not systematized and their reading can only be memorized or checked in the dictionary.

Other currently important difficulties of the Chinese language, which were not revealed during the experiments, but exist depending on the situation.
1. There are 7 dialect groups in Chinese, which means that residents of different provinces often even can not support a conversation on everyday topics because dialects imply significant differences at the phonetic and lexical levels.

2. Traditional hieroglyphics, which is still used in the territory of Hong Kong, Macau, Taiwan and Singapore.

3. The presence of more than one hundred words in the Chinese language, which imply the use of a certain classifier after each numeral that refers to a particular noun. Sometimes native speakers find it difficult to use this or that counter word and that is why they use certain universal counter words.

4. Strong adaptation of proper names to their almost complete unrecognizability and dissimilarity in sounding with the original name. For example, names from the text **山本麻由** Shānběn máyóu Mayu Yamamoto, **哈佛** Hāfó Harvard, etc ..

5. A huge number of words that have a large number of meanings based on the situation and context. This applies especially to verbs, for example, the verb 打 dǎ, which means "to hit, to beat," can be translated into Russian by many other words: 打杯子 break a glass, 打鼓 hit / beat a drum, 打针 make an injection, etc. [3].

Other difficulties of the Russian language, which were not revealed during the experiments, but exist depending on the situation:

1. In the Russian language there is a large number of long terms, which are very difficult not only to use in speech, but even simply to write down. Let us give an example of several chemical terms that are extremely long:

– "Tetrahydropyranylcyclopentyltetrahydropyridopyridine" (55 letters, chemical substance).

– "Hydrazinocarbonylmethylbromphenyldihydrobenzdiazepine" (50 letters, tranquilizer Gidazepam).

– "Cocamidopropylpropyleneglycoldimmoniumchloridephosphate" (48 letters, chemical substance).

– "methoxychlorodiethylaminomethylbutylaminoacridine" (44 letters, chemical substance, another name – acrichine) [4].

2. In Russian, the word stress is very difficult. When the case and number are changing the stress can shift and it causes certain difficulties. In many languages the stress is almost always put on the last syllable (Turkish, Tatar), or on the first syllable (Czech, Latvian), or on the penultimate syllable (Polish). However in Russian there are no clear rules and the stress varies in different forms of the word. For example, *golová – gólovu – golóv – golovámi* (different forms of the word "head"). In a long word it can be somewhere at the very end, for example: pereraspredelenó (redistributed). The chemical word "výkristallizovavshiesya" (crystallized) is a perfectly correct Russian word with an initial stress [5].

3. A large number of polysemantic words, this is especially the case of nouns (and of Chinese verbs as it was mentioned above). For example, the word "order"", which was noted by one of the Chinese students participating in the experiment. If you look into the explanatory dictionary, you will see seven different meanings of the word: 1. Well-organized state 2. Control system, structure 3. Lifestyle, way of life 4. Certain sequence, course 5.

Military organization, formation. 6. Way, method 7. Variation, type of something 8. Conversation word for assessing the situation [6].

Table 1. Total calculation of all the advantages and disadvantages that were revealed during the experiment.

	The Chinese language	The Russian language
First experiment	Disadvantages 7 Advantages 2	Disadvantages 4 Advantages 4
Second experiment	Disadvantages 4 Advantages 0	Disadvantages 1 Advantages 2
Other cases	Disadvantages 5 Advantages 0	Disadvantages 3 Advantages 0
Total	Disadvantages 16 Advantages 2	Disadvantages 8 Advantages 6

3 Conclusion

It is obvious that it is not possible to analyze and mention all the disadvantages and advantages of the Chinese and Russian languages as these language systems are rather complicated and extensive. However, the data that were collected during the experiments and theoretical knowledge allow us to see that the Russian language has more advantages that the Chinese language. Although grammatical categories of the Russian language are difficult to use [7], they nevertheless contribute to the most detailed description of a phenomenon, which is very important in the most specific areas of expertise, where it is necessary to describe every minor detail of what is happening, for example, during experiments on chemistry or surgical operations. The Chinese language is easier for understanding the meaning of individual words, but not of the whole sentence or a text. This is especially the case of any specialized terms in certain areas, for example, in chemistry [8]. The Russian language is harder for understanding technical terms, since most of the words were borrowed from Latin and Greek.

References

1. A. Piperski, Language Complexity, URL: https://postnauka.ru/video/54750 [in Russian]
2. A. Berdichevskiy, Language Complexity. Topics in the Study of Language, **5** (2012) [in Russian]
3. U. Lizhu, Young Scientist **7**, 1012 (2015) [in Russian]
4. Interesting Russian Words, URL: https://orfogrammka.ru
5. A. A. Zaliznyak, About Russian accent, URL: http://elementy.ru/nauchno–populyarnaya_biblioteka/432371/Iz_russkogo_udareniya [in Russian]
6. S. Ozhegov, Dictionary of Russian Language, URL: http://www.ozhegov.org [in Russian]
7. M. A. Kulkova, E. A. Denisova, Th. H. Zinecker, MJLTM 202 (2016) [in Russian]
8. T. A. Martseva, A. Yu. Snisar, Yu. V. Kobenko, K. A. Girfanova, International Conference on Linguistic and Cultural Studies **677**, 266 (2018)

Ordinary differential equations for the dynamic characteristics of heating boilers

Sergei Khaustov[1,*], *Olga* Guk[1], and *Igor* Razov[2]

[1]National Research Tomsk Polytechnic University, 634050 Tomsk, Russia
[2]Tyumen industrial University, 625003 Tyumen, Russia

Abstract. The paper presents ordinary differential equations for the dynamic characteristics of a solid fuel boiler that can be combined into one-dimensional nonstationary mathematical model for simulating long-term dynamics of a solid fuel boiler. This model requires less computational time for a qualitative simulation of boiler's operation than known CFD-solutions. The long-term dynamic model presented can help to determine annual costs, to set up automatic control systems and to detect dangerous deviations in the project.

1 Introduction

With the development of computer technology, mathematical models have become widely used. These models enable us to correct technical decisions even while the design. Highly detailed CFD-models (Computational Fluid Dynamics) in a steady state [1-7] are the most commonly used method for heating boiler gas-dynamics evaluation. However, the thermal load on heating boilers varies considerably throughout the year followed by the changes of outdoor temperature [8-10]. In addition, during the long-term operation of solid fuel boilers, the fuel characteristics generally change, which also requires an adjustment of the operating mode [8, 11]. Detailed CFD modeling of physical processes requires obtaining a stable solution of a system of complex differential equations by a resource-intensive finite element method [12-14]. It makes CFD models ineffective in solving practical-oriented non-steady-state heat exchange problems at long time intervals [15].

In general, the dynamic characteristics are the dependence of the change in time of productivity and efficiency of the boiler when it is varied by the regime parameters [16]. Dynamic simulation is practically used to determine the dynamic characteristics of the boiler [8-10, 15, 17, 18]. The aim of this paper is to develop and present practically applicable equations for the dynamic characteristics of heating boiler.

We failed to find any mathematical models for simulating long-term dynamics of a solid fuel boiler. However these models can be useful for determining annual costs depending on meteorological and technological features. Moreover the long-term dynamic model of a boiler can help to set up automatic control systems and to detect dangerous deviations in the project [4, 18].

* Corresponding author: khaustovsa@tpu.ru

2 Mathematical apparatus

Suppose that there is a diffusion combustion regime and the reaction rate is controlled by changing the inlet pressure. Let's denote: F is the area (m^2) of the grate on which the fuel is combusted, m is the mass (kg) of fuel on the grate at an arbitrary time t. As already noted, m is a variable that varies in proportion to the flow rate of the combustion air. The amount of fuel consumption that determines the boiler's thermal output is the first derivative of m in time t:

$$\dot{m} = \frac{\partial m}{\partial t}.$$

The heat output of a hot-water boiler is usually defined as the amount of heat (kW) transferred to the coolant (water) during the combustion. According to the known values of the calorific value of fuel Q (kJ/kg) and the efficiency of the boiler, this amount of heat (kW) can be estimated as:

$$P = Q \cdot \eta \cdot \dot{m}.$$

However due to the thermal inertia during transient operation of a boiler, the amount of heat transferred to the consumer differs from this value:

$$P = \dot{m}_w \cdot c_w \cdot (T_{out} - T_{in}),$$

where m_w - mass flow of boiler water (kg/s), c_w - specific heat of water ($kJ \cdot kg^{-1} \cdot K^{-1}$), Tout and T_{in} - temperatures of direct and reverse supply water, respectively, °C.

The filtration rate – the air flow per second referred to the total cross-sectional area of the layer – is often used in investigations of layer combustion. Using this concept, we write the volumetric air flow, where ω is the average filtration rate at an arbitrary moment of time t, m/s. The fuel consumption (kg/s) will be written by the equation:

$$\dot{m} = \frac{F \cdot \varpi}{V_0 \cdot \alpha}. \tag{1}$$

The process of coal burning on the grate is controlled with forced and induced draft fans. The control parameters are the air pressure in the discharge air duct and the pressure of the flue gases behind the boiler (Pa). It is possible to control the sum of the pressure losses in the gas and air tracts of the boiler (Pa):

$$p_{in} - p_{out} = \Delta p_g + \Delta p_w.$$

Let us express the pressure loss in the air path through the average velocity of filtration with the Darcy-Weisbach equation:

$$\Delta p_w = p_{in} - p_{out} - \Delta p_g = (\zeta + \lambda \cdot \delta)\frac{\varpi^2 \rho_w}{2} \tag{2}$$

where ω is an average filtration rate, ρ_w is an air density, kg/m^3, and ζ is an pressure loss coefficient, λ is the drag coefficient of the layer, m^{-1}; δ is the thickness of the coal layer (m) at an arbitrary time t:

$$\delta = \frac{m}{\rho_c \cdot F}$$

where ρ_c is a coal density.

Expressing the value of the filtration rate from formula (2) and substituting it in (1), we obtain the final formula for fuel consumption:

$$\dot{m} = \frac{F \cdot \sqrt{\frac{2}{\rho_w} \cdot \frac{p_{in} - p_{out} - \Delta p_g}{\zeta + \lambda \cdot \delta}}}{V_0 \cdot \alpha} \quad (3)$$

Despite the assumption (1), the analytical solution of equation (3) is a complicated problem because of its nonlinearity. The most universal and simple numerical method for solving ordinary differential equations is the Euler method. According to this method, we replace the derivative entering into equation (3) with a finite difference analogue in such a way that the time axis is sampled at intervals within which the fuel consumption is assumed constant.

To simulate the possible unevenness in the thickness or density of the coal layer, it can be sampled at N sections with a fuel mass m_i and a lattice area F_i, where the index i is the number of the section in the range 1.N. The passage cross-section of the H_i lattice can also be set unevenly, which makes it possible to model the burning of coal on complex geometry grates:

Air flowrate: $\sum_{i=1}^{N} F_i \cdot \varpi_i$ Fuel massflow: $\sum_{i=1}^{N} \dot{m}_i$ $\dot{m}_i = \frac{F_i \cdot \sqrt{\frac{2}{\rho_w} \cdot \frac{p_{in} - p_{out} - \Delta p_g}{\zeta_i + \lambda \cdot \delta_i}}}{V_0 \cdot \alpha}$

This approach enables us to predict the distribution of air filtration rates and coal burnup in a layer and simulate changes of the layer geometry in time (Fig. 1).

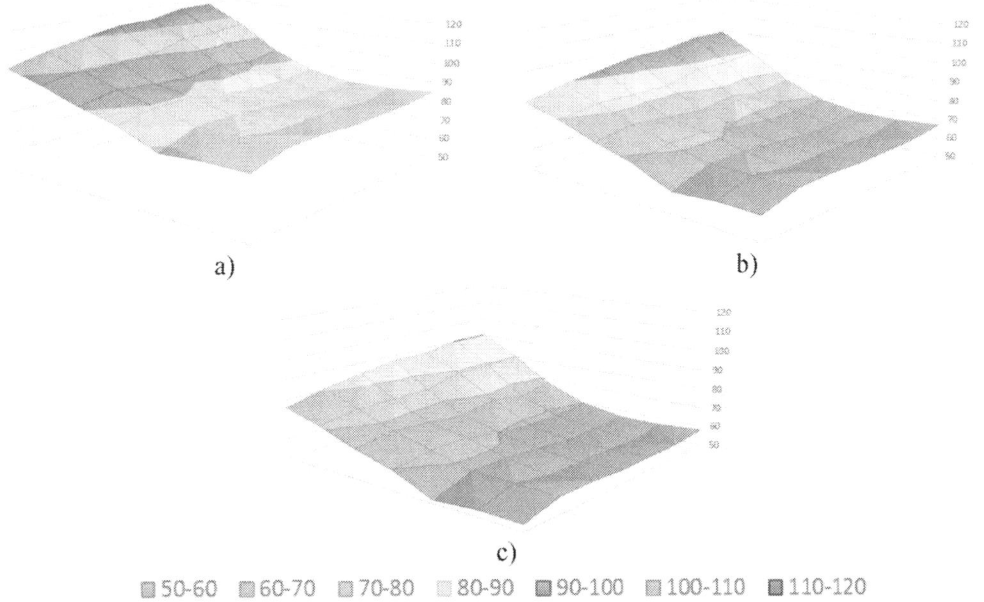

Fig. 1. Coal layer (mm) thickness distribution at different time points: a) start up; b) after 30 minutes; c) after one hour.

The rate of filtration and the burning rate of fuel (Fig. 2) are redistributed in inverse proportion to the drag of parallel sections. Even with a slight unevenness in the thickness or density of the coal layer, it burns down in places of least drag, tending to form sections without fuel – craters. The formation of craters, in turn, leads to a useless breakthrough of the air into the furnace through the bare sections of the grate. At the same time, the filtration rate sharply decreases through the areas covered with a layer of coal.

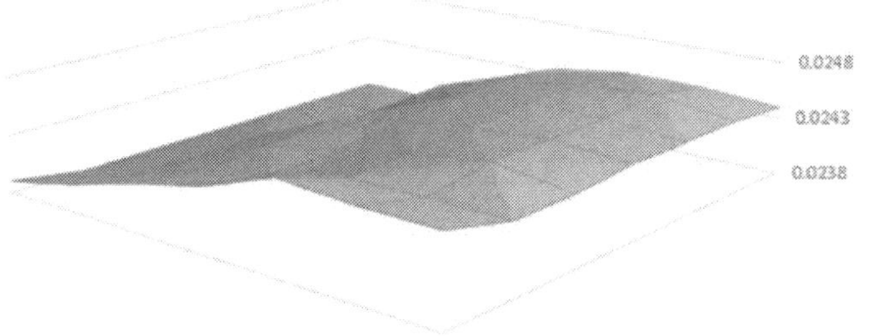

Fig. 2. Coal burning rate (kg·s^{-1}·m^{-2}) distribution.

According to the known values of the calorific value of the fuel Q (kJ / kg) and the efficiency of the boiler, the amount of heat (kW) that is transferred to the coolant during the combustion of the fuel can be written as

$$Q \cdot \eta \cdot \sum_{i=1}^{N} \dot{m}_i$$

In order to take into account the thermal inertia of the boiler's water volume, we assume that it has an average temperature T_{out}. Then from the coolant heat balance equation we write the expression how T_{out} changes in time:

$$\frac{\partial T_{out}}{\partial t} = \frac{Q \cdot \eta \cdot \sum_{i=1}^{N} \dot{m}_i + \dot{m}_w \cdot c_w \cdot (T_{in} - T_{out})}{m_w \cdot c_w}$$

where $m_{в}$ is the mass of water in the boiler, kg.

3 Conclusion

Given ordinary differential equations for the dynamic characteristics of a solid fuel boiler can be combined into one-dimensional nonstationary mathematical model of the fire-tube boiler. This model requires less computational time for a qualitative simulation of boiler's operation than known CFD-solutions. With the Euler method the proposed model can be easily applied to solve engineering problems.

At the design stage, the dynamic model of the boiler is indispensable for detecting dangerous deviations in the heat engineering characteristics of fuel and operating parameters, as well as for the development and adjustment of automatic control systems. In addition, the dynamic characteristics of the boiler are decisive in the development of various simulators for training operators and boiler room staff.

References

1. H. Kim, S. Choi, Int. Commun. Heat Mass. **32**, 786 (2005)
2. P. J. Coelho, P. A. Novo, M. G. Carvalho, J. Super Comput. **13**, 211 (1999)

3. A. A. Bhuiyan, J. Naser, Fuel **159**, 150 (2015)
4. S. A. Khaustov, Y. A. Belousova, K. V. Buvakov, A. Y. Dolgih, R. N. Kulesh, 11th International Forum on Strategic Technology (IFOST - 2016) Proceedings **2**, 548 (2016)
5. S. A. Khaustov, A. S. Zavorin, K. V. Buvakov, N. A. Zakharushkin, MATEC Web Conf. **19**, 01020 (2014)
6. S. A. Khaustov, A. S. Zavorin, K. V. Buvakov, L. D. Kudryashova, A. V. Tshelkunova, EPJ Web Conf. **82**, 01041 (2015)
7. S. A. Khaustov, A. S. Zavorin, K. V. Buvakov, V. A. Sheikin, EPJ Web Conf. **82**, 01039 (2015)
8. S. A. Khaustov, O. V. Khaustova, A. N. Ermolaev, Bulletin of the Tomsk Polytechnic University. Geo Assets Engineering **329** (2), 16 (2018)
9. A. N. Subbotin, Izvestiya Tomsk Polytechnic University **4**, 23 (2008)
10. R. N. Kulesh, A. S. Mazanik, A. N. Subbotin, Izvestiya Tomsk Polytechnic University **4**, 25 (2014)
11. J. A. Oraz, I. E. Korzilova, A. N. Ermolaev, S. A. Khaustov, A. Y. Dolgih, MATEC Web Conf. **92**, 01001 (2016)
12. A. Gómez, N. Fueyo, L. I. Díez, Appl. Therm. Eng. **28**, 532 (2008)
13. M. Pezo, V. D. Stevanovic, Z. Stevanovic, Int. J. Heat Mass Tran. **49**, 1214 (2006)
14. A. Habibi, B. Merci, G. J. Heynderickx, Comput. Chem. Eng. **31**, 1389 (2007)
15. C. K. Weng, A. Ray, X. Dai, Appl. Math. Model. **20**, 501 (1996)
16. V. R. Sabanin A. A. Starostin, A. I. Repin, A. I. Popov, Therm. Eng. **64** (2), 151 (2017)
17. A Behbahaninia., M. Bagheri, R. Bahrampoury. Appl. Therm. Eng. **30**, 2378 (2010)
18. P. S. Gergelizhiu, S. A. Khaustov, R. B. Tabakaev, P. Y. Novoseltsev, A. V. Kazakov, A. S. Zavorin, *Proc. of 2014 Int. Conf. on Mechanical Engineering, Automation and Control Systems*, 109804 (2014)

Simulating a convectional heat transfer in buildings with radiant gas heating

Anton N. Ermolaev[1,*], *Olga* V. Khaustova[2], and *Ilya* A. Turaev[2]

[1]Tyumen industrial University, 625000 Tyumen, Russia
[2]National Research Tomsk Polytechnic University, 634050 Tomsk, Russia

Abstract. The paper presents the results of the simulation of convectional heat transfer in buildings with radiant gas heating. The study is provided by numerical researches based on existing buildings and numerical studies with the use of modern software ANSYS Multiphysics Fluent, Autodesk Inventor. The data reliability is ensured by the use of fundamental laws in regards to numerical researches and approved mathematic models application; certificated measuring equipment; licensed software products; matching of numerical and experimental researches; matching of the simulation with other works. The simulation of luminous infrared gas emitters of all thermal powers (5, 10, 15, 20, 30, 40 kW) in pre-wall and central placements at heights of 4, 5, 6, 7, 8, 9, 10 m was conducted. Gas-fired infrared heat emitters with heat output of 5, 10, 15, 20, 30, 40 kW were studied. Obtained function of IR-heated zone diameter is close to logarithmic function of thermal output. Effective emitter placement heights and least horizontal pitch for the given room volume were determined.

1 Introduction

It's essential to know the thermal performance [1,2] when designing and operating a radiant gas heating that is the radiant heat flow during the operation of infrared gas emitters [3 – 5]. The aim of the work is to identify the dependences, which describe the thermal rate and heat flow density distribution during the operation of infrared gas burners. Another aim is to determine the effective spots for emitters in terms of the room space.

Together with the fact that buildings with radiant gas heating have high thermal performance [7, 8], they also have a number of drawbacks [9, 10]: their upper zone is overheated; their walls are often covered by smut leading ultimately to subsequent destruction. To solve the problems assigned, several numerical, experimental and analytic researches were executed.

Problem solving is provided by numerical researches based on existing buildings and numerical studies with the use of modern software ANSYS Multiphysics Fluent [10 – 15], Autodesk Inventor.

The data reliability is ensured by the use of fundamental laws in regards to numerical researches and approved mathematic models application [3 – 5, 10 – 13]; certificated

* Corresponding author: ermolaevanton03@gmail.com

measuring equipment; licensed software products; matching of numerical and experimental researches; matching of the simulation with other works.

There are several heat flow control methods: the infrared gas emitter design, its relative arrangement in the room, materials of all system elements, optic properties of environment. Consequently, there is a large number of variables in parametric model. We carried out a series of CFD-experiments to obtain temperature fields formed in a computational domain. Computational domain was convectional heat transfer in buildings with radiant gas heating system (Fig. 1).

2 Method of the research

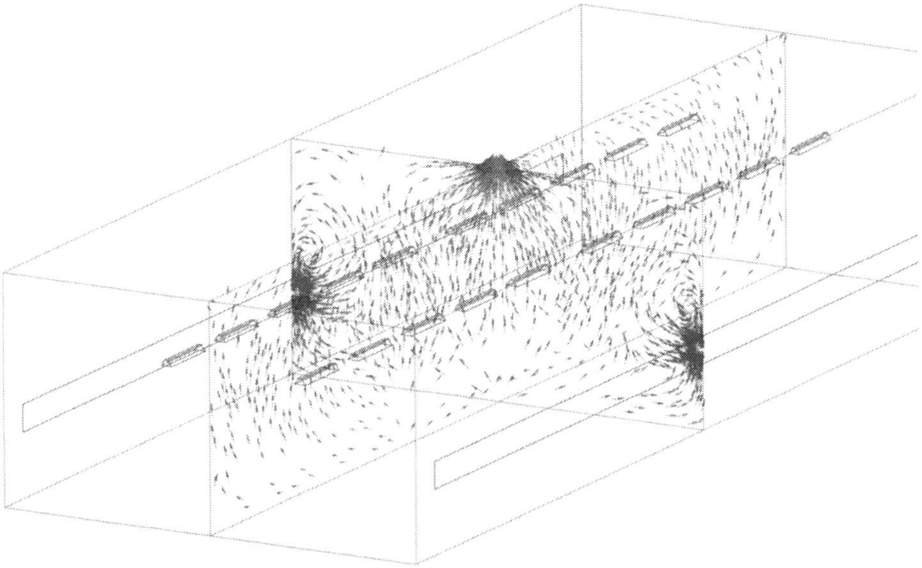

Fig. 1. Isometric view of computational domain.

The analysis of differentiated parameter evaluation possibility of microclimate was executed by simulation modeling of environment inside the modular prefabricated industrial building (workshop) with 60x24x12 structure gauge and radiant gas heating. The most typical and practically significant sets of boundary conditions were considered. To identify the most effective way of managing the air and heat flows inside the industrial building, 125 most typical arrangements of radiant gas heating system were considered.

The simulation of luminous infrared gas emitters of all thermal powers (5, 10, 15, 20, 30, 40 kW) in pre-wall and central placements at heights of 4, 5, 6, 7, 8, 9, 10 m was conducted [3 – 5]. Gas-fired infrared heat emitters with heat output of 5, 10, 15, 20, 30, 40 kW were studied. The gas pressure before the inlet nozzle was 1270 Pa. The air-fuel ratio (AFR) was 9.996. All high-temperature gas-fired infrared emitter's tech specs are taken according to nominal data. Heat exchange with the environment is restricted by four external walls and light holes with the conditions arranged for the third type of heat transfer. Outside environment temperature of the coldest five days – 35 °C was taken. The wall material is sandwich-panel made of polyurethane (λ = 0.019 W/m·K, C = 1.47 kJ/kg·K). The glass-material with reduced thermal conductivity of λ = 1 W/m·K and width of 0,27 m was set for the light holes as it matches the window thermal insulance of 0.42 m^2·K/W. Air exchange in the building is provided by general ventilation, which implies air induction in lower part of the building and air intake in the upper. Mass air flow determines by volume of the room and air change rate of 0.5.

All geometric model elements of calculated area (walls, light holes, vent holes, heating devices) were modeled in Ansys DesignModeller. There is a calculation subarea inside the area, which includes flame chamber and reflector. Heat exchange and emitter calculation area are plotted as conjugated gas-dynamic areas with doubled walls in the place of contact. Discretization was made uneven to obtain the acceptably accurate and efficient calculations. The grid is more accurate in the area of the highest gradient. Primary calculations with the use of grids with different topology and number of cells have shown that approved grid with $3{\cdot}10^6$ control volumes provides low results sensitivity to the further mesh grinding.

3 Results and Conclusion

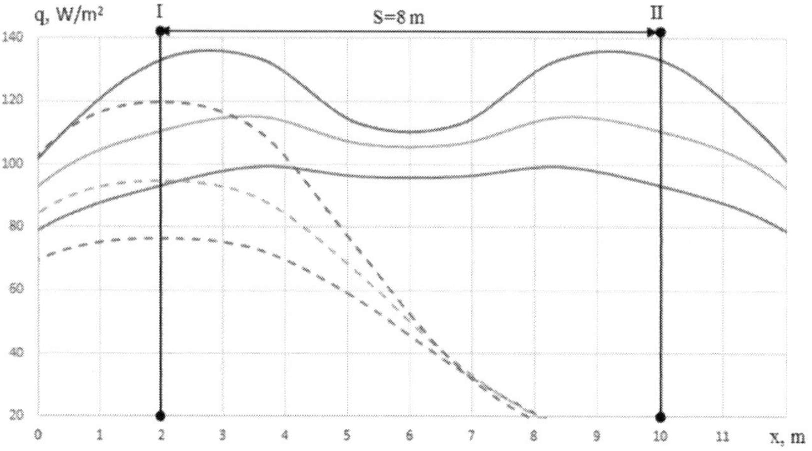

Fig. 2. Heat flow intensity in the working area at one-row(dotted line) and multi-row(continuous line) placements of emitters with the power of 20 W(– 8 m, – 9 m, – 10 m) I, II – placements of first and second emitters; S – emitters placement pitch, m.

To verify the heat comfort task fulfillment, thermal exposure intensity distribution of working area (1 m above the floor), depended on the parameters researched, was calculated. According to Russian set of rules, infrared radiation intensity of working area mustn't exceed 150 W/m^2. The infrared gas emitter placement heights of 4 m (5 kW, 119.3 W/m^2); 6 m (10 kW, 106.1 W/m^2); 7 m (15 kW, 116.9 W/m^2); 8 m (20 kW, 119.3 W/m^2); 10 m (30 kW, 114.6 W/m^2); 10 m (40 kW, 148.2 W/m^2) match this requirement. It was derived that the higher infrared gas emitters are placed, the more uniform local heat flow distribution and the lower amplitude in the IR-heated zone center are. IR-heated zone center implies the heating surface area onto which it didn't fall less than 10 W/m^2 from the infrared gas emitter researched. Obtained function of IR-heated zone diameter is close to logarithmic function of thermal output in the researched range of variables (Fig. 2).

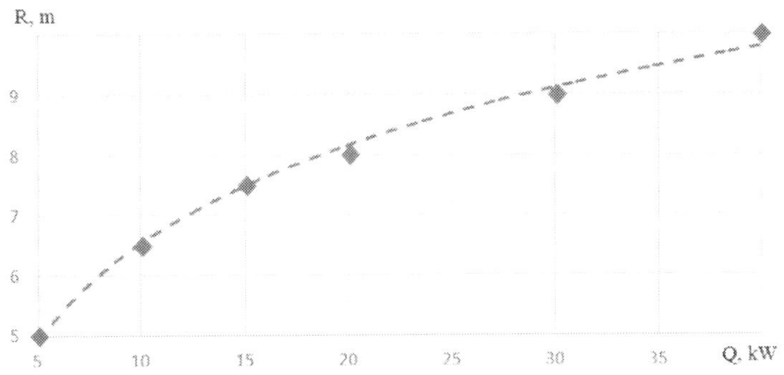

Fig. 3. Impact of gas-fired infrared emitter's thermal output on the radius of IR-heated zone.

Practically applicable dependences were derived. They can be used in gas heating system designing as they describe thermal rate and heat flow intensity distribution in the working area of the building. Effective emitter placement heights and least horizontal pitch for the given room volume were determined.

It was proved in the research that the most uniform working area radiation intensity can be obtained by setting high-temperature gas-fired infrared heat emitter of the same thermal output at a distance, shown in the Fig. 3.

References

1. Y. M. Luo, L. Chevalier, F. Utheza, Int. Polym. Process. **30**, 554 (2015)
2. M. Ostafin, M. Balkenhol, J. Erlemeyer, M. Köhler. Mater. Werks. **41**, 11 (2010)
3. A. N. Ermolaev, O. V. Khaustova, MATEC Web Conf. **141**, 01019 (2017)
4. A. N. Ermolaev, S. A. Khaustov, MATEC Web Conf. **110**, 01024 (2017)
5. A. N. Ermolaev, S. A. Khaustov, MATEC Web Conf. **110**, 01038 (2017)
6. P. S. Gergelizhiu, S. A. Khaustov, R. B. Tabakaev, P. Y. Novoseltsev, A. V. Kazakov, A. S. Zavorin, *Proc. of 2014 Int. Conf. on Mechanical Engineering, Automation and Control Systems* (Tomsk, 2014)
7. D. Patino, B. Crespo, J. Porteiro, J. L. Miguez. Appl Therm Eng. **100**, 849 (2016)
8. S. A. Khaustov, A. V. Kazakov, G. A. Cherkashina, L. A. Sobinova, EPJ Web Conf. **110**, 01028 (2016)
9. K. A. Al-attab, J. C. Ho, Z. A. Zainal, Exp. Therm Fluid Sci. **62**, 1–8 (2015)
10. S. A. Khaustov, A. S. Zavorin, K. V. Buvakov, and V. A. Sheikin, EPJ Web Conf. **82**, 01039 (2015)
11. S. A. Khaustov, A. V. Kazakov, G. A. Cherkashina, L. A. Sobinova, EPJ Web Conf. **110**, 01028 (2016)
12. J. A. Oraz, I. E. Korzilova, A. N. Ermolaev, S. A. Khaustov, A. Y. Dolgih, MATEC Web Conf. **92**, 01001 (2016)
13. S. A. Khaustov, Y. A. Belousova, K. V. Buvakov, A. Y. Dolgih, R. N. Kulesh, IFOST 2016 Proceedings **2**, 548 (2016)
14. S. A. Khaustov, A. S. Zavorin, K. V. Buvakov, N. A. Zakharushkin, MATEC Web Conf. **19**, 01020 (2014)
15. S. A. Khaustov, A. S. Zavorin, K. V. Buvakov, L. D. Kudryashova, A. V. Tshelkunova, EPJ Web Conf. **82**, 01041 (2015)

Levitation of liquid microdroplets over a dry heated substrate near triple-phase contact line

Dmitry Kirichenko[1,2,*], *Dmitry* Zaitsev[1,2], and *Oleg* Kabov[1,2]

[1]Kutateladze Institute Institute of Thermophysics, 630090 Novosibirsk, Russia
[2]Novosibirsk State University, 630090 Novosibirsk, Russia

Abstract. Levitating droplets of liquid condensate are known to organize themselves into ordered arrays over hot liquid-gas interfaces. We report experimental observation of similar behaviour over a dry heated substrate in the vicinity of the triple-phase contact line with high local evaporation rate. It was found that there is an area near the contact line, where no levitating droplets are observed. The width of this area increases with the substrate temperature. Also, it was found that the distance between adjacent levitating droplets increases with the temperature.

1 Introduction

One of effective solutions for removing high heat fluxes are systems based on two-phase flows and sprays [1-3]. The effectiveness of these methods can be impaired by the Leidenfrost effect [4]. Liquid may not reach the heating surface, which can affect the cooling capacity. The behavior of liquid microdroplets falling on a solid surface heated below the Leidenfrost temperature (about 200 °C for water) is not well understood. Droplet levitation over hot liquid-gas interfaces [5-6] can have a significant effect on the performance of two-phase cooling systems. Experimental studies of microdroplets levitation over a dry solid surface with relatively low temperature have been conducted in the works [7-10].

A number of applications such as spray cooling [1], heat pipes [11], film rupture [12] involve contact lines; that is, lines where the liquid–vapor or liquid–air interface comes into contact with the solid wall. We note that macroscale observations do not always allow one to determine whether the solid wall is actually dry or covered by an ultrathin adsorbed film. The term "contact line" is often used for the latter situation even though the true contact of the fluid interface with the solid wall is not achieved; some authors prefer the term "apparent contact line" to describe this situation. In [13-14], the behavior of levitating droplets was studied when they transit from hot liquid interface to dry area. It was found that when approaching the contact line, the droplet levitation height increases several times, which indicates the existence of intense evaporation in the region of the contact line.

* Corresponding author: Dmitriy_Kirichenko@mail.ru

2 Experimental technique

A sketch of the experimental setup is shown in Fig. 1. The test section is a stainless steel plate of 40×40×5 mm^3 in size with a flush-embedded 10×10 mm^2 copper rod, electrically heated by a nichrome wire. According to the measurements, the condition of constant temperature, denoted by T_W, is satisfied along the copper surface. The test section is installed horizontally and open to the atmosphere; the temperature of ambient air is 25±2°C. Optical recording was made with a high-speed camera FAST- CAM SA1.1 (5.400 fps, resolution of 1.024 × 1.024 pixels) coupled with a microscope objective through a set of extension tubes and bellows.

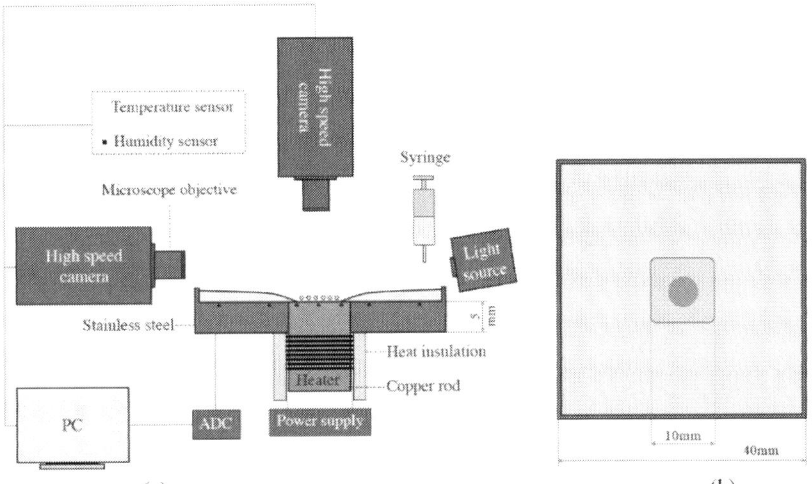

(a) (b)

Fig. 1. (a) Sketch of the side view of the experimental set-up; (b) sketch of the top view of the test section.

Working liquid in the experiments is ultra-pure degassed distilled water (Merck Millipore) with the initial temperature of 25°C. Wetting properties of the surface are important for obtaining the steady configuration needed for our experimental study (a dry patch in a liquid layer, as described in more detail below) so particular attention was paid to control and characterization of surface properties. To obtain the desired wetting characteristics, the working surface of the test section was rough polished prior to the start of the experiment. The morphology of the surface was analyzed using scanning electron microscope (JEOL JSM6700F) and atomic force microscope (Solver Pro NT MDT). The root mean square (RMS) surface roughness was found to be 0.50 μm. The equilibrium contact angle for water on the copper surface was measured at different points by the method of a sessile liquid drop at the room temperature of 25 ± 2°C. To obtain the profile of the drop surface, the shadow method was used with a collimated light source and digital camera Nikon D800 having resolution of 6 micron/pixel. The resulting images were processed using the software by KRUSS. The advancing static contact angle measured at different points over the surface of the copper heater was 74±9°, whereas the receding contact angle was close to zero (less than 10°).

3 Results and discussions

At the start of the experiment, the substrate is covered by a uniform layer of ultrapure water of thickness of 0.4 mm. A dry patch, shown in the sketch in Fig. 2a, is formed in the liquid layer by a downward pulsed jet of air. The heater surface is sufficiently rough to ensure pinning of the contact line, so the dry patch created by the air jet persists throughout the duration of the experiment. Levitating droplets are shown near the right boundary of the Fig. 2a. These droplets are part of a larger ordered array. As they approach the contact line, the local gas flow velocity increases due to stronger evaporation and the resulting force on the droplet becomes higher than its weight, resulting in an upward deflection of the droplet trajectory. However, as the droplet finds itself over the dry patch, it starts moving downward until it reaches a location very close to the solid surface. These processes were recorded by the high-speed camera; typical video frames are shown in Fig. 2b (the side view and the top view correspond to different experimental runs because the camera position had to be changed to obtain the top view Fig. 2a). As the droplets find themselves in the region above the hot dry surface, they disappear by evaporation within about a second. We have conducted experimental investigations of interaction of arrays of levitating droplets with a region of high evaporation near the contact line on a heated surface. Droplets are seen over the solid surface except in the vicinity of the contact line, where there is a narrow region near the contact line where no levitating droplets are observed. This forbidden zone is clearly seen in both the side-view and top-view photographs in Figure 2a. Its width is between 10 and 50 μm and typically increases with temperature, which can be explained by the higher evaporation rate. Fig.3 shows the dependence of width of "forbidden zone" on temperature of the substrate. Dependence of interdroplet distance on the substrate temperature have also been obtained, Fig. 4. Interdroplet distance (distance between centers of two adjacent drops) have been measured using top view experimental photos and also was found to increase with temperature.

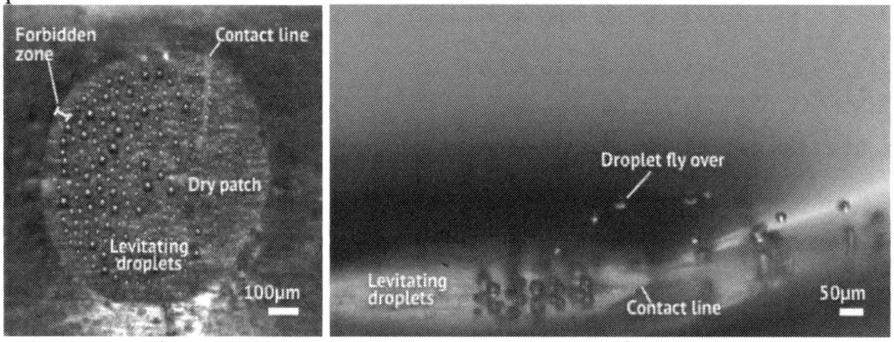

(a) (b)

Fig. 2. Experimental photo: (a) top view of a dry patch and an array of ordered levitating droplets for the substrate temperature 90°C; (b) side view of the dry patch.

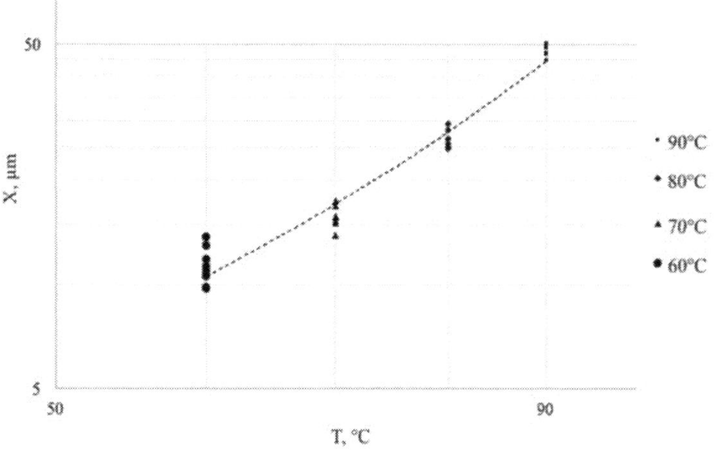

Fig. 3. Dependence of width of "forbidden zone" on substrate temperature. Dashed line – generalization of all data.

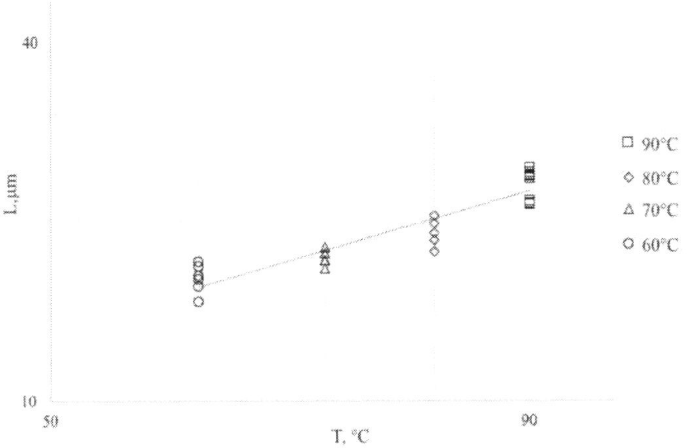

Fig. 4. Dependence of interdroplet distance on substrate temperature. Dashed line – generalization of all data.

We have conducted experimental investigations of interaction of arrays of levitating droplets with a region of high evaporation near the contact line on a heated surface. It was found that there is an area near the contact line, where no levitating droplets are observed. The width of this area increases with the substrate temperature. Also, it was found that the distance between adjacent levitating droplets increases with the temperature.

The study was financially supported by the grant of Russian Foundation for Basic Research (agreement No. 17-38-50222).

References

1. J. Kim, Int. J. Heat Fluid Flow **28**, 753 (2006)
2. R. Srikar, T. Gambaryan-Roisman, C. Steffes, P. Stephan, C. Tropea, A. L. Yarin, Int. J. Heat Mass Transfer **52**, 5814 (2009)
3. O. A. Kabov, D. V. Zaitsev, V. V. Cheverda, A. Bar-Cohen, Exp. Therm. Fluid Sci. **35**, 825 (2011)
4. M. Shirota, M. A. J. van Limbeek, C. Sun, A. Prosperetti, D. Lohse, Phys. Rev. Lett. **116**, 064501 (2016)
5. A. A. Fedorets, JETP Lett. **79**, 372 (2004)
6. A. A. Fedorets, I. Marchuk, and O. Kabov, Tech. Phys. Lett. **37**, 116 (2011)
7. D. V. Zaitsev, D. P.Kirichenko, V. S. Ajaev, O. A. Kabov, Phys. Rev. Lett. **119**, 94503 (2017)
8. D. P. Kirichenko, D. V. Zaitsev, O. A. Kabov, MATEC Web Conf. **72**, 01046 (2016)
9. D. V. Zaitsev, D. P. Kirichenko, MATEC Web Conf. **84**, 00044 (2016)
10. D. P. Kirichenko, D. V. Zaitsev, O. A. Kabov, J. Phys. Conf. Ser. **754**, 042008 (2016)
11. A. Chatterjee, J. L. Plawsky, P. C. Wayner, Adv. Colloid Interface Sci. **168**, 40 (2011)
12. D. V. Zaitsev, A. A. Semenov, O. A. Kabov, T and A **23**, 625 (2016)
13. O. A. Kabov, D. V. Zaitsev, D. P. Kirichenko, V. S. Ajaev, Interfacial Phenomena and Heat Transfer **4**, 207 (2016)
14. O. A. Kabov, D. V. Zaitsev, D. P. Kirichenko, V. S. Ajaev, Nanoscale and microscale thermophysical engineering **21**, 60 (2017)

Study of in-furnace gas-dynamic processes with different design of vortex burners

Oleg M. Koksharev[1,*], and *Andrey* V. Gil[1]

[1]National Research Tomsk Polytechnic University, 634050 Tomsk, Russia

Abstract. The paper considers the combustion chamber of the boiler unit designed for burning Ekibastuz coal by vortex burners with opposed wall firing. Since Ekibastuz coal is not highly reactive, and also has a high content of the mineral part, then vortex burners are used for its combustion. The use of standard two-channel burners leads to increased generation of nitrogen oxides, therefore, in this article the combustion processes of burning Ekibastuz coal using multi-channel burners are considered. The studies were carried out using numerical methods.

1 Introduction

In the Russian energy sector large boiler units are used mainly flare polyfractional burning of pulverized coal with the application of various types of burners and their layout. The choice of burners and combustion scheme, as well as the type of combustion chamber, is determined by the thermal characteristics of the project fuel from the condition of its rational use.

The specifics of the thermal properties of Ekibastuz coal is a combination of explosion safety, low sulfur content, good flowability and low humidity with high ash content, which is an average of $A^r = 32{,}9 - 41{,}9\%$. In this regard, in the boiler unit there are such negative processes as abrasive wear of the heating surfaces, and in the case of a significant change in the content of the mineral part of the fuel, the instability of the combustion process is observed. It is also possible to distinguish a sufficiently high generation of nitrogen oxides, since vortex burner devices are used in the organization of burning of Ekibastuz coal [1, 2].

2 Problem statement

The open type combustion chamber with solid slag removal and water-walled furnace by 60×5.5 mm pipes with a step of 64 mm. The side screens at the bottom ash removal form the slopes of the "cold hopper". The ceiling of the furnace is water-walled by tubes of the radiation superheater. The volume of the combustion chamber (1957 m^3) is structurally divided into two parts: the upper part in the horizontal section along the axes of the pipes of opposite screens has dimensions of 15744×3776 mm, the lower – 15744×9024 mm (Fig. 1).

* Corresponding author: omk2@tpu.ru

Vortex burners are located at 10800 mm with an opposite-fired arrangement in one storey of 6 pieces on the side walls of the furnace, with alternating left and right twist of the flow, which provides additional mixing of the adjacent jets of each pair of burners. For the boiler heating oil nozzles of steam spraying are provided, built-in pulverized coal burners. The diameter of the burners on the external secondary air supply channel is 1032 mm, and on the internal air mixture supply channel-530 mm. The diameter of the fuel oil nozzle is 218 mm.

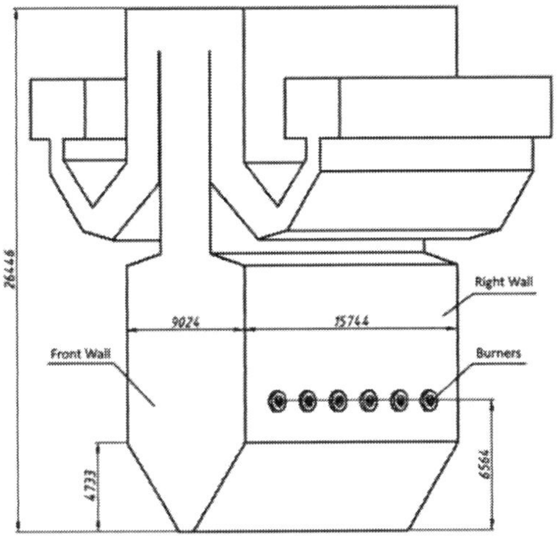

Fig. 1. Geometrical model of the boiler furnace chamber.

For the reconstruction of the boiler was applied a four-channel vortex burners. In the central part there is a fuel oil nozzle with a diameter of 209 mm, then the channel of the dust-air mixture with a diameter of 490 mm and three channels for supplying secondary air.

During the numerical experiment, the rates of fuel supply and secondary air through the channels were taken to be the same. The axial speed of the fuel-air mixture was 20.0 m/s, and the secondary air was 25.4 m/s.

Numerical studies are carried out using Euler-Lagrange approach. Turbulence characteristics are described using the k-ε standard model. The calculation is carried out by the second order of accuracy. To perform the numerical simulations commercial software ANSYS FLUENT V. 12.1 was used [3, 4].

3 Results and discussion

Fig. 2 presents the results of numerical simulation of velocities in the horizontal cross section of the combustion chamber in the center of the burners. In both figures, the formation of a vortex swirling flow of burner jets against each burner is observed. Also in the central part of the coaxially spreading swirling streams ejection zone is visible. These results indicate the correctness of the initial and boundary conditions for the combustion of fuels using vortex burners.

On average, the results in both versions are quite similar in absolute values, there are differences in the location of the areas with the minimum speed values.

Analyzing fig. 2 and 3, it can be noted that at the end of the flow of fuel-air mixture and secondary air from the burners with two channels, the mixing intensity is higher than at the end of the flows from the vortex burners with four channels.

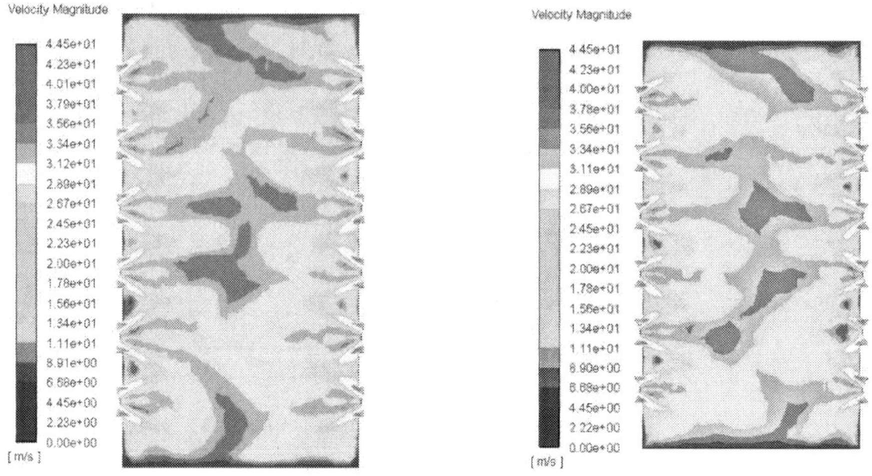

(a) (b)

Fig. 2. Velocity: a – two-channel vortex burners; b – four-channel vortex burners.

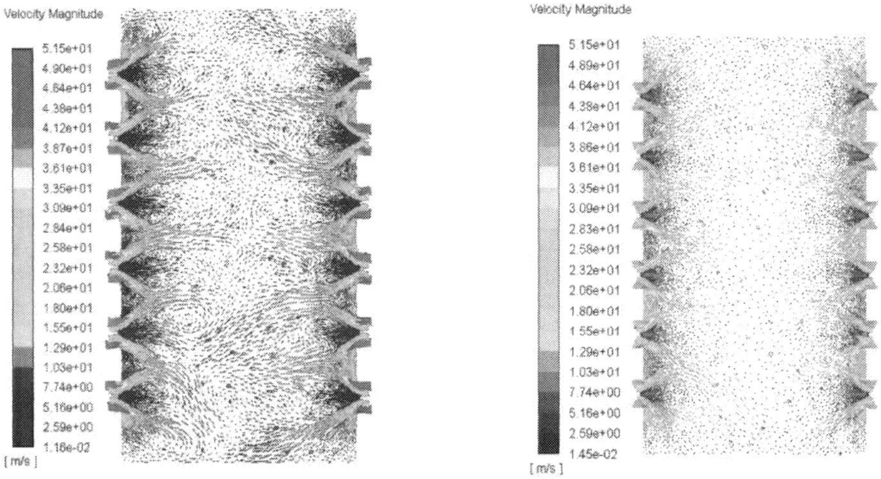

(a) (b)

Fig. 3. Velocity vectors (m/s): a – two-channel vortex burners; b – four-channel vortex burners.

At the specified speeds and the angle of twist of the primary and secondary air, it can be noted that the process of ignition of the fuel occurs in the central part of the swirling burner jets (Fig. 4). The maximum temperature values reach 1700 K when coal is burned in two-channels burners. When burning coal in four-channel burners the temperature in the horizontal section of the burner devices center is lower and about 1550 K. This fact is associated with the intensity of mixing and oxidation of combustion products with oxygen, respectively.

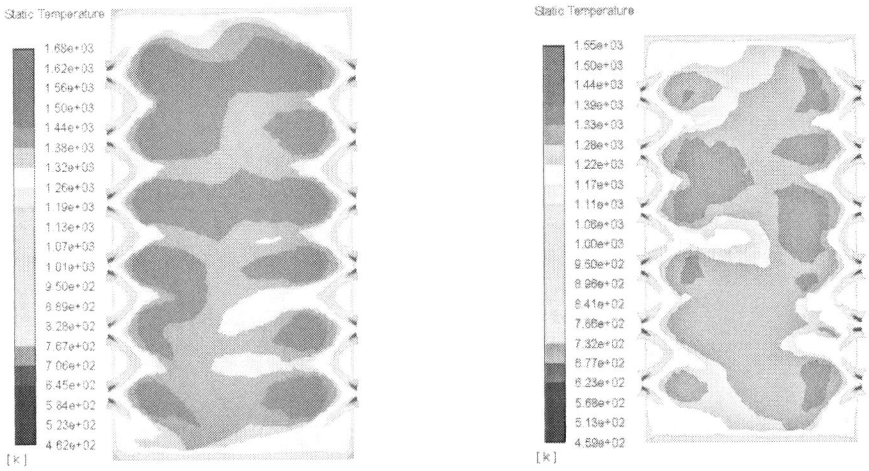

a) b)

Fig. 4. Flue gas temperature (K): a – two-channel vortex burners; b – four-channel vortex burners.

4 Conclusions

This paper presents a numerical study of combustion processes, and in particular the structure of aerodynamic flows and temperature values in the region of flow of burner jets from vortex burners. The numerical simulations performed by commercial software ANSYS FLUENT. The use of such software systems allows to effectively evaluate the alternate methods in the reconstruction or modernization of the boiler units combustion chambers. Based on the results it can be noted that the use of multi-channel burner devices contributes to the reduction of the temperature level in the combustion core and thus contributes to the reduction of the generation of nitrogen oxides.

The studies were carried out with the financial support of the RFBR (project No. 18-38-00775_mol_a).

References

1. A. V. Gil, A. S. Zavorin, A. V. Starchenko, S. V. Obukhov, Power Technology and Engineering **45**, 42 (2011)
2. N. V. Vizgavljust, A. V. Starchenko, A. V. Gil, T. S. Taylasheva, Eur.Phys. J.Web Conf. **82**, 01040 (2015)
3. P. S. Gergelizhiu, S. A. Khaustov, R. B. Tabakaev, P. Y. Novoseltsev, A. V. Kazakov, A. S. Zavorin, Proc. of 2014 Int. Conf. on Mechanical Engineering, Automation and Control Systems, 109804 (2014)
4. S. A. Khaustov, Y. A. Belousova, K. V.Buvakov, A. Y. Dolgih, R. N. Kulesh, 11th International Forum on Strategic Technology (IFOST - 2016) Proceedings, **2** (2016)

Influence of thermo-gravitational convection to temperature fields in small surroundings of the heat source

Alexander Kondakov[1,*]

[1]National Research Tomsk Polytechnic University, 634050 Tomsk, Russia

Abstract. The mathematical modeling of temperature and velocity fields in the system "the heat source - environment - the object of heating" was conducted. The impact assessment of thermo-gravitational convection to the temperature field in comparison with the model of conductive heat transfer was done.

1 Introduction

When developing technologies of formation of thermal conditions of rooms, as a rule, methods which are based on relatively simple balancing equations, are used. When outputting the last ones it was supposed that temperature in the room and conditions of heat transfer between the object and the heat source do not change in space and time.

In the real practice temperature differentials can obtain a few degrees in the concerned conditions, and heat removal capacity from the heating object is higher in two-three times even in the conditions of natural convection than in the conditions of conduction. This depends on geometrical and operating characteristics of the system "the heat source - environment - the object of heating".

Because of these reasons, enhancement of advantageous thermal operation ensuring systems can succeed when describing spatial temperature fields and the air velocity during the operation of thermo-gravitational convection in the concerned system. These methods have not being developed till now. In this regard the purpose of the work is numerical study of heat transfer in the system "the heat source - environment - the object of heating" and getting of temperature and velocity fields for evaluation of heat transfer characteristics change in the natural convection.

2 Formulation of the problem

The problem of heat transfer in the conditions of natural convection between heat source and the heating object is considering.

* Corresponding author: kondakov-alexandr@rambler.ru

Two parallel vertical plates with constant temperatures T_1 and T_0, the height H and which are disposed in the distance L one to another are accepted as the source and the object (Fig. 1).

There is gas with constant viscosity, heat conduction and specific heat capacity between the source and the object. When formulating the problem, assumption about constancy of liquid density except lift force estimation was accepted. Here linear dependence of density from temperature is used.

The movement of the gas between the source and the heating object is considered as laminar. This takes a place at relatively small differences of temperatures and moderate object's sizes.

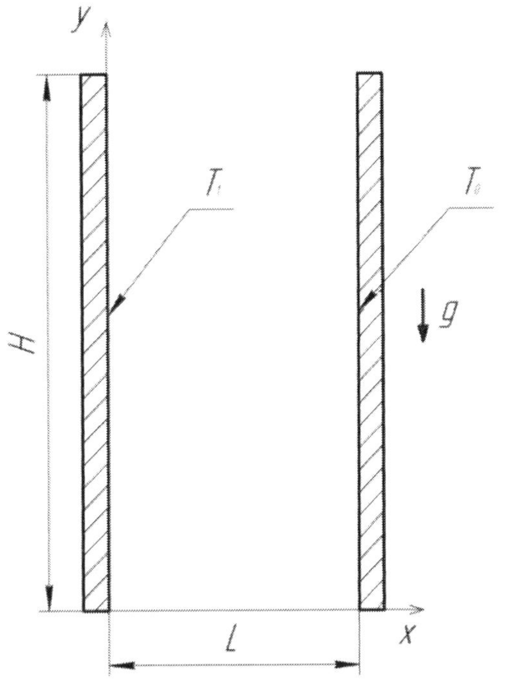

Fig. 1. Formulation of the problem.

The movement of the gas between the source and the heating object is considered as laminar. This takes a place at relatively small differences of temperatures and moderate object's sizes.

3 Mathematical model and study technique

The problem was solved by analogy with [3]. Two-dimensional unsteady system of differential Navier-Stokes equations in variables eddy (w) - streamline function (ψ) in Boussinesq approximation was used.

The following equations was accepted to lead the system to nondimensional form:

$$x = \frac{X}{L} \tag{1}$$

$$y = \frac{Y}{L} \tag{2}$$

$$u = \frac{U}{V_0} \tag{3}$$

$$v = \frac{V}{V_0} \tag{4}$$

$$\theta = \frac{T - T_0}{\Delta T} \quad \psi = \frac{\Psi}{V_0 L} \tag{5}$$

$$w = \frac{\Omega}{V_0 / L} \tag{6}$$

$$V_0 = \sqrt{g\beta\Delta TL} \tag{7}$$

$$\Delta T = T_1 - T_0 \tag{8}$$

Here: X, Y – coordinates, *m*; x, y – nondimensional coordinates; L – width of the calculation area *m*; U, V – horizontal and vertical components of the velocity vector, *m/s*; u, v – nondimensional components of the velocity; V_0 convection velocity, *m/s*; θ – nondimensional temperature; T – temperature, *K*; g – acceleration of gravity, m/s^2; β – volume expansivity, K^{-1}; Ψ – stream function, m^2/s; ψ – nondimensional stream function; Ω – eddy, s^{-1}; w – nondimensional analogue Ω; σ – iteration parameter, *s*.

$$\begin{cases} \dfrac{\partial w}{\partial t} + u\dfrac{\partial w}{\partial x} + v\dfrac{\partial w}{\partial y} = \dfrac{1}{\sqrt{Gr}}\left(\dfrac{\partial^2 w}{\partial x^2} + \dfrac{\partial^2 w}{\partial y^2}\right) + \dfrac{\partial \theta}{\partial x} \\ \dfrac{\partial \psi}{\partial \sigma} = \dfrac{\partial^2 \psi}{\partial x^2} + \dfrac{\partial^2 \psi}{\partial y^2} - w \\ \dfrac{\partial \theta}{\partial t} + u\dfrac{\partial \theta}{\partial x} + v\dfrac{\partial \theta}{\partial y} = \dfrac{1}{\Pr\sqrt{Gr}}\left(\dfrac{\partial^2 \theta}{\partial x^2} + \dfrac{\partial^2 \theta}{\partial y^2}\right) \end{cases} \tag{9}$$

Initial and boundary conditions for the system:

$$t = 0: w(x, y, 0) = \psi(x, y, 0) = \theta(x, y, 0)\ ; \tag{10}$$

$$x = L, 0 < y < H : \psi(L, y, t) = \theta(L, y, t) = 0, \left.\frac{\partial \psi}{\partial x}\right|_{x=L} = 0\ ; \tag{11}$$

$$x = 0, 0 < y < H : \psi(0, y, t) = \theta(0, y, t) = 0, \left.\frac{\partial \psi}{\partial x}\right|_{x=0} = 0\ ; \tag{12}$$

$$y = H, 0 < x < L : \left.\frac{\partial^2 \psi}{\partial y^2}\right|_{y=H} = \left.\frac{\partial^2 \theta}{\partial y^2}\right|_{y=H} = 0\ ; \tag{13}$$

$$y=0, 0<x<L: \left.\frac{\partial^2 \psi}{\partial y^2}\right|_{y=0} = \left.\frac{\partial^2 \theta}{\partial y^2}\right|_{y=0} = 0 \ . \tag{14}$$

The discretization of differential equations was conducted by finite difference method on the constant mesh with the number of nodes 100×200 on the base of second-order Taylor's approximation of the required functions (the choice of computation mesh dimension is determined by approximation scheme [3]).

The system of algebraic equations was calculated by the method of alternating directions [4] on each step by time.

Permanence of temperature and velocity fields on two adjacent time layers was taken as the convergence criteria (with the divergence no more than 10^{-8} for all variables).

4 Results of numerical modelling

Calculations was conducted in the computational domain size $H \times L = 2 \times 1$. Air is accepted as working fluid between the heating object and the source (*Pr=0.71*).

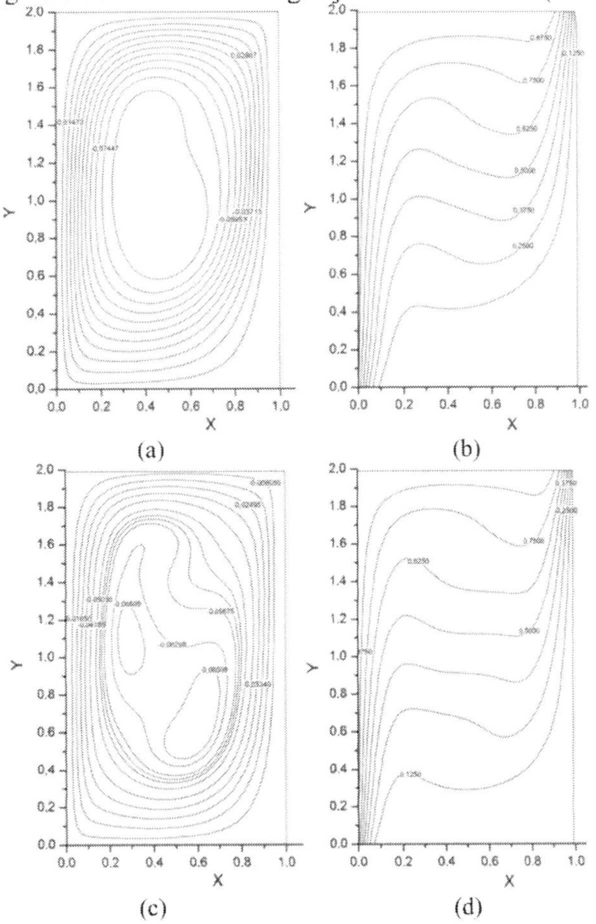

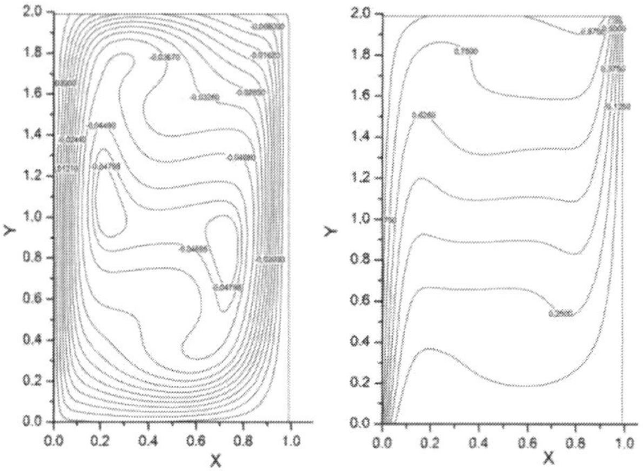

(e) (f)

Fig. 2. Streamline functions (*a. c. e.)* and temperature (*b. d. f.*) at different Grashof numbers: $Gr = 10^5$ (*a. b*); $Gr = 2 \cdot 10^5$ (*c. d*); $Gr = 5 \cdot 10^5$ (*e. f*).

Results of numerical modelling of air movement between the object and the source at numbers $Gr = 10^5; 2 \cdot 10^5; 5 \cdot 10^5$ (steady flow) are shown on the figure 2. They are presented as isotherms (*a. c. e*) and streamline functions (*b. d. f*).

When gas is near with the source, it heats. As the result, temperature gradient in the crosscut direction of the calculation area is appeared. Nonuniformly heated liquid starts to move under the influence of gravity; lift movement takes place along the left bound and the lowering movement - along the right bound (the heating object).

It was established that circulating gas motion in the considering area occurs at Grashof number $Gr = 10^5$.

Growth of Grashof (difference raising of the object's and the source's temperatures) number leads to intensification of heat transfer; secondary streams start to forming inside the main eddy (*c. e*). These processes also influence on the behavior of isotherms when the gas moves in the considering system. The increase of temperature gradients' value on the whole calculation area is shown in the figure 2 (*d. f*).

Temperature distributions between the object and the source in the midsection ($y = 1$) are presented on the figure 3. The air movement is absent in case of heat conduction and the steady temperature distribution has linear dependence. This corresponds to the model of air heat conduction with constant physical features.

The influence of natural convection on the temperature distribution can be valued according the results of numerical modeling which are represented on the picture 3 (*Gr>0*).

Arithmetic mean of temperature deviation in the similar space points from the linear profile grows from *0.048* at *Gr = 5000* to *0.1899* at *Gr = 200 000* in the absolute values. Extremal deviation meanings reach *0.0903* at *Gr = 5000* and *0.4201* at *Gr = 200 000*. The further growth of Grashof number leads to the greater deviation value of temperature from linear distribution.

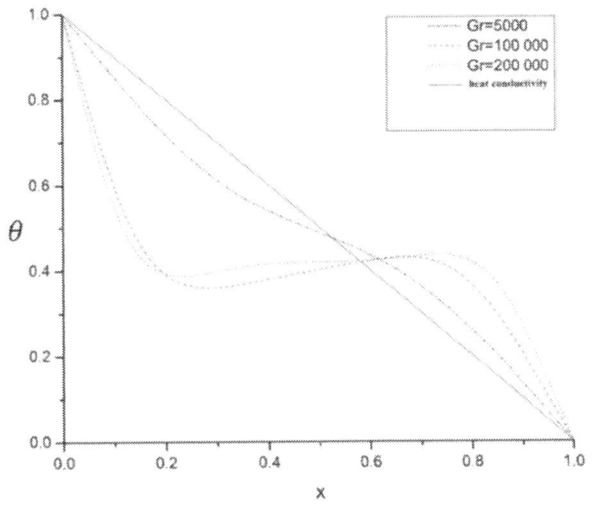

Fig. 3. Temperature distributions in the midsection (*y=1*) at various Grashof numbers.

The received results are shown that significant temperature distributions' deviations from linear profile appear when considering the system "the heat source - environment - the object of heating". Therefore, the natural convection component of heat transfer when designing of microclimate's thermal conditions in rooms with heat sources should be taken into account.

5 Conclusion

Mathematical modeling of temperature fields in the system "the heat source - environment - the object of heating" was carried out.

The problem of heat transfer between the object and the source in conditions of thermo-gravitational convection at small Grashof numbers was solved.

The natural convection influence to temperature fields in the system "the heat source - environment - the object of heating" comparing with heat conduction model was evaluated.

References

1. P. J. Roache, *Fundamentals of Computational Fluid Dynamics* (Hermosa Publishers, Albuquerque, 2003)
2. A. A. Samarskii, *The Theory of Difference Schemes* (Marcel Dekker, New York, 2001)
3. H. K. Versteeg, W. Malalasekera, *An introduction to computational fluid dynamics* (Longman group, New York, 2007)
4. S. V. Patankar, *Numerical Heat Transfer and Fluid Flow* (CRC Press, Florida, 1980)
5. A. I. Vinogradova, V. G. Zubkov, Mathematical simulation E **6**, 3 (2002)

Features of thermal behavior and ignition of HEM with bimetal powders

Alexander Korotkikh[1,2,*], *Ivan* Sorokin[1], and *Ekaterina* Selikhova[1]

[1]National Research Tomsk Polytechnic University, 634050 Tomsk, Russia
[2]National Research Tomsk State University, 634050 Tomsk, Russia

Abstract. The use of metal fuel in high-energy materials (HEM) for propulsion is the most energy efficient method allows to the combustion characteristic increase for solid propellants and specific impulse. Aluminum powder is used in original HEM. To improve the ignition characteristics of HEM advisable to use the catalyst (nonmetals, metals or their oxides). Paper presents the experimental data of the thermal behavior and ignition for HEM based on AP and butadiene rubber, containing bimetal powders: aluminum/iron and aluminum/boron. The use of Alex/Fe powder in HEM decreases the ignition time by 1.3–1.9 times under initiation by CO_2 laser in air at the range of heat flux density of 55–220 W/cm^2 and increases of the recoil force of gasification products outflow from burning surface by 27 % during combustion of propellant due to possible of the catalytic effect, which reduces the beginning temperature of AP high-temperature decomposition by ~20 °C, and interaction of thermite mixture of aluminum and iron particles in the reaction layer of propellant. Then the use of Alex/B in HEM the ignition time is decreased by 1.2–1.4 times, the recoil force of gasification products outflow from burning surface is slightly increased by 9 %.

1 Introduction

According to the study [1] on the aluminum oxidation at combustion of composite solid propellants considerable influence renders the presence on the surface particles of refractory layer of aluminum oxide. Melting point of alumina is significantly above the aluminum melting point. The combustion aluminum particles is possible at the high temperature gradient near the reaction layer of high-energy materials (HEM) near the combustion surface with occurrence of cracks and destruction of the oxide layer resulting in oxidation of the active metal. It is well known [1, 2] the destruction of the aluminum oxide layer of particles is possible in cooperation with the carbon particles to form aluminum carbide.

Analysis of obtained results [3] showed that the main role of iron powder additive is reduced to acceleration of thermal decomposition process of oxidizer and combustible binder occurring in condensed phase of HEM. Furthermore, the iron and titanium powders used for the intermetallic compounds by heating with aluminum powder reacts with high

* Corresponding author: korotkikh@tpu.ru

exothermic effect. This indirect to confirming is the reduction of the plate surface temperature at the same value of the ignition time of sample under introduced into the HEMs these metal additives [4]. Additive of boron powder can change the agglomeration of aluminum nanoparticles on the HEM reaction layer surface then combustion composite solid propellants [5]. Moreover according to the thermodynamic calculations and experimental data [6] the additives of boron and iron powders allows to reduce of the mass fraction of condensed combustion products of HEM.

This paper presents the experimental data of thermal decomposition and the ignition characteristics for the HEM samples based on ammonium perchlorate (AP) and butadiene rubber, containing metal: aluminum, aluminum/iron and aluminum/amorphous boron.

2 Experimental

2.1 The HEM samples

We studied the HEM samples on the basis of bidispersed AP (fraction less than 50 μm and 160–315 μm in the ratio of 40/60), inert combustible-binder (19.7 wt. %) – butadiene rubber plasticized by transformer oil, and ASD-4 micron or Alex aluminum NP (15.7 wt. %) obtained in argon using electric explosion of conductors. In the second and third compositions 2 wt. % of the Alex nanopowder (NP) was replaced by aluminum/iron and aluminum/amorphous boron NP in the ratio of 87/13. According to the measurements on the BET analyzer Nova 2200e in nitrogen the specific surface area of Alex amounts to 7.04 m^2/g, iron – 1.08 m^2/g, amorphous boron – 8.63 m^2/g. Micrographs of scanning electron microscope Jeol JSM-7500F of aluminum, iron and boron nanopowders are shown in Fig. 1. According to the manufacturers data the metal content in the tested NP Alex comprised 85 wt. %, iron – 92 wt. %, and boron – 99.5 wt. %.

Fig. 1. SEM images of tested aluminum Alex (a), iron (b) and amorphous boron (c) NP.

The studied cylindrical samples of HEM in diameter 10 mm and height 30 mm produced in the laboratory by extrusion pressing with the subsequent curing. The density of solid samples was in an interval of 1.53–1.59 g/cm^3 depending on the component composition.

2.2 Ignition of HEM

The ignition study of tested HEM was carried using the setup of the radiant heating on the basis of CO_2 laser with the wavelength of 10.6 μm and power of 200 W (Fig. 2). Prior to testing the samples were cut into tablets of 5 mm in height. The HEM samples were placed in the hollow cylinder of 10 mm depth made of ebonite for inhibiting the lateral surface and making one-dimensional flow of gasification products.

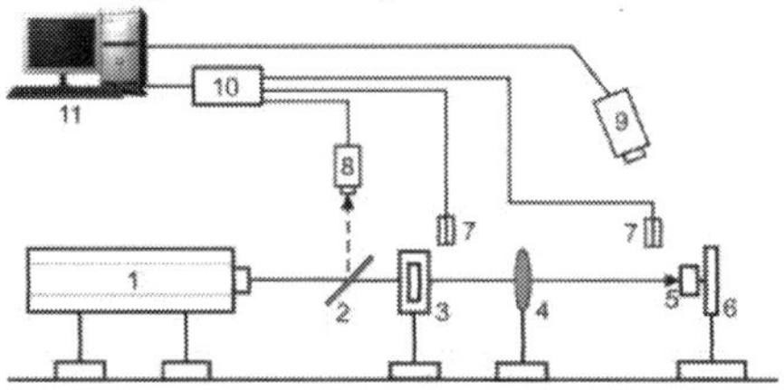

Fig. 2. The scheme of experimental setup based on CO_2 laser: 1 – CO_2 laser; 2 – beam-splitting mirror; 3 – shutter; 4 – lens; 5 – HEM sample; 6 – holder; 7 – photodiodes; 8 – thermoelectric sensor of radiation power; 9 – video camera; 10 – ADCs; 11 – PC.

The HEM sample (6) was attached to the substrate of recoil force transducer (8) to register the gasification products outflow from the burning surface. When opening the shutter (4) the radiation was focused by the sodium chloride lens (5) to the HEM sample (6). Signals from the recoil force transducer (8) and photodiodes (7) were transmitted to the L-card-E 14-440 ADC (9) and recorded in the personal computer (10), and then processed with the software application LGraph2. The time delay of start gasification t_{gas} of HEM sample was determined as time interval between the moments of signals change of photodiode near the shutter (7) (or a thermocouple installed in the laser beam behind the shutter), and the recoil force transducer (8). Photodiode (7) registered the moment of opening the shutter, transducer (8) recorded the appearance of recoil force signal of gasification products flowing from the front (irradiated) sample surface. The ignition time t_{ign} of HEM was determined by difference between the moments of signals from two photodiodes (7), one of which registered the appearance of flame near the end surface of HEM sample. The relative error of delay times measuring of t_{gas} and t_{ign} was equal to 5–12 % at the value of confidence probability of 0.95.

The values of recoil force of gasification products outflow from the end surface of HEM sample during the heating of reaction layer, ignition and combustion of HEM were determined using recoil force transducer [7].

The radiation power and heat flux density of CO_2 laser beam was measured by the thermoelectric sensor of radiation power (3). The maximum radiation power was defined in the center of the laser beam.

3 Results and discussion

3.1 Thermal decomposition of HEM

Thermal analysis of HEM samples containing metal ultrafine powders was carried out using a Netzsch Simultaneous Analyzer STA 449 F3 Jupiter at the heating rate range of 2–

10 °C/min. A set of TG measurements performed for the tested HEM samples in argon at the heating rate of 10 °C/min. is presented in Fig. 3.

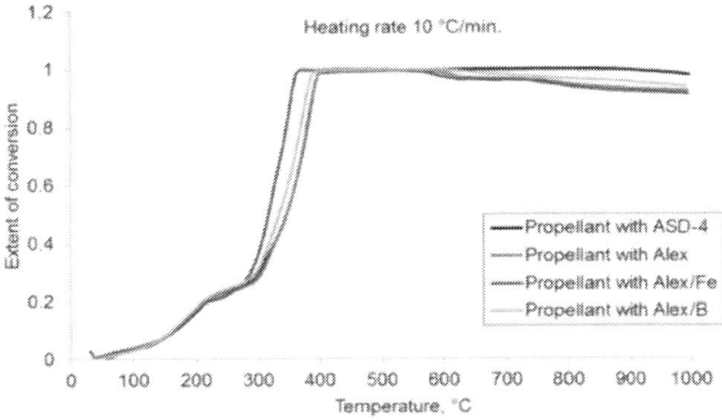

Fig. 3. The extent of conversion for tested HEM samples vs. the temperature.

The results of DSC-TG measurements show that the tested HEM samples have two stage decomposition. The use of Alex/Fe in HEM reduces the starting temperature of intensive decomposition by ~20 °C and increases the rate of intensive decomposition of the sample due to a possible catalytic effect [8], which decreases the starting temperature of high-temperature decomposition of ammonium perchlorate. Using of Alex/B UFP in HEM does not lead to change the starting temperature and the rate of intensive decomposition.

3.2 Ignition parameters of HEM

The values of ignition and gasification times, recoil force of gasification products outflow for the HEM samples in atmospheric conditions were determined. Processing of signal recording from the recoil force transducer, the time delays of ignition and gasification, the recoil force outflow of gasification products from the burning surface of HEM samples have been got. The ignition time of HEMs dependences vs. the maximum value of the heat flux density (in the center of the laser beam) were determined (Fig. 4).

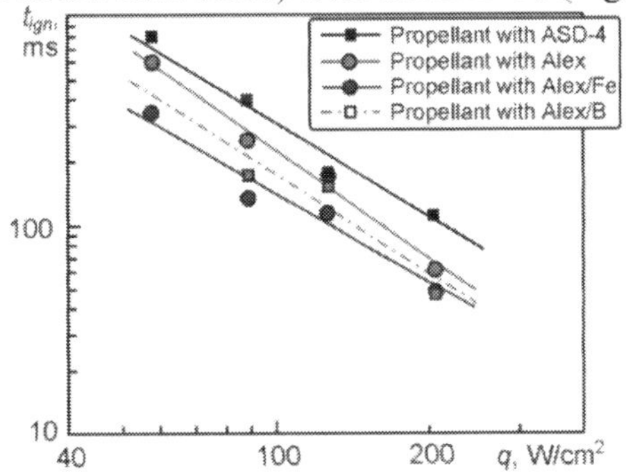

Fig. 4. The ignition time vs. the heat flux for tested HEM: ASD-4 – $t_{ign}=2.97\cdot10^5q^{-1.49}$; Alex – $t_{ign}=4.37\cdot10^5q^{-1.65}$; Alex/Fe – $t_{ign}=0.74\cdot10^5q^{-1.37}$; Alex/B – $t_{ign}=1.91\cdot10^5q^{-1.53}$.

The use of Alex/Fe powder in HEM decreases the ignition time by a factor of 1.3–1.9 in the heat flux range of 55–220 W/cm^2 in comparison with the base HEM sample with Alex and increases the recoil force of the gasification product outflow from the burning surface by 27 % at stationary combustion of HEM with the increase of the surface temperature on the HEM reaction layer (Table 1). The use of Alex/B powder in HEM causes 1.2–1.4 fold decrease in the ignition time. The recoil force of gasification product outflow from the burning surface slightly increases by 9 %.

Table 1. The ignition surface temperature for HEM samples in the flame appearance moment.

q, W/cm^2	T_{ign} of HEM sample, °C		
	with Alex	with Alex/Fe	with Alex/B
35	590 ± 50	720 ± 70	790 ± 80
80	720 ± 80	950 ± 110	770 ± 90

4 Conclusions

The effect of aluminum/iron and aluminum/amorphous boron fuels on ignition characteristics was studied in the HEM samples on the basis of AP and butadiene rubber. It was found that the use of Alex/Fe powder in the HEM sample decreases the ignition time by 1.3–1.9 times under initiation by CO_2 laser in air at the range of heat flux density of 55–220 W/cm^2 and increases of the recoil force of gasification products outflow from burning surface by 27 % during combustion of propellant due to possible of the catalytic effect, which reduces the beginning temperature of AP high-temperature decomposition by ~20 °C, and interaction of thermite mixture of aluminum and iron particles in condensed phase with the increase of temperature on the reaction layer surface of HEM.

The use of Alex/B powder in the HEM sample causes 1.2–1.4 fold decrease in the ignition time, the recoil force of gasification products outflow from burning surface is slightly increased by 9 % while the beginning temperature of high-temperature decomposition of HEM does not change and equals to ~310 °C.

The reported study was supported by RFBR according to the research project No. 16-03-00630 a and by The Tomsk State University competitiveness improvement programme.

References

1. P. F. Pokhil, A. F. Belyaev, Yu. V. Frolov, V. S. Logachev, A. I. Korotkov, *Combustion of powdered metals in active media* (Nauka, Moscow, 1972)
2. A. Abraham, H. Nie, M. Schoenitz, A. Vorozhtsov, M. Lerner, A. Pervikov, N. Rodkevich, E. Dreizin, Combust. Flame **173** (2016)
3. A. A. Gromov, A. G. Korotkikh, A. P. Il'in, L. T. DeLuca, V. A. Arkhipov, K. A. Monogarov, U. Teipel, *Energetic nanomaterials: synthesis, characterization, and application* (Elsevier, Amsterdam, 2016)
4. V. A. Arkhipov, A. G. Korotkikh, V. T. Kuznetsov, A. A. Razdobreev, I. A. Evseenko, Russian Journal of Physical Chemistry B **5**, 4 (2011)
5. O. G. Glotov, V. V. Perov, V. E. Zarko, G. S. Surodin, 57th Israel Annual Conference on Aerospace Sciences (2017)
6. A. G. Korotkikh, O. G. Glotov, V. A. Arkhipov, V. E. Zarko, A. B. Kiskin, Combust. Flame **178** (2017)
7. V. A. Arkhipov, A. B. Kiskin, V. E. Zarko, A. G. Korotkikh, Combust. Explos. Shock Waves **50**, 5 (2014)
8. M. K. Berner, M. B. Talawar, V. E. Zarko, Combust. Explos. Shock Waves **49** (2013)

The effect of the liquid layer thickness on the evaporation intensity

Aleksei Kreta[1,2,3*], and *Vyacheslav* Maksimov[3]

[1]Novosibirsk State University, 630090 Novosibirsk, Russia
[2]Kutateladze Institute of Thermophysics, 630090 Novosibirsk, Russia
[3]National Research Tomsk Polytechnic University, 634050 Tomsk, Russia

Abstract. An experimental study of the influence of thermo-capillary forces and shear stresses with the side of the gas flow to the evaporation flow rate has been made. The experiments were carried out at various thicknesses of the liquid layer and constant gas velocity. The influence of the thickness of the liquid layer on the evaporation flow rate (the intensity of evaporation) has been analyzed. It is shown that the thermocapillary forces have a direct effect on the evaporation flow rate of the liquid layer.

1 Introduction

One of the main directions of modern technology development is the miniaturization of electronic equipment [1]. As a result, with decreasing the size of electronic devices, the specific and volumetric density of the heat flux increases greatly. For removing high heat fluxes, promising two-phase cooling system in which a thin layer of liquid moves in a flat mini or micro channel under the action of a gas stream is used. In such systems, the relationship between evaporation and convection in the liquid layer has a direct effect on heat and mass transfer. Intensive evaporation from the free surface of a liquid under the action of a gas flow causes different convection currents in the liquid layer. The shape and structure of these flows are quite complex and can have a significant effect on the intensity of evaporation and, consequently, on the efficiency of heat exchangers.

Convective flows of liquids, accompanied by evaporation at the interface, are actively studied experimentally and theoretically [2-6]. The space experiment "Evaporative convection and interphase heat and mass transfer" is devoted to the study of thermocapillary convection in a liquid layer due to intensive evaporation under the action of a gas flow [2]. This is a joint experiment of the European Space Agency and Russian Space Agency, which will be planned to carry out on aboard the International Space Station. Numerical study of the development of convective currents in horizontal liquid layers of different thickness evaporating under the action of a gas flow was carried out in [4]. It is shown that in time both the cross-section and longitudinal dimensions of the vortex structures increase. Also it is shown that an increase in the thickness of the liquid layer also contributes to the growth of the dimensions of the vortex structures. The development of mathematical models of

* Corresponding author: alexy.slav@yandex.ru

convective motions with allowance for the processes of heat and mass transfer at the interface and the formulation of conditions on the interface on the basis of exact solutions of the Navier-Stokes equations in the Oberbeck-Boussinesq approximation are reviewed in [5, 7]. In article [7], a comparison of the analytical and experimental results [2] on the evaporation of liquid from the thermocapillary interface was carried out and their qualitative coincidence was revealed.

The aim of the work is an experimental study of the interaction of various mechanisms of convective currents arising due to thermogravitational, thermocapillary convection, as well as shear stresses, on the intensity of evaporation of a horizontal moving liquid layer into a gas stream.

2 Experimental setup

The experimental investigations were conducted on the setup shown in Fig. 1. The setup consists of the following components: test section, gas and liquid loops, data acquisition system, thermal stabilization system, and optical techniques.

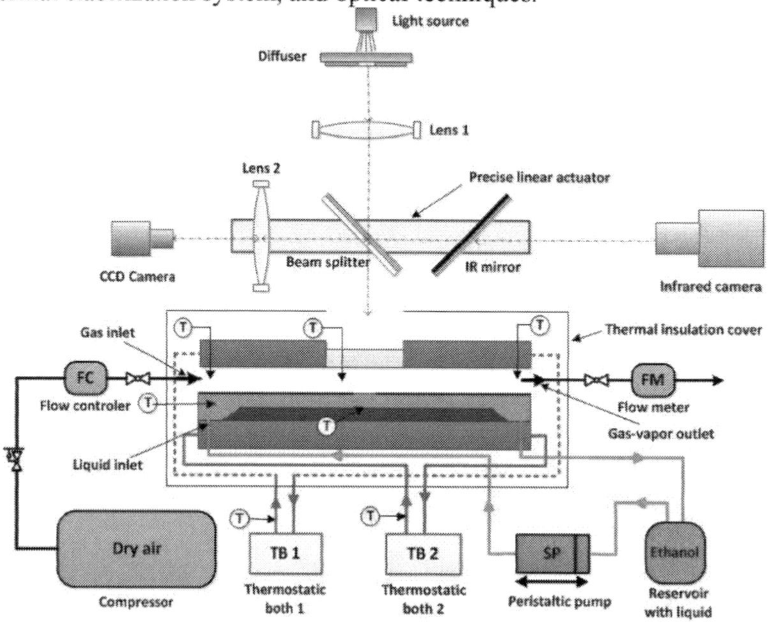

Fig. 1. Scheme of the experimental setup.

In the centre of the installation there is a working section. The working area consists of gas and liquid channels, which are separated by a thin metal plate having a thickness of 200 microns. The metal plate has a square cut-out in the centre of 10x10 mm. It is at this point that the liquid contacts the gas. The height of the liquid channel is changed by means of metal plates. Thermostabilization of the working section was carried out by two thermostats TB1 and TB2. Gas enters to the test section from the compressor. The gas flow rate at the inlet of the gas channel was set by means of a mass flow regulator. The liquid was injected into the liquid channel by a peristaltic pump. The liquid flow rate in the channel was changed by changing the flow rate at the peristaltic pump. The working fluid evaporated under the action of the gas flow and the vapor-gas flow moved to the outlet of the gas channel of the working section. The flow of the vapor-gas mixture at the outlet of the working section was measured with a mass flow meter. The position of the interface level was controlled with an accuracy of 10 micrometres using a Schlieren technique and the peristaltic pump.

According to the experimental conditions, the interface between the liquid and the gas must be maintained in a flat state throughout the entire experiment. The position control of the level of the interphase boundary was carried out using the method Schlieren and the peristaltic pump. At the same time, it was necessary to measure the temperature distribution on the interphase surface. Measurements of the temperature distribution on gas-liquid interface were conducted with the help of Titanium 570M IR camera. Infrared imaging of liquid surface was performing through quartz optical window that implemented in the top cover of the test cell. IR camera was placed on the optical tabletop horizontally and measuring of the surface temperature of the liquid was contacted with the help of the IR mirror. The IR mirror was inclined to 450 to infrared camera and liquid surface. Optical components of the Schlieren technique and IR mirror were mounted on an optical platform of the linear actuator. Linear actuator is controlled from personal computer and by special software. The optical platform moved in the range of 50 mm and speed of 104 mm/s in horizontal direction. Initially, the platform was set in position where optical axes of Schlieren technique coincide and the interface visualized. When a flatness of interface and the stationary regime of the experiment were established the linear actuator moved the platform in position where optical axes of the IR camera and mirror also coincide and measurement of surface temperature was running.

2.1 Calculation of shear stresses

Fig. 2 shows a thermogram of the liquid surface at the intensive evaporation. The IR image snapshot does not explain all dynamic of the process. Analyzing of the video the movement of gas-liquid interface in counter-current direction to the gas flow is observed.

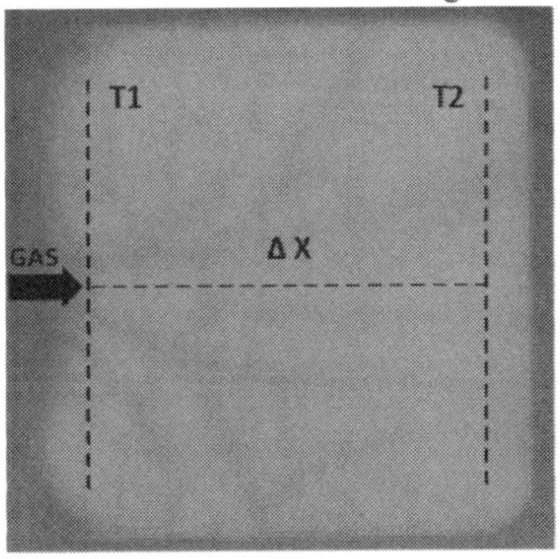

Fig. 2. The thermogram of the evaporating liquid layer.

Convective flows in the liquid layer propagate under the action of thermocapillary forces and shearing stresses caused by friction of the gas flow on the interphase surface. Thermocapillary stresses on the surface of the liquid layer caused by the action of the thermocapillary effect are calculated by the formula:

$$\tau_{sur} = \frac{\delta\sigma}{\delta T}\frac{\delta T}{\delta x} \qquad (1)$$

where, $\sigma_T = \frac{\delta\sigma}{\delta T}$ - temperature coefficient of surface tension is equal 1.14 x 10^{-4} N/mK.

The temperature gradient is defined as: $\frac{\delta T}{\delta x} = \frac{T_2 - T_1}{\Delta x}$. Points of the temperature measuring and length x=0.008 m are shown in Fig. 2.

Shear stresses on the liquid surface induced by gas flow can be estimated as [8]:

$$\tau_g = \eta \frac{dV}{dy} = -\frac{H}{2}\frac{\delta p}{\delta x} \tag{2}$$

where, H= 0.005 m is height of gas channel; η – dynamic viscosity.

Pressure gradient is calculated from the formula: $V_g = -\frac{H^2}{12\eta}\frac{\delta P}{\delta x}$, where average gas velocity V_g, m/c is known value. It is assumed that in the gas channel the Poiseuille flow is realised. The gas-liquid interface is considered as a rigid surface. In more detail, the method of calculating tangential stresses is described in the article [9].

3 Results and discussion

The experiments were conducted at atmospheric pressure in a test section. The thickness of the liquid layer in the experiments was 3, 5 and 8 mm, respectively. Ethanol liquid was used as the working liquid. The evaporation surface area was 100 mm^2 with the corresponding 10x10 mm cut-out in the plate. The temperature of the "liquid-gas" system has changed from 20 °C to 40 °C, with the step of 10 °C. The difference between the temperature of the gas and liquid does not exceed 0.1 °C. The gas flow rate was a constant and equaled 1000 ml/min that corresponds to gas average velocities 0.138 m/s. Reynolds fluid throughout the experiment was constant and was equal to 10 (Re=10).

Dependency of the evaporation flow rate on the depth of the liquid layer at different temperatures is shown in Fig. 3

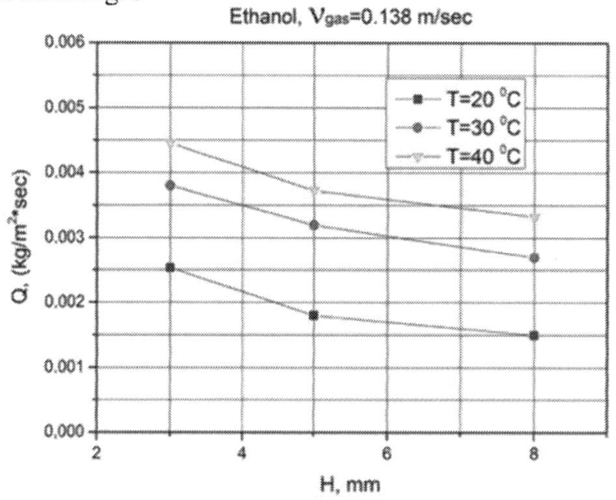

Fig. 3. Evaporation flow rate.

It can be seen from the graph that the maximum evaporation rate is reached with a liquid layer thickness equal to 3 mm. With increasing thickness of the liquid layer, the evaporation flow rate decreases substantially. On average, the evaporation rate decreases by 30%. It is shown that this recession is caused by a sharp drop in thermocapillary forces on the interphase surface of the phase section. It is supposed that this is connected with the complication of the structure of convective currents in the liquid layer with increasing thickness. Using a thermal imager, it was noted that as the thickness of the layer increases, the structure of the flow in the liquid layer changes, which directly affects the evaporation flux rate. With increasing thickness of the liquid layer, the number of convective structures in the liquid layer increases, and as a result, the mass flow through the interfacial surface decreases. A similar phenomenon in the case of a stationary liquid layer was observed by Lyulin and Kabov [2].

Fig. 4 shows the dependence of the thermocapillary stresses on the thickness of the liquid layer at various temperatures of the "liquid-gas" system.

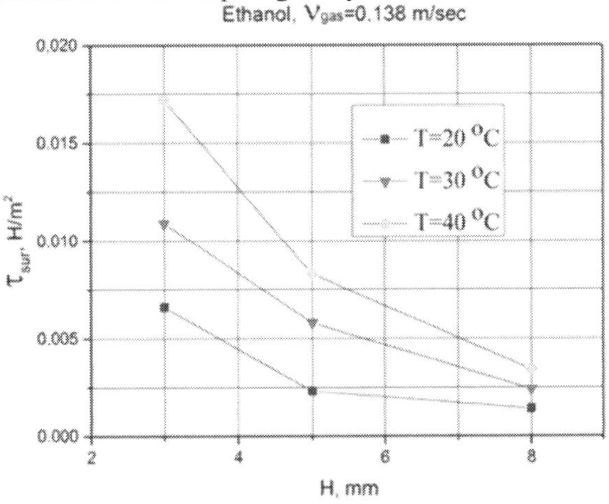

Fig. 4. Thermocapillary stresses.

It can be seen from the graph that the thermocapillary forces on the interphase surface decrease significantly with increasing thickness of the liquid layer. First of all, this drop is due to the fact that the influence of gravitational forces increases with increasing layer of liquid. The influence of gravitational forces significantly reduces the temperature gradient on the interphase surface, which leads to a sharp decrease in thermocapillary forces. Secondly, as already mentioned, when the thickness of the liquid layer is increased, the structure of the convective currents in the layer becomes much more complicated, which also leads to a drop in the gradient of the temperature on the interphase surface. Such a drop in thermocapillary forces has a direct effect on the rate of evaporation.

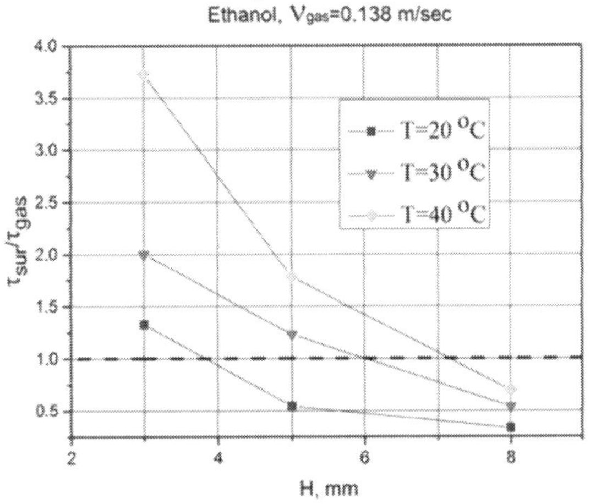

Fig. 5. The ratio of shear stresses.

It was also noted that with increasing thickness of the layer, the thermo-capillary shear stress to the gas shear stress are decreased (Fig. 5). With a maximum layer thickness in the experiment of 8 mm for all diapasons of the "liquid-gas" temperatures, the shear stresses on the gas flow side are greater than the thermocapillary stresses. This correlation of forces directly affects the character and rate of development of convective currents in the liquid layer and, as a result, on the intensity of evaporation.

4 Conclusions

Experiments were conducted to study the influence of the liquid layer thickness on the evaporation flow rate at various temperatures of the "liquid-gas" system and at a constant gas flow rate. Infrared study and measurements of the temperature field on the surface of horizontal liquid layer evaporating into gas flow has been performed. Temperature gradient of the gas-liquid interface has been measured with the help of Titanium 570M IR camera. Shear stresses on gas-liquid interface induced by thermocapillary effect and inert gas flow have been defined. It is shown that the thermocapillary stresses have a direct effect on the evaporation flow rate of the liquid layer. It is found that as the thickness of the liquid layer increases, the evaporation intensity decreases as a result of a sharp decrease in thermocapillary forces on the interphase surface.

The work was financially supported by Russian Foundation for Basic Research (Project identifier No. 17-38-50223 "mol_nr").

References

1. E. A. Chinnov, V. V. Guzanov, O. A. Kabov, Letters to the Journal of Technical Physics **35,** 14 (2009)
2. Yu.V. Lyulin, O. A. Kabov, Int. J. Heat Mass Transf. **70**, 599 (2014)
3. B. Scheid, J. Margerit, C. S. Iorio, L. Joannes, M. Heraud, P. Queeckers, P. C. Dauby, P. Colinet, Experiments in Fluids **52**, 1107 (2012)
4. H. Machrafi, C. S. Iorio, P. C. Dauby, Interfacial Phenomena and Heat Transfer **2**, 3 (2014)
5. N. A. Goncharova, O. A. Kabov, Int. J. Heat Mass Transf. **53**, 2795 (2010)
6. N. A. Goncharova, M. Hennenberg, E. V. Rezanova, O. A. Kabov, Interfacial phenomena and heat transfer **1**, 3 (2013)
7. N. A. Goncharova, E. V. Rezanova, Yu. V. Lyulin, O. A. Kabov, Thermophys. Aeromech. **22**, 5 (2015)
8. L. Landau, E. Lifshitz, *Hydrodynamics* (Science, Moscow, 1986)
9. A. S Kreta, Yu. V. Lyulin, MATEC Web Conf. **92**, 01048 (2017)

Calculation of temperature distribution and effective temperature in high burnup fuel of WWER-1000

Alexander Kudrov[1*], *Arian* Kuzmin[1], and *Denis* Savitsky[1]

[1]National Research Tomsk Polytechnic University, 634050 Tomsk, Russia

Abstract. The problem of finding an effective temperature is highly important. This characteristic provides calculation of the power effect reactivity (PER), which defines safe and stable operation of nuclear reactor. The numerical experiment, which is considered on average and maximum heat load conditions, is put to find out changing of the effective temperature in the high burnup fuel of WWER-1000. In the course of the experiment mathematical statement of the problem is made, numerical solution of which is found with using the finite differences approximation of both control equations and border conditions. Also, the method of simple iteration is used for calculating temperature distributions, according to determination of the effective temperature. The linear approximation was obtained basing on effective and maximum temperatures depending on the burnup.

1 Introduction

One of the most important characteristics of a nuclear reactor is the power effect of reactivity (PER), which is defined with the effective temperature and actual amount of ^{235}U in fuel. The effective temperature is an average temperature of a fuel pellet; it can be determined with the exact integral formula [1]:

$$T_{eff} = \frac{\int_0^{r_T} T_T(r) 2\pi r dr}{\int_0^{r_T} 2\pi r dr} \tag{1}$$

where
$T_T(r)$ – temperature distributions over the fuel pellet;
T_{max} – maximum temperatures on the axis of the fuel block;
T_c – temperatures on the surface of the fuel block.

* Corresponding author: alex12kudrow@gmail.com

Due to the Doppler effect, rising temperature leads to broadening of resonance levels of ^{238}U, which increases neutron absorption, thereby it reduces reactivity. PER has to be negative during the reactor company for providing safe and stable operation.

The rim-effect is a process of forming a thin layer around a periphery of the fuel pellet, it has high porosity and has increased heat release. The rim-layer is always associated with the high burnup fuel ($\overline{B} > 40MWd / kgU$). According WWER-1000 design, maximum burnup, that can be reached, is $70MWd / kgU$. It means that calculations of the effective temperature must be provided with considering of the rim-effect.

2 Mathematical formulation of the problem

Calculation of the effective temperature requires the known temperature distribution, which can be found from following equations:

$$\frac{1}{r}\frac{d}{dr}\left(\lambda_0(T_1)r\frac{dT_1(r)}{dr}\right)+q_v(r)=0, \tag{2}$$

$$\frac{1}{r}\frac{d}{dr}\left(\lambda_{rim}(T_2)r\frac{dT_2(r)}{dr}\right)+q_v(r)=0, \tag{3}$$

where

$\lambda_0(T_1)$ is the thermal conductivity of the inner area, $W/(m\cdot{}^\circ C)$;

$T_1(r)$ is the temperature distribution in the inner area, ${}^\circ C$;

$q_v(r)$ is radial distribution of heat release in the fuel pellet, W/m^3;

$T_2(r)$ is the temperature distribution in the rim-layer, ${}^\circ C$;

$\lambda_{rim}(T_{rim})$ is the thermal conductivity of the rim-layer, $W/(m\cdot{}^\circ C)$.

Boundary conditions:

$$r=0, \left.\frac{dT_1(r)}{dr}\right|_{r=0}=0; \tag{4}$$

$$r=R_1, T_1(r=R_1)=T_2(r=R_1); \tag{5}$$

$$r=R_1, -\lambda_1(T_1)\left.\frac{dT_1}{dr}\right|_{r=R_1}=-\lambda_2(T_2)\left.\frac{dT_2}{dr}\right|_{r=R_1}; \tag{6}$$

$$r=r_p, T_2(r=r_p)=T_c, \tag{7}$$

where

R_1 is the radius of inner area, mm;

T_c is temperature on a the surface of the fuel pellet, ${}^\circ C$.

Due to a non-linearity of (2) and (3), the analytical solution cannot be found, therefore the numerical method of finite differences and method of the simple iteration was used.

3 Influence of radiation on properties of the fuel rod

Degradation of thermal conductivity may be expressed with following formulas [2]:

$$\lambda_{95} = [(0.1148 + 3.0864 \cdot 10^{-3}\overline{B}) + 2.475 \cdot 10^{-4}(1 - 2.95 \cdot 10^{-3}\overline{B})T]^{-1} + 0.0132\exp(1.88 \cdot 10^{-3}T), \quad (8)$$

where

$\overline{B}$ is an average burnup in the fuel pellet, MWd / kgU;

T is temperature, $^\circ C$.

$$\lambda_0 = \lambda_{95} / [1 - (2.58 - 0.58 \cdot 10^{-3}T) \cdot 0.05]; \quad (9)$$

$$\lambda_{rim} = \lambda_0 \left\{1 - P^{2/3}\left[1 - 1/(1 + P^{1/3}(\lambda_0 / \lambda_{Xe} - 1))\right]\right\}, \quad (10)$$

where

$P = N\pi d^3 / 6$ is porosity of the rim region (d – diameter of the pore, N – volume density of pores, m^{-3});

λ_{Xe} is thermal conductivity of gas in the pores, $W / (m \cdot {}^\circ C)$.

The expression (8) describes degradation of the thermal conductivity due to microstructural changes caused by the radiation effect, (9) takes account of accumulation of solid fission products; (10) is thermal conductivity of rim region.

The high burnup fuel rod experiences swelling, which causes absence of a central hole in pellet and gas gap between the pellet and the fuel cladding. Thickness of the rim-layer can be calculated with the following formula [2].

$$\delta_{RIM} = -32.5 + 2.34 \cdot \overline{B}. \quad (11)$$

The radial distribution of the heat release is represented on figure 1 [3]. It has the same form as the radial distribution of ^{239}Pu in the fuel pellet [4,5], and does not depend on burnup.

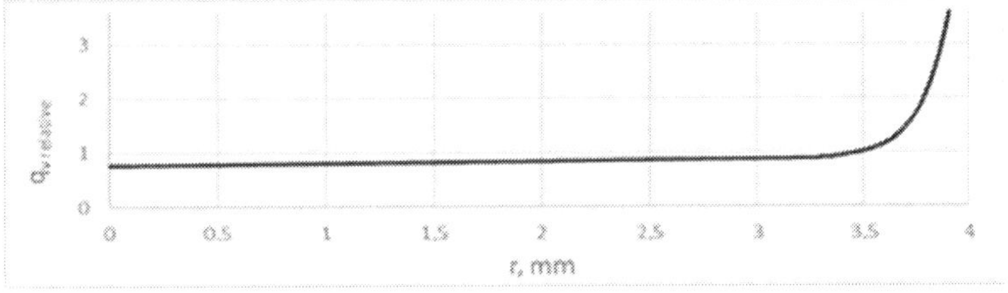

Fig. 1. Radial distribution of relative heat release.

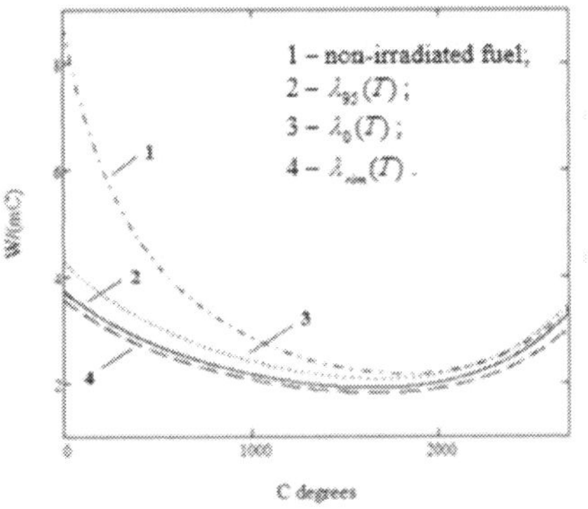

Fig. 2. Dependence of thermal conductivity on temperature.

4 Results of calculations

Temperature distributions were found in interval of burnups from $40MWd/kgU$ to $100MWd/kgU$. Dependence of maximum and effective temperatures is represented in the Table 1.

Table 1. Maximum and effective temperatures on different burnup levels.

Average heat load							
Burnup, MWd/kgU	40	50	60	70	80	90	100
T_{max}, °C	1379.5	1455.0	1527.1	1595.6	1660.6	1722.2	1798.7
T_{eff}, °C	833.6	872.4	910.2	946.9	982.7	1017.4	1066.4

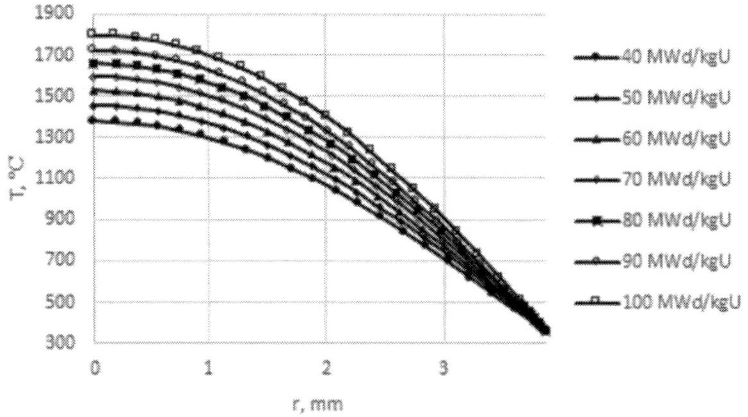

Fig. 3. Temperature distributions in the fuel pellet on different burnup levels.

Dependence of effective and maximum temperatures from burnup was approximated by linear functions. For average heat load those functions look like:

$$T_{eff} = 3.7889 \cdot \overline{B} + 681.9; \quad (11)$$

$$T_{\max} = 6.8767 \cdot \overline{B} + 1109.9. \quad (12)$$

These expressions have error less than 1%.

References

1. G. G. Bartholomew, G. A. Bat', V. D. Bajbakov, M. S. Alhutov, *Fundamentals of the Theory and Methods for Calculating Nuclear Power Reactors* (YULAND Publ., Yekaterinburg, 2016)
2. B. H. Lee, Y. H. Koo, D. S. Sohn, JNST **38**, 45 (2010)
3. S. U. Kurchatov, V. V. Likhansky, A. A. Sorokin, O. V. Khoruzhy, Atom. Energy **92**, 317 (202)
4. K. Lassmann, C. O'Carroll, J. van de Laar, C. T. Walker, J. Nucl. Mater. **208**, 223 (1994)
5. L. D. Palmer, K. Hesketh, P. A. Jackson, IAEA **321** (1982)

Internal stress distribution in weld-affected zone under the effect of constrained loads

Nikita V. Hapov[1], *Aigul* S. Kulesh[1,*], and *Anton* A. Gerasimov[1]

[1]National Research Tomsk Polytechnic University, 634050 Tomsk, Russia

Abstract. The paper presents the experimental results of identify the redistribution pattern under the influence of mechanic cycle loads. The subject of the research was the weld metal zone of the uniform joint made of steel X12Cr1MoV. The research method was to organize the mechanic cyclic deformation of sample and to increase the load in each regular load cycle. The result of the work is the determination of sign-variable pattern of internal stress changes under deformation leading to propagation of fatigue and destruction.

1 Introduction

There is a major number of emergencies connected with heat-affected zone fatigue as well as with weld-affected zone. It is proportional to the increase of welded seam use.

Durability of welded seams can be obtained by rational selection of constructional, metallurgical and technological means. Nevertheless, the working capacity and welded seam functional qualities required are not always provide

Among the broad range of tasks solved in the framework of the Russian energy development, among the first in importance is the problem of ensuring the reliability and efficiency of the thermal power plants equipment.

This applies in particular to those components where in manufacture materials different in their mechanical, chemical, strength and thermal characteristics are used. It is obvious that the most dangerous elements in this sense are welded joints which transitional zone is a kind of stress concentrators.

The durability calculation is based on admissible stress in welded seam design. In its turn, admissible stress is based on the material properties, type of effort (stretching, bending, shearing) and type of load. They are determined as a quotient of conventionally disrupting load by factor of safety.

If a welded design works under the repeated loads and it experiences the cycle fatigue and fatigue damage, admissible loads should be directly reduced in relation to factored stress.

However, admissible loads cannot match the actual design stresses by an order of magnitude. It refers to the hard-to-predict internal stress field redistribution in particular. This is what may cause the destruction by itself.

* Corresponding author: ronikul@tpu.ru

In this regard, the aim of the work was to identify the redistribution pattern under the influence of mechanic cycle loads.

2 Materials. Samples. Experimental techniques

The research sample was cut off the pipe of a superheater (Fig. 1) in the brick envelope zone (point B, Fig. 1). The pipes were made of steel X12Cr1Mo) in 1999. The weld assembly was made in 2005. The pipes in the boiler were operating for 16500 hours. Working temperature in point B was 480°C. The pipe was taken out of service due to the fracture in point A, where the flue gas flow temperature was up to 540 – 590 °C.

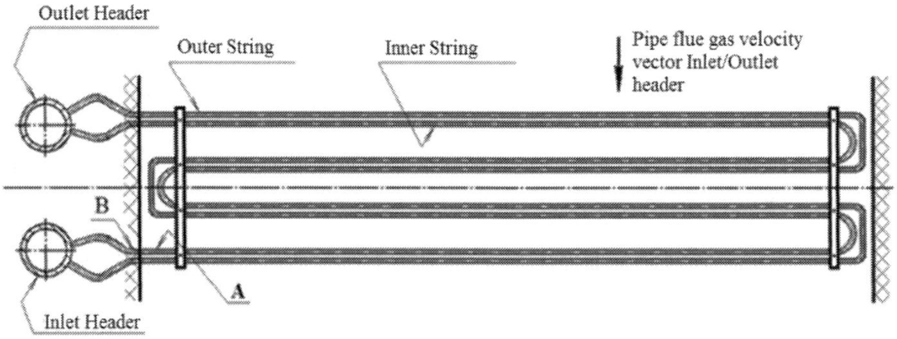

Fig. 1. Top view of the superheater.

The subject of the research was the weld metal zone of the uniform joint made of steel X12Cr1MoV. The research sample was cut off at a distance of 7,5 mm from the welding joint axis (Fig. 2).

The research method was to organize the mechanic cyclic deformation of sample and to increase the load in each regular load cycle.

The rate of macrostress of the first kind was implemented with the Röntgen's method for the crystal lattice parameter alteration based on Hooke's law.

$$\frac{\Delta\ell}{\ell} = \frac{\sigma}{E} \quad \text{or} \quad \frac{\Delta a}{a} = \frac{\sigma}{E},$$

where σ is the internal tensile stress of the first kind, $\Delta a/a$ is the relative elongation of lattice cell parameter during the stretching, E is the Young's modulus.

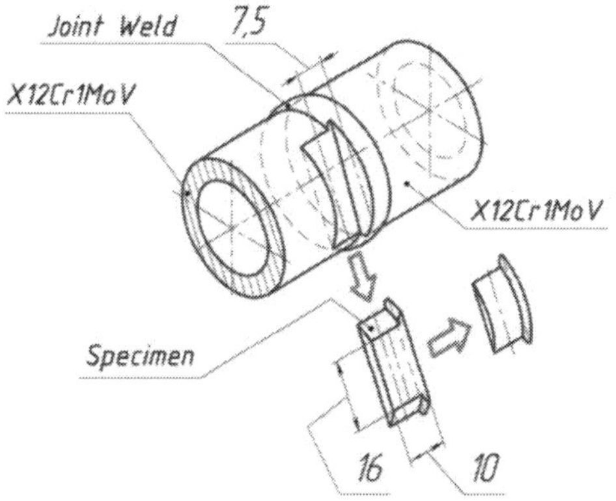

Fig. 2. Sample made of X12Cr1MoV steel for the research of circumferential zone.

3 The experimental facts. Discussion

As a result of extrenal pulse disturbance application, shear strain tracks can be observed (Fig. 3). Plastic shear band spreads from the mid-point to the edge of the sample, then dissipates on the unevennesses (grains) and goes around them. The waves interfere beyond the unevennesses and transform a developed surface relief into interferential rings shape (Fig. 3a). The deformable sample surface takes a pitting form (Fig. 3b, 3c). The important consistent pattern about the plastic shearing process is the fragmentation and localization tendency at all stages.

The area in front of shear wave is subjected to deformation hardening. At a certain moment microband can't overcome the hardening zone and slows down emerging on the surface as a wave with peak and deepening, determined by material capacity for plastic flow (Fig. 3d). Probably, the shear wave slowing down moment fixates the fluidity limit of the material and is a pre-destruction sign.

Subsequent load application leads to the emergence of individual destructions that are represented as rather big pores (Fig. 4). In the next load cycles separated pores on the sample's surface merge and form main cracks (Fig. 3e).

So, it makes possible to see the interaction of two separate processes at the same time in alternate cycle load conditions. These are creep and fatigue, which lead to the destruction stage. The impact of cycle load on the destruction is the wave periodic deformation hardening known as annealing.

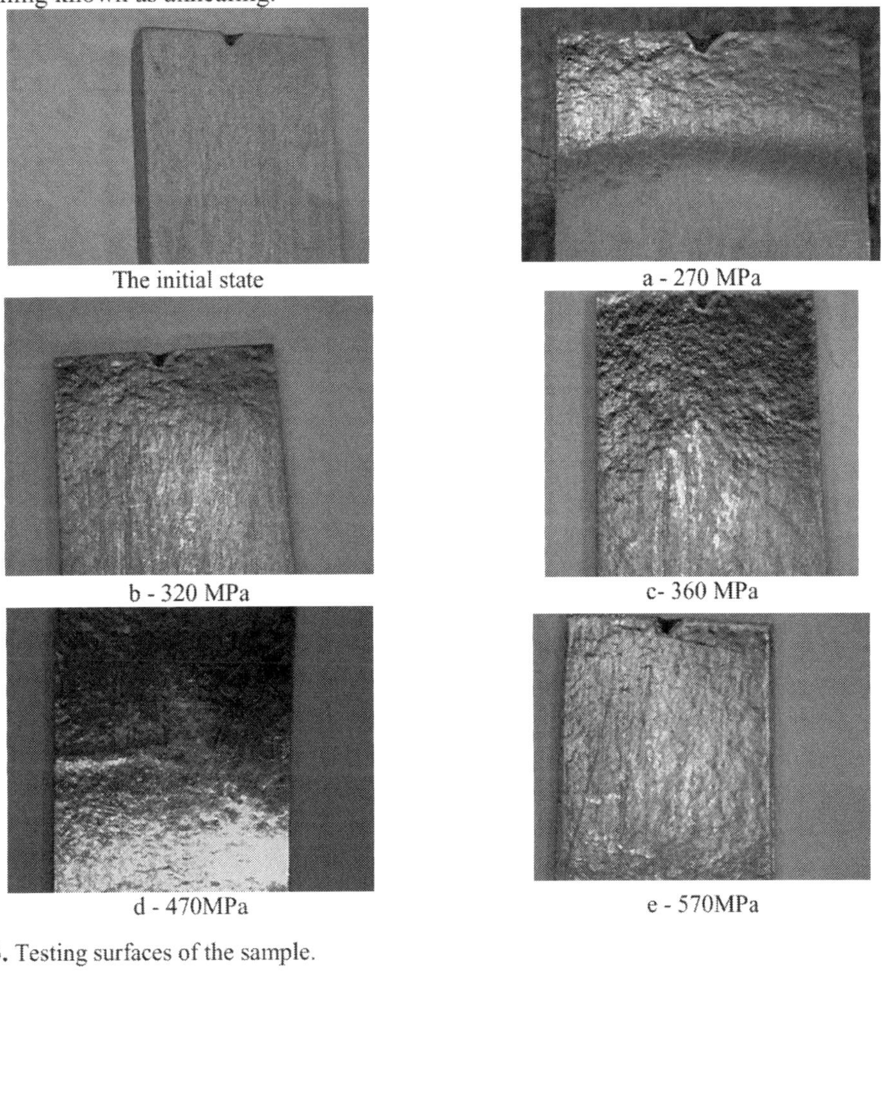

Fig. 3. Testing surfaces of the sample.

Fig. 4. Porosity and Crack formation of the sample.

It is verified by the results of cyclic internal stress changes. According to Fig. 5 there are two types of interacting processes run coherently in deformation section. On one hand, these are relaxation processes that emerge as a result of dislocation movements, i.e. plastic relaxation forming. On the other hand, every plasticity act is accompanied by acoustic wave generation, which is connected with elastic deformation redistributing during the relaxation acts of deformation.

Internal stress relaxation is a typical process during deformation (Fig. 5). It matches the annealing and marks the forming moment of microstructurally short cracks [1-5]. Interatomic bonds are strained in the grain boundaries. And atomic bonds will break when the interatomic distance increases by more than 10-15% [1]. The bond breaking process is the leading process at all stages of destruction. Its result is the stress relaxation and crack forming [1-5].

According to A. Griffith's notion, the crack disclosure critical stress for plane strain is written as [6]:

$$\sigma = \sqrt{\frac{2E\cdot\gamma}{\pi\cdot(1-\mu)\cdot l}} \tag{1}$$

where E is the Young's modulus, l is the crack's length, μ is the Poisson's ratio, γ is the free surface energy density. ($\gamma = 0{,}1E\cdot b; b = 2{,}48\cdot 10^{-8}$ cm is the interatomic distance).

It's on this basis that the length of implied crack can be evaluated with Fig. 5a,b. When deep (zero) relaxation occurs crack length under load of P = 280 MPa (Fig. 5) will be:

$$l = \frac{2E\cdot\gamma}{\pi\cdot(1-\mu)\cdot\sigma^2} = \frac{2\cdot(205\cdot10^3\cdot0.1\cdot205\cdot10^3\cdot2.48\cdot10^{-8})}{\pi\cdot(1-0.27)\cdot280^2} = 1.16\cdot10^{-3}cm \approx 12\mu m. \tag{2}$$

In this case, internal pipe surface relaxation occurs earlier than the external one (75 MPa), and the implied crack length is:

$$l = \frac{2E\cdot\gamma}{\pi\cdot(1-\mu)\cdot\sigma^2} = \frac{2\cdot(205\cdot10^3\cdot0.1\cdot205\cdot10^3\cdot2.48\cdot10^{-8})}{\pi\cdot(1-0.27)\cdot75^2} = 16.16\cdot10^{-3}cm \approx 162\mu m. \tag{3}$$

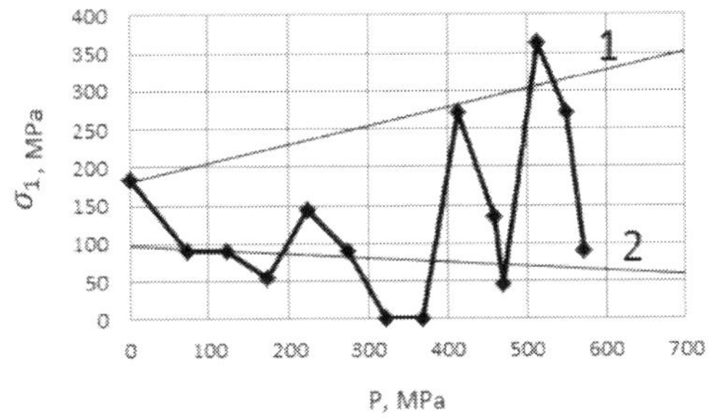

(a)

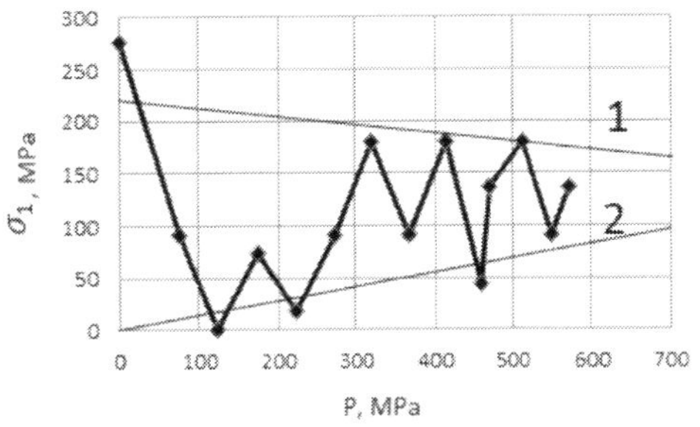

(b)

Fig. 5. Distribution of internal stresses in circumferential zone of joint weld: a – external surface; b – internal surface (1 - hardening line; 2 - annealing line).

4 Conclusion

The result of the work is the derived plastic flow localization tendency, which starts up with the plastic flow and ends up with the fusion of macroscopical plastic localization areas in the moment of destruction. It enables us to determine propagation features of plastic deformation localization during the material pre-destruction stage.

Another result is the determination of sign-variable pattern of internal stress changes under deformation leading to propagation of fatigue and destruction.

This work was supported by RFBR grant "Fundamentals of engineering sciences" No. 15-08-99544a in 2014, No. 18–08–01265 in 2018).

References

1. J. Friedel, B. D. Gullity, C. Crussard, Acta Met **1**, 79 (1953)
2. L. L. Lyubimova, A. A. Tashlykov, R. B. Tabakaev, A. A. Levin, A. Y. Popov Engineering Failure Analysis **81**, 45 (2017)
3. L. L. Lyubimova, R. N. Fisenko, A. A. Tashlykov, R. B. Tabakaev, Met. Mater. Int. **24**, 121 (2018)
4. L. L. Lyubimova, R. N. Fisenko, R. B. Tabakaev, A. A. Tashlykov, A. S. Zavorin, Mater. Sci. Eng. A **682**, 248 (2017)
5. L. L. Lyubimova, R. B. Tabakaev, A. A. Tashlykov, A. S. Zavorin, V. Y. Zyubanov, EPJ Web Conf. **110**, 1037 (2016)
6. A. A. Griffith, Phil. Trans. Roy. Soc. London A **221**, 163 (1920)

Numerical investigation of localization and suppression of thermal decomposition of forest combustible materials using specialized water supply

Dmitrii Antonov[1], *Geniy* Kuznetsov[1], and *Alena* Zhdanova[1,*]

[1]National Research Tomsk Polytechnic University, 634050 Tomsk, Russia

Abstract. The danger of forest fires and their large-scale consequences are becoming ever more complex problems for mankind every year. The results of numerical studies of heat transfer processes under suppressing the thermal decomposition reaction of forest combustible materials by films and groups of water droplets are presented. We used forest combustible materials, typical for the Siberian region. We have established the relationship between the dimensions of the heated near-surface layer of the material and the thickness of the film layer of the liquid that evaporated when the thermal decomposition reaction was suppressed in forest combustible materials. Typical times of suppression of the thermal decomposition reaction of typical forest combustible materials are given when the water film on the surface evaporates. The influence of the mutual arrangement of droplets on the surface of the forest combustible material and the temperature of the outer vapor-gas mixture on the times of suppression of its pyrolysis is revealed. The results can be used to develop technologies for extinguishing fires, containment of flame sources and combustion front.

1 Introduction

Over the past five years, wildfires have claimed the lives of thousands of people around the world, the number of injured people exceeds tens of thousands [1–3]. In addition to thousands of hectares of burnt forest massifs, numerous cases of suffocation and poisoning of people with combustion products of forest combustible materials should be noted. The only possible way to solve the problem of forest fires is to create effective extinguishing technologies that will ensure the suppression of flame in the shortest possible time with the rational use of extinguishing agent [4].

Numerical [5] and experimental [6] studies have shown that in a typical local aviation discharge of water into the flame zone in the process of extinguishing forest fires, as a rule, only a small part (less than 10%) of the liquid is used (directly evaporated). It was shown [7] that for a substantial (several hundred degrees) reduction in the flame temperature with

* Corresponding author: zhdanovaao@tpu.ru

minimal water consumption, it is advisable to use a fine dispersion of the liquid. After the spraying time has elapsed, liquid films can be formed on the surface of the thermally decomposable material. It should be noted that the components of the forest massif (bark, leaves, and branches) can reach the surface of a burning forest material together with the liquid. In this case, an inhomogeneous system is formed on the surface of the thermally decomposable forest material, which includes water, leaves, needles, and bark. This system is a "buffer layer" between the surface of the thermally decomposable material and the environment. It should also be noted that only some of the droplets of the extinguishing liquid can reach the surface of the forest combustible material during movement through the flame [7]. In such a case, the formation of monolithic liquid films on the surface of the reacting material is unlikely with fine spraying (droplets fall at a certain distance between each other).

The purpose of this study is to numerically study the processes of heat transfer under conditions of endothermic phase transformations during the evaporation of a water film on the surface of typical forest combustible materials.

2 Problem statement

In the formulation of the heat transfer problem under consideration (Fig. 1 *a*), it was assumed that the "water shell" (large set of water droplets) passed through the flame and the water was evenly distributed over the surface of the forest combustible material. A liquid film of fixed dimensions (thickness $L_{w\text{-}f}$) was formed. We assumed that the leaves, tree branches, bark, and needles together with the liquid reached the surface of the forest combustible material. Thus, a heterogeneous system (water with solid inclusions) is formed between a layer of forest combustible material with temperature $T \gg T_d$ and a trace of a "water shell". Steam formation is intensified at the boundary of "forest combustible material – water with solid inclusions" when the phase transition conditions are reached. It was believed that the liquid was heated by heat exchange with the thermally decomposable forest material. At the boundary of the "water film – forest combustible material", vaporization occurs when the phase transition conditions are reached. The temperature of the forest combustible material is reduced due to the heat of the endothermic phase transition. During intensive vaporization, the thickness of the "water shell" (liquid film) decreases.

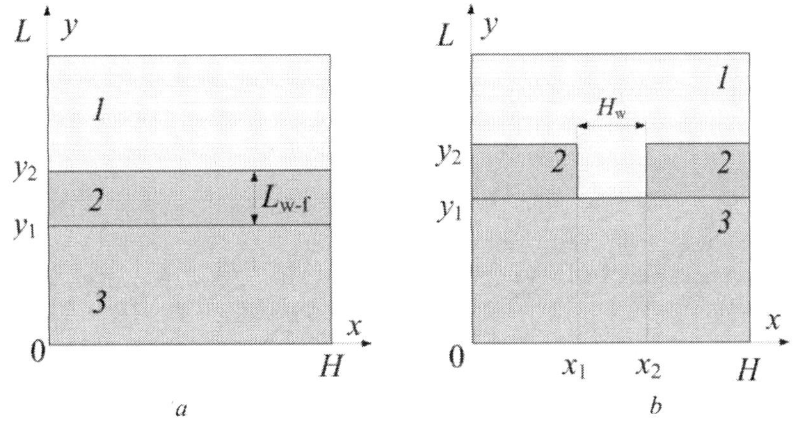

Fig. 1. Scheme of the solution domain for the heat transfer problem ($0<t<t_d$): *a* – upon evaporation of a water film with solid inclusions – ("buffer layer") on the surface of a forest combustible material; *1* – vapor-gas mixture, *2* – water, *3* – forest combustible material; *b* – upon evaporation of water droplets on the surface of a forest combustible material; *1* – vapor-gas mixture, *2* – water droplets, *3* – forest combustible material.

In the formulation of the problem (Fig. 1 *b*), it was assumed that a small group of drops had reached the surface of the thermally decomposable forest material. In general, different distances between individual water drops are possible on the surface of a forest combustible material under intense pyrolysis. From the whole variety of possible locations of the droplet group on the surface of the forest combustible material, the most significant for the subsequent analysis is the variant of two drops (Fig. 1 *b*), located at a certain distance from each other ($H_w=X_2–X_1$). It was believed that the liquid is heated by heat exchange with a thermally decomposable forest material heated to high temperatures. Evaporation occurs at all boundaries of water droplets when the corresponding conditions are reached. The temperature of the forest combustible material is reduced due to the heat of the endothermic phase transition and conductive heat transfer.

Based on the results of numerical simulation, the time (t_d) of the decrease of the forest combustible material temperature to the temperature at which the thermal decomposition began (T_d). The value of t_d illustrates the time of suppression (termination) of the thermal decomposition reaction of a forest combustible material.

3 Mathematical models and methods of solution

When carrying out numerical studies, we used two software complexes – Ansys Fluent and Matlab. Mathematical statements for the systems (Fig. 1) are given in [8].

The problems were solved in the Matlab environment by nonstationary differential equations of heat transfer in partial derivatives for systems "vapor-gas mixture – water film with solid inclusions ("buffer layer") – forest combustible material" and "forest combustible material – water droplets – vapor-gas mixture" respectively. The system of nonstationary differential equations was solved by the method of finite differences [9]. The difference analogues of differential equations were solved by a locally one-dimensional method [10]. To solve one-dimensional difference equations, we used the sweep method using the implicit four-point scheme [9]. Nonlinear equations were solved by the iteration method [10]. Uneven time steps ($10^{-4}÷10^{-2}$ s) and coordinate increments ($10^{-8}÷10^{-6}$ m) were used to improve the accuracy of solving differential equations. Near the boundaries of the phase transition, the difference grid was condensed according to an algorithm similar to [5]. The method for evaluating the reliability of the theoretical studies results, based on the verification of the conservativeness of the difference scheme, is analogous to that used in [5].

When solving problems in Ansys Fluent, the finite element method was used. This method was originally included in this commercial software package. The solution of the heat and mass transfer problem assumed the developing a geometry of the tasks (Fig. 1), the formation of a grid area and a description of models of evaporation, combustion, thermal decomposition, turbulence, etc.

Numerical studies were performed with the following values of the parameters [9]: the initial temperature of water and gas-vapor mixture T_n=300 K; the temperature of the onset of thermal decomposition of forest combustible material T_d=500 K; the parameters of the chemical reaction were assumed to be identical for the considered types of forest combustible material k_3^0 =3.63·10^4 s^{-1}, E_3=78.114 kJ/mol, Q_3 =1·10^3 J/kg; the dimensionless coefficient of condensation (evaporation) β=0.1; molar mass of water M=18 kg/kmol; thermal effect of water evaporation Q_e=2.26 MJ/kg. The thickness of the near-surface layer of forest combustible material was varied in the range L_f=0.02–0.06 mm. The thickness of the liquid film was assumed equal to L_w=0.01 mm. The size of the solution area was changed depending on the values of L_f in the range L=0.02–0.2 m. The longitudinal and transverse dimensions of the droplets were varied in the range H_{dr}=0.01–0.05 m at L_{dr}=0.01 m. The distance between droplets was varied in the range H_w=0.001–

0.09 m. The characteristic dimensions of the solution region were varied depending on the values of H_{dr}, H_w and L_f in the ranges H=0.05–0.1 m, L=0.05–0.1 m. Thermophysical characteristics of forest combustible materials (birch leaves, pine needles and spruce needles), of pyrolysis products, of water and water vapor were taken according to [11, 12].

4 Results and Discussion

Fig. 2 shows the characteristic times of suppression of the thermal decomposition of forest combustible material under the impact of a water film with solid inclusions. It was found that the inertia of the process of suppressing the thermal decomposition reaction increases substantially with changing $L_{w\text{-}f}$ in the range from 0.002 m to 0.004 m. With further growth of $L_{w\text{-}f}$ the times t_d almost do not change (deviations less than 1%). This result can be explained by the small mass of water evaporating upon suppression of the decomposition reaction.

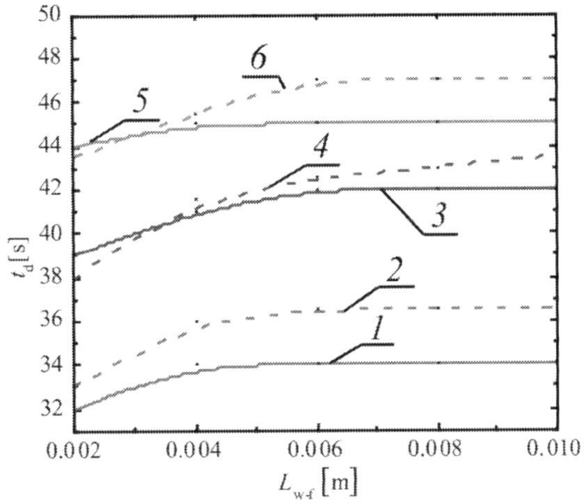

Fig. 2. Characteristic times of suppression of the thermal decomposition reaction of the forest combustible material depending on the thickness of the water film $L_{w\text{-}f}$: *1*, *3*, *5* – results of numerical studies in Matlab; *2*, *4*, *6* – results of numerical studies in Ansys Fluent; *1*, *2* – pine needles, *3*, *4* – birch leaves, *5*, *6* – spruce needles.

We have established that the considered “buffer layer” is heated to a depth of less than 0.005 m and the thickness of the evaporated water layer is about 0.001 m. Therefore, the possible increase of $L_{w\text{-}f}$ relatively to 0.005 m almost does not affect the cooling conditions of the forest material and, as a consequence, the conditions for suppressing the reaction of its decomposition. Fig. 3 illustrates the dependence of the times of suppression of the thermal decomposition of forest combustible material on the distance between two water droplets (H_w) at the temperatures in the trace of the “water shell” 300–350 K. It can be noted that with decreasing H_w, the times of the thermal decomposition reaction termination t_d are decrease nonlinearly.

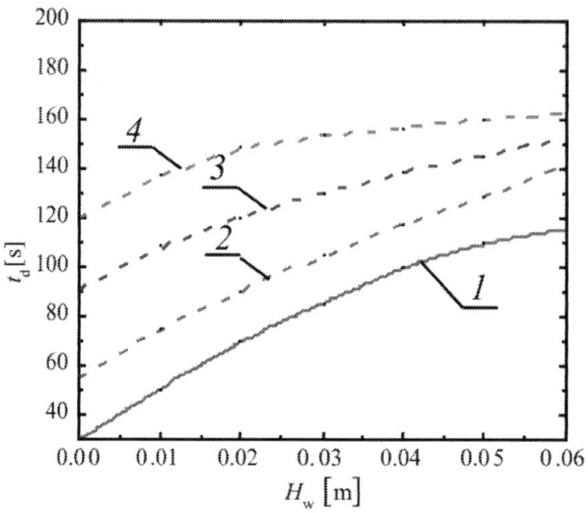

Fig. 3. Characteristic times of suppression of the thermal decomposition reaction of the forest combustible material depending on the distance between water droplets H_w when extinguishing by single water droplets: *1* – results of numerical studies in Matlab; *2–4* – results of numerical studies in Ansys Fluent; *1*, *2* – pine needles, *3* – birch leaves, *4* – spruce needles.

The established (Fig. 3) dependences $t_d=f(H_w)$ can be explained by the fact that the temperature of the vapor-gas mixture in the region between the droplets increases significantly under the thermal decomposition of the forest material due to intense heat release. A large proportion of this heat is consumed for the implementation of phase transformations on the lateral boundaries of droplets ($X=X_1$, $X=X_2$, $Y_1<Y<Y_2$) when the distances between the droplets is less than 0.02 m (due to the large thermal effect of liquid evaporation, Q_2=2.26 MJ/kg). In this case, the area of heating of the vapor-gas mixture is limited by the thickness of the liquid droplets L_{dr} (Fig. 1). A significant increase in the size of the heating region of the vapor-gas mixture relatively to L_{dr} was established (Fig. 1) with increasing the values of H_w.

Deviations of the numerical studies results in Matlab and Ansys Fluent are primarily due to the fact that the rate of water evaporation is given by a constant in Ansys Fluent, but is determined by the experimental approximations, if Matlab is used.

5 Conclusion

Excessive flooding of the surface of forest combustible material with water is wasteful to reduce the possibility of its thermal decomposition. A special spraying of the liquid with the formation of a small amount of water on the forest combustible material surface is recommended ($L_{w\text{-}f}$=1–3 mm). Even in the presence of "buffer layers" (water films with solid inclusions) considered in this work, it is possible to stop forest material pyrolysis in sufficiently short (less than 2 minutes) times.

However, with specialized spraying, the probability of forming relatively large "buffer layers" is significantly reduced due to low water consumption and a small fraction of leaves, branches and bark carried away by the liquid.

Numerical studies have shown that the distance between water droplets (H_w) significantly affects the termination of the reaction of thermal decomposition of forest combustible material (birch leaves, pine needles and spruce). It was shown that the times t_d can differ by 10–40% for identical values of H_w as a function of the temperature of the vapor-gas mixture at the surface of the forest combustible material and, consequently, between the droplets of the liquid.

The research was funded by the Russian Science Foundation (project 18-19-00056).

References

1. A. Dimitrakopoulos, C. Gogi, G. Stamatelos, I. Mitsopoulos, Pol. J. Environ. Stud. **20**, 327 (2011)
2. F. X. Catry, F. C. Rego, F. Moreira, F. Bacao, 1st International Conference on Modelling, Monitoring and Management of Forest Fires **119**, 213 (2008), DOI: 10.2495/FIVA080221
3. D. H. Klyde, D. J. Alvarez, P. C. Schulze, T. H. Cox, M. Dickerson, AIAA Atmospheric Flight Mechanics Conference 97625 (2010)
4. A. O. Zhdanova, R. S. Volkov, I. S. Voytkov, K. Y. Osipov, G. V. Kuznetsov, Int. J. Heat Mass Transfer **126**, 703 (2018), DOI: 10.1016/j.ijheatmasstransfer.2018.05.085
5. R. S. Volkov, G. V. Kuznetsov, P. A. Strizhak, Int. J. Heat Mass Transfer **79**, 838 (2014), DOI: 10.1016/j.ijheatmasstransfer.2014.09.006
6. R. S. Volkov, G. V. Kuznetsov, P. A. Strizhak, Int. J. Heat Mass Transfer **96**, 20 (2016), DOI: 10.1016/j.ijheatmasstransfer.2016.01.029
7. R. S. Volkov, G. V. Kuznetsov, P. A. Strizhak, Int. J. Heat Mass Transfer **85**, 1 (2015), DOI: 10.1016/j.ijheatmasstransfer.2015.01.078
8. A. O. Zhdanova, G. V. Kuznetsov, J. C. Legros, P. A. Strizhak. Therm. Sci. **21**, 2565 (2017), DOI: 10.2298/TSCI151006121Z
9. A. A. Samarskij, *Teorija raznostnyh shem* [The theory of difference schemes] (Nauka, Moskow, 1983) [in Russian]
10. L. A. Kozdoba, *Metody reshenija nelinejnyh zadach teploprovodnosti* [Methods for solving nonlinear heat conduction problems] (Nauka, Moskow, 1975) [in Russian]
11. N. B. Vargaftik, *Spravochnik po teplofizicheskim svoistvam gazov i zhidkostey* [Gases and liquids thermophysical properties guide] (Stars, Moskow, 2006) [in Russian]
12. V. N .Yurenev, P. D. Lebedev, *Teplotechnicheskiy spravochnik* [Thermophysical guide] (Energiia, Moskow, 1975) [in Russian]

Influence of $Cu(CH_3COO)_2$ promoting additive on bituminous coal oxidation process

Darya Bolgova[1], *Kirill* Larionov[1], *Andrey* Zenkov[1], and *Stanislav* Yankovsky[1,*]

[1]National Research Tomsk Polytechnic University, 634050 Tomsk, Russia

Abstract. The process of coal oxidation with applied $Cu(CH_3COO)_2$ additive was studied by the capillary incipient wetness impregnation method with 5% mass concentration. The experiment was conducted by thermogravimetric analysis at a heating rate of 2.5°C/min to a maximum temperature of 600°C in atmospheric air. It was established that application of the initiation additive leads to a significant reduction in the initial temperature of sublimation and active oxidation of volatile compounds (ΔTi = 78°C) and the oxidation end temperature (ΔT_f = 64°C). It was established that in the presence of copper acetate the nature of coal oxidation reaction significantly changes (DTG data). The parameters of the coals oxidation process in the presence of copper acetate were determined, and an assumption was made about the presence of a composite catalytic effect.

1 Introduction

Coal is one of the most common types of fuel in the Russian energy industry [1]. The most simple and widely used method of its transformation is direct combustion in boiler furnaces [2]. Combustion of coal is associated with a number of serious problems: the emission of flue gases, the inertia of burning fuel, the incompleteness of its combustion [2]. It is possible to increase the coefficient of fuel use due to application of catalytic combustion technology, which has a significant effect on the kinetics of the combustion process [3]. As a catalytic additive, as a rule, oxides of alkali and rare-earth metals are used, which accelerate the oxidation process [3-4]. In turn, a comparative analysis of the effect of oxide additives and their precursors in the form of metal nitrates and metal chlorides on the process of coal oxidation showed that the precursors contribute to a greater decrease in the initial temperature of active oxidation of volatile substances and an increase of the reaction rate, shortening the residence time of the sample in its active oxidation area [4]. The mechanism of coal catalytic combustion is explained by the fuel reactivity activation, resulting from the close contact of the coal particles with the catalyst surface [2]. The paper presents the results of an experimental study of the coal oxidation process with $Cu(CH_3COO)_2$ initiation additive.

2 Experimental setup and study technique

* Corresponding author: jankovsky@tpu.ru

The coal of the Allardinskoye deposit of the Kuzbass coal basin was taken as the initial sample. The large-dispersed sample of the initial coal (particle diameter d = 5÷10 mm) was ground in a drum mill with an equal ratio of the mass of the milling bodies and the material for 8 hours. After grinding, the sample was fractionated on sieves with a mesh size of not more than 80 μm. The obtained micro-dispersed coal powders were previously dried at a temperature of 105°C to a constant weight. The physical and chemical characteristics of the prepared samples were determined using standard techniques [5] and are presented in Table 1.

Table 1. Physical and chemical characteristics of a coal sample and initiation additive.

Characteristics of bituminous coal sample	
Ash content, %wt.	17.0
Moisture content, %wt.	0.5
Volatile compounds, %wt.	14.0
Carbon, %wt.	68.5
Humidity ratio, ml/g	2.9
Characteristics of $Cu(CH_3COO)_2 \cdot H_2O$ initiation additive	
Molar mass, g/mol	181.63
Decomposition temperature, °C	~290
Solubility in water, g/ml	0.10

Application of the initiation additive was conducted by the incipient wetness impregnation method [6]. To overcome the hydrophobicity of the coal powder, a water-alcohol solution with a volume ratio of C_2H_5OH/H_2O = 50/50 was used. The prepared solution was applied to the previously prepared coal powder by means of a mechanical dispenser. The impregnated coal powder was kept in a drying oven at a temperature of 105°C for 20 hours. The mass fraction of $Cu(CH_3COO)_2$ in the modified sample (in terms of dry salt) was 5%. For comparative analysis, a reference sample containing no initiation additive was prepared which was subjected to exactly the same treatment procedures. The prepared samples had the following designation: the reference sample was BC, the modified sample was BC/Cu.

Study of the coal samples oxidation (modified and initial samples) was conducted using STA 449 C Jupiter synchronous thermal analyzer (Netzsch, Germany). The experiments were performed under the same conditions, at a heating rate of 2.5°C/min in a corundum crucible with a perforated cover (sample mass ~ 15 mg) in the temperature range 45-600°C. A mixture of air (60 ml / min) and nitrogen (10 ml / min) was used as an oxidizing medium. Nitrogen was used as a shielding gas to ensure reliable operation of the analyzer and correct registration of the obtained data. A qualitative determination of the oxidation products composition was also performed using the STA attachment in the form of QMS 403 D Aeolos quadrupole mass spectrometer (Netzsch, Germany). All experiments were conducted at atmospheric pressure.

Parameters of the coal oxidation process were determined graphically (Fig. 1) with the help of the obtained TG and DTG curves. The points A and B on the TG curve characterize the onset temperature of the released volatiles active oxidation (T_i) and the end temperature of the combustible mass loss of the sample (T_f). The characteristic point C, lying on the DTG curve, reflects the maximum value of the reaction rate (w_{max}) and the temperature at a certain moment (T_1).

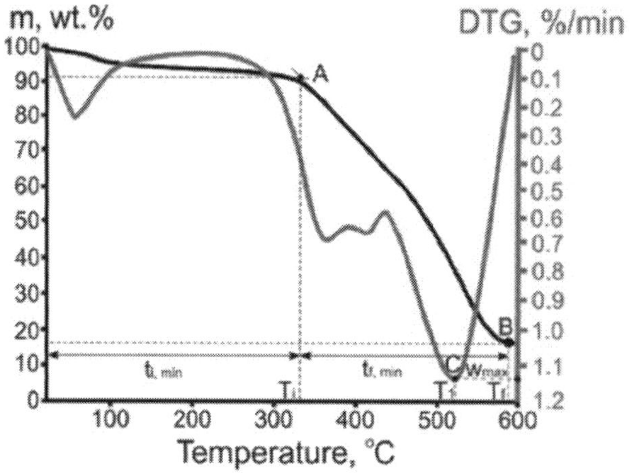

Fig. 1. TG- and DTG-curves for oxidation process of the initial (unmodified) sample of bituminous coal.

3 Experimental results

The TG and DTG curves for oxidation process of the researched coal samples are presented in Fig. 2. It can be seen that application of $Cu(CH_3COO)_2$ initiation additive leads to a significant change in the nature of the process.

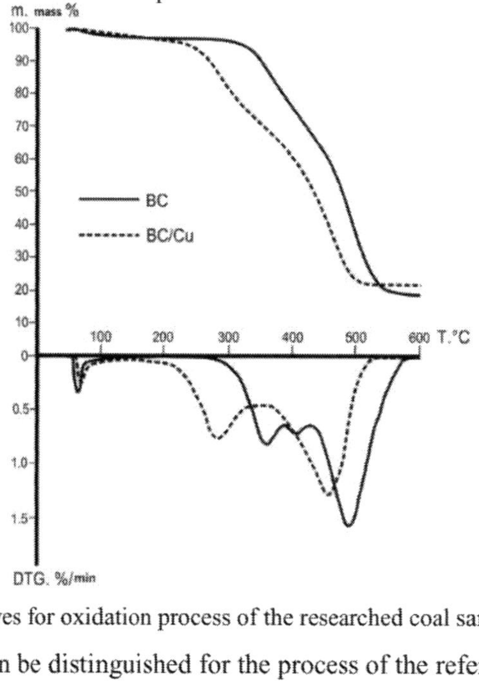

Fig. 2. TG- and DTG-curves for oxidation process of the researched coal samples.

Four main stages can be distinguished for the process of the reference sample oxidation. Moisture evaporates in the first stage (up to 100°C). In the second - coal particles undergo thermal destruction, which is accompanied by sublimation of volatile compounds (140-360°C). The third and fourth stages are characterized by active oxidation of the released volatile compounds and the obtained coke residue (360-600°C).

The main parameters of the oxidation process of the researched coal samples were calculated based on the TGA data. The results of the calculations are presented in Table 2.

Table 2. Parameters of oxidation process of the researched coal samples.

Parameter	BC	BC/Cu
Active oxidation onset temperature of volatile compounds, T_i (°C)	361	283
Time of the sample heating to the active oxidation onset of volatile compounds, t_e (min)	126	94
Oxidation end temperature of the combustible mass of the sample, T_f (°C)	583	519
Temperature at the point of maximum oxidation rate, T_l (°C)	527	493
Maximum oxidation rate, w_{max} (%wt./min)	0.49	0.40

As can be seen from Table 2, modification of a coal sample by addition of $Cu(CH_3COO)_2$ leads to a significant decrease in the onset temperature of active coal oxidation (ΔT_i = 78°C), a decrease in the oxidation end temperature (ΔT_f = 64°C), and a reduction of the sample warm-up time before the start of volatile compounds oxidation (Δt_e = 32 min).

The decrease of the onset temperature of active oxidation is accompanied with an increase of the average process flow rate (Δw = 0.023%/min) and a decrease in the maximum reaction rate (Δw_{max} = 0.09%/min). At the same time, the temperature corresponding to the maximum oxidation reaction rate of the sample decreases (ΔT_{max} = 34°C).

The mass spectrometric curves corresponding to the appearance of CO_2 (m/z=44) in the composition of gas phase oxidation products of the researched coal samples are presented in Fig. 3. It can be seen that in the presence of $Cu(CH_3COO)_2$ initiation additive, a considerable shift of the emission peaks to the low-temperature region is observed, which agrees well with the data of DTG (Fig. 2). In this case, a significant change in the shape of the curves also occurs.

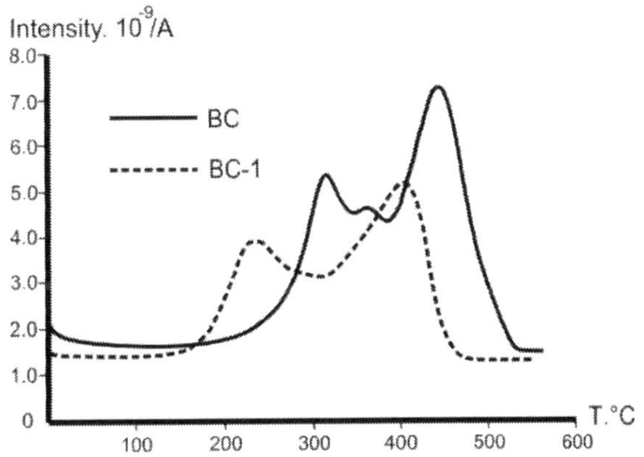

Fig. 3. MS-curves of CO_2 emission in the composition of oxidation products of the researched coal samples.

Thermal destruction of $Cu(CH_3COO)_2 \cdot H_2O$ copper acetate monohydrate in an oxygen-containing medium is accompanied by a loss of water at ~ 130°C (stage 1), followed by salt decomposition in the 220-300°C range [7]. The primary gas-phase decomposition products of $Cu(CH_3COO)_2$ (acetone, acetaldehyde and acetic acid vapor) are oxidized to carbon dioxide and water (step 2) when reacted with air oxygen. Despite the complex mechanism of destruction and a large set of intermediate products of copper acetate decomposition [7], the main reaction (stage 2) can be presented in a simplified form:

$$Cu(CH_3COO)_2 \cdot H_2O \xrightarrow{<130\,^{\circ}C} Cu(CH_3COO)_2 + H_2O \tag{1}$$

$$Cu(CH_3COO)_2 \cdot 4O_2 \xrightarrow{<220\,^{\circ}C} CuO + 4CO_2 + 3H_2O \tag{2}$$

The oxidation reactions of the gas-phase decomposition products of copper acetate are exothermic, as a result of which additional heat is released contributing to the earlier thermal destruction of the carbon particles, which initiates the process of volatile compounds sublimation.

It is important to note that the solid phase product formed as a result of $Cu(CH_3COO)_2$ salt decomposition can be presented by a set of different oxide phases: CuO, Cu_2O and Cu_4O_3 [7]. Nevertheless, with a further rise of the temperature to 400°C and higher in oxygen-containing atmosphere leads to stabilization of the only phase of copper oxide (II). At the same time, it is known [8] that the copper oxide formed as a result of the reaction is an active catalyst of the complete oxidation processes in heterogeneous catalysis. Thus, the appearance of dispersed CuO particles on the surface of the samples, apparently, contributes to the further acceleration of the coals oxidation process.

4 Conclusion

The results of the study showed that application of $Cu(CH_3COO)_2$ initiation additive in the composition of coal in an amount of 5% by weight has a positive effect on the oxidation process nature. The presence of the copper acetate promoter in the coal sample leads to a decrease of the temperature corresponding to the onset of release and active oxidation of the volatile compounds, a reduction of the sample warm-up time prior to the active interaction onset of the volatile compounds and the blown medium, and also to a shift to the lower temperature region of the process in general. The temperature range of volatile compounds oxidation for the modified sample corresponds to the decomposition process beginning of $Cu(CH_3COO)_2$ initiation additive. Copper oxide (II) formed as a result of copper acetate decomposition catalyzes the further process of complete coal oxidation. The revealed patterns of the initiated coals oxidation in the presence of copper acetate can be further used in the development of more efficient methods for solid fuels combustion.

The work was conducted with the financial support of the Ministry of Education and Science of the Russian Federation within the framework of the project No. 13.7644.2017/BCH.

References

1 Russian Federation Federal State Statistics Service: extraction of fuel and energy minerals, www.gks.ru/free_doc/new_site/business/prom/natura2g.html, 2017, [accessed 12.03.2018]

2 J. Cheng, F. Zhou, X. Xuan, J. Liu, J. Zhou, K. Cen, Fuel **187**, 398 (2017) DOI: 10.1016/j.fuel.2016.09.083

3 V.N. Parmon, A.D. Simonov, V.A. Sadykov, S.F. Tikhov, Comb. Explos. Shock Waves **51**, 143 (2015) DOI: 1134/S001050821502001X

4 G.W. Zhao, W.Q. Yu, Y.H. Xiao, Adv. Mater. Res. **236**, 660 (2011) DOI: 10.4028/www.scientific.net/AMR.236-238.660

5 D.V. Miroshnichenko, Y.S. Balaeva, Coke and Chemistry **56**, 85 (2013) DOI: 10.3103/S1068364X1303006X

6 I.V. Tokareva , I.V. Mishakov, D.V. Korneev, A.A. Vedyagin, K.S. Golokhvast, Nanotechnologies in Russia **10**, 158 (2015) DOI: 10/1134/S199507801501019X

7 Z. Lin, D. Han, S. Li, JTAN **107**, 471 (2012) DOI: 10.1007/S10973-011-1454-4

8 Y. Wang, J. Wang, H. Chen, M. Yao, Y. Li, CES **135**, 294 (2015) DOI: 10.1016/j.ces.2015.03.024

Physical simulation of heat exchange between 6(10) kV voltage instrument transformer and its environment with natural convection and insolation

Sergey N. Litvinov[1], *Vladimir* D. Lebedev[1], *Nikolay* N. Smirnov[1,*], *Vladimir* V. Tyutikov[1], and *Ilkhom B.* Makhsumov[2]

[1]Ivanovo State Power Engineering University, 153003 Ivanovo, Russia
[2]National Research South Ural State University, 454080 Chelyabinsk, Russia

Abstract. This study examines the results of thermal tests on a 6(10) kV digital combined current and voltage transformer conducted in an environmental chamber. This measuring instrument consists of current and voltage transformers, featuring a resistive divider, and is used for commercial and technical electric power accounting. Ambient temperatures and levels of insolation on the transformer surface were set for the environmental chamber, with simulation of voltage transformer functioning in normal and emergency modes. We determined the time needed for the thermodynamic system to move to steady heat exchange mode, and also the final temperatures in the lower resistors and on the surface of the voltage transformer insulation cover. The results of our study have been used in developing algorithms for self-diagnostics of the thermal state of a digital combined transformer.

1 Introduction

An important task in the functioning of digital combined current and voltage transformers in commercial and technical electric power accounting in summer conditions with high ambient temperatures and levels of insolation and the possibility of grid voltage exceeding rated values is to examine the exchange of heat between a transformer and its environment.

Methods of thermal protection against overheating for voltage power transformers are widely known [1-2]. Unfortunately, thermal tests during operation of voltage power transformers, and even more so in the case of current and voltage instrument transformers are often performed according to maintenance schedules [3], which negatively impacts the efficiency of such electrical equipment [4]. Developing new systems of self-diagnostics checks for the thermal state of digital instrument transformers that determine temperature according to the most heated up electrical equipment part in real time is therefore an important task, especially in the development of "smart" power grids. Developing algorithms for thermal self-diagnostics requires data relating to physical and mathematical

* Corresponding author: nsmirnov@bk.ru

simulation of the thermal environment in which transformers operate under the influence of various unfavourable factors.

An important indicator in research conducted with the aid of software enabling simulation of heat exchange between a transformer and its environment following changes in initial conditions is the accuracy of the resulting data. Laboratory research conducted with physical models of digital transformers is one of the best methods of verifying such models.

We conducted a series of experiments in the high voltage laboratory of our High Voltage Equipment Department to study the influence of voltage, high ambient temperatures and insolation on the operating temperature of a digital combined transformer. We also studied the dynamics of the processes involved.

The digital instrument transformer [5] consists of a resistive voltage divider and current transformer (Fig. 1), several resistors being contained within the body of the instrument transformer.

Fig. 1. External view of a 6 (10) kW digital combined transformer: 1 – primary voltage converter; 2 – primary current power converter; 3 – low voltage electronic unit.

Electrical processes occurring in resistive voltage dividers comprising groups of resistors are related to heat emission. Resistor overheating may result in measuring equipment failure.

2 Description of tests conducted inside a high voltage environmental chamber

A part of our research was conducted in an environmental chamber designed for high-voltage testing. The diagram and general view of this chamber with installed equipment are shown in Fig. 2. The environmental chamber comprises a closed, isolated structure consisting of two compartments: a warm compartment and one for a step-up voltage transformer. The warm compartment is shielded by means of PVC sheeting, and features a heating element used to maintain constant ambient temperature, creating the necessary temperature conditions corresponding to the worst transformer operating conditions, taking into account the environmental version and placement category.

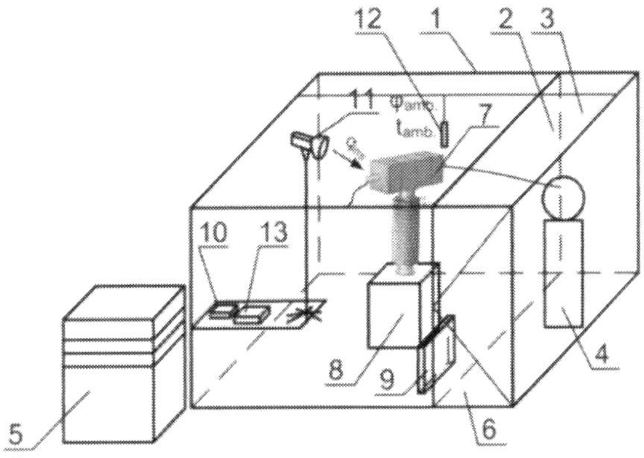

Fig. 2. Diagram of an environmental chamber in a high-voltage test laboratory: 1 – warm compartment; 2 – partition; 3 – step-up transformer compartment; 4 – step-up transformer; 5 – control panel; 6 – compartment door; 7 – test item; 8 – earthed prop for the test item; 9 – heating element; 10 – temperature measuring and controlling device; 11 – incandescent-filament lamp for simulating insolation; 12 – primary humidity and temperature converter; 13 – thermal hygrometer.

An electrode was introduced into the chamber to supply high voltage to the test item from the laboratory UIV- 150kV unit, consisting of a control panel and step-up transformer. Ambient temperature and humidity, also temperature at the resistors, were determined by means of sensors. A TESTO-875-1i thermal imager was also used to determine transformer surface temperature fields. Insolation on the collecting surface of the warm compartment in the chamber was simulated using a 250W incandescent filament lamp (Fig. 3) located on a device enabling us to change and fix lamp angle of elevation. Preliminary experiments were conducted to calibrate the values of radiation energy emitted from the incandescent filament lamp onto the collecting surface.

Fig. 3. Simulated insolation on a transformer using an incandescent filament lamp.

In accordance with electric power safety requirements, temperature sensors were placed on the lower contact cap of the lower resistors. Sensor resistance was R_s=10 kOhm at 25°C, within 1%.

A resistive voltage divider functions at various thermal conditions, and the temperature (as well as the number of temperature cycles) significantly influences both the metrological properties of the resistor, and its performance.

For a 6(10) kV voltage instrument transformer we simulated normal operating mode, with nominal voltage U_{nom}=10kV (phase voltage $U_{nom.phase}$=5.77 kV), and also emergency mode. In emergency mode, voltage increased 2.1 times in comparison with the nominal value. Voltage was changed only after the system moved into steady heat exchange mode. Thermal conditions inside the environmental chamber were maintained with an ambient temperature of t_{amb}=44°C, normal atmospheric pressure and natural convection [6] near the transformer surface.

3 Results of tests

Thermographs for transformer surface (Fig. 4) show that most heating occurred in the middle part of the resistor divider. Thus, insulation surface temperature in the area of the middle resistor was t_{surf}=109 °C, and 101°C at the upper resistor. The temperature was also lower in the area of the lower resistor.

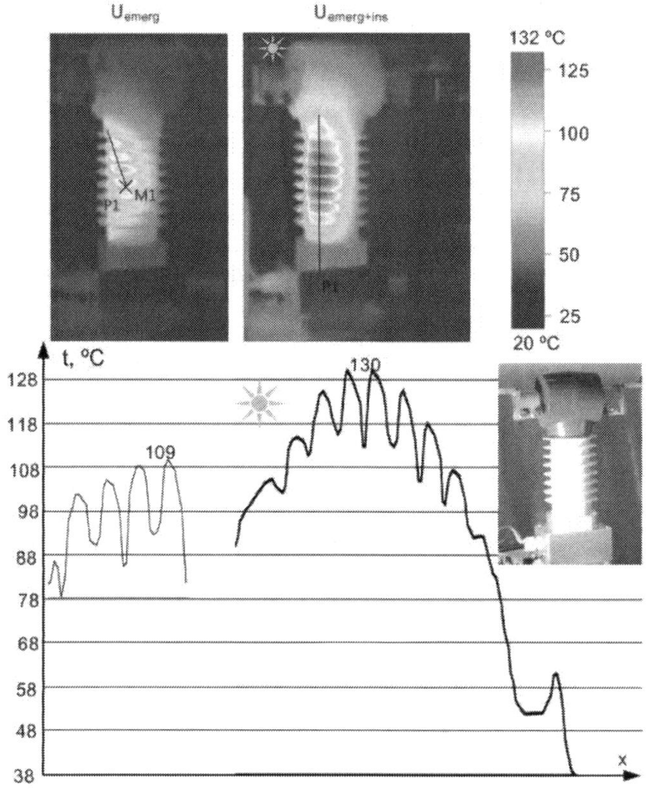

Fig. 4. Thermograph for voltage transformer surface during emergency mode in the absence of/with insolation, and also temperature charts along transversal P1.

Therefore, because of the cumulative effect of heating in the upper and lower resistors, the middle resistor warmed up more than the others, and placing the temperature sensor on the lower resistor cannot unequivocally reflect thermal conditions at the middle resistor; adjustment must be made for the cumulative effect.

Insolation in emergency operating mode intensified heating of the middle resistor even further, the temperature on the surface of the transformer cover in the area of the most heated resistor increasing to t_{surf}=130°C. Hence, the temperature on the most heated resistor, located inside the transformer, will be somewhat higher than 130°C. Here adjustment must also be made for the heat characteristics of the incandescent filament lamp (the heat stream along the lamp axis being higher than in the peripheral area).

As shown by experimental data for examination of the thermal condition of resistor dividers in a voltage instrument transformer, supplied voltage and simulated insolation have a significant impact on the thermal condition of the controlled facility (Fig. 5).

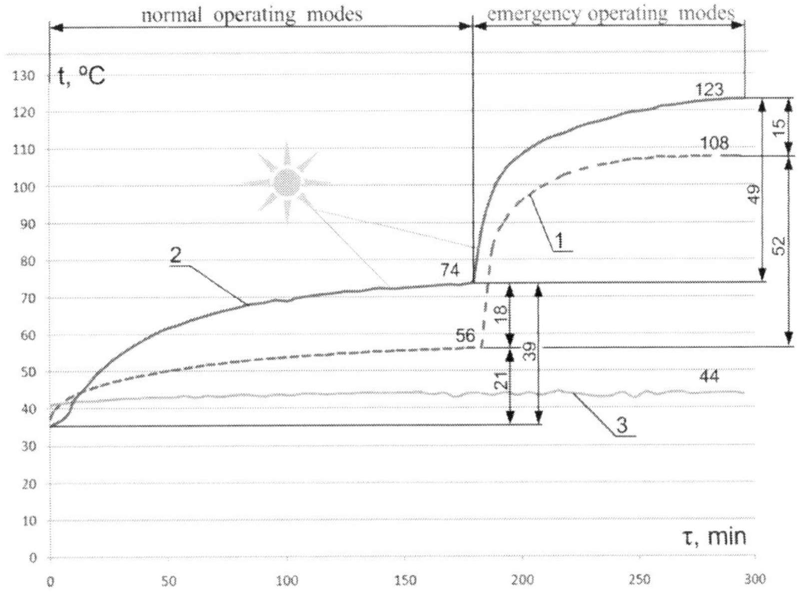

Fig. 5. Charts showing dependence of temperatures at the lower contact cap of the lower resistor and ambient temperature on time in normal and emergency transformer operating modes: 1 – without insolation; 2 – with insolation; 3 – ambient temperature.

Since the transformer being tested was energized immediately after the ambient temperature tamb in the warm compartment had reached 41°C, we may observe in the chart that air warmed up more rapidly than the transformer body, therefore resistor temperature at the moment of transformer energization was somewhat lower than the ambient temperature due to a certain heat retention in the materials of the above electrical equipment. However, self-warming of the resistors resulted in quite a rapid increase in temperature in the area of the resistors' location.

With normal voltage U_{norm} without insolation the resistor heated up to a temperature of t_r=56 °C (increasing by Δt_r=21 °C). Resistor warming from the initial state to steady thermal conditions occurred within 3 hours.

Simulating insolation on the transformer surface significantly increased the above temperature parameters. Thus, the resistor heated up to a temperature of t_r=74 °C (increasing by Δt_r=39 °C). In this case, insolation caused the resistor to warm up by a further Δt_{ins} =18 °C. Warming of the resistors from the initial state to steady thermal conditions in our experiment also occurred within 3 hours.

The simulated emergency operating mode for electric power grids logically resulted in significantly increased temperatures. In the absence of insolation, resistor temperature increased to t_r=108 °C (or by Δt_{emerg} =52 °C, that is, 2.5 times more than the increase in temperature with normal transformer operating mode). The system changed over to steady mode within 2.2 hours, with a significant rise in temperature observed during the first 20 minutes of our experiment with simulated emergency voltage.

Insolation further increased the impact of equipment operation in emergency mode. Resistor temperature rose to t_r=123 °C (increasing by Δt_{emerg}=49 °C, that is, 1.3 times more than the increase in temperature with normal equipment operating mode. In these emergency conditions, insolation caused the resistor to further warm up by Δt_{ins}=15 °C, which is comparable with the effect of insolation in normal equipment operating mode.

It may be noted that the increase in resistor temperature as a result of voltage increasing to emergency levels (Δt_{emerg}=52 °C) was 2.9 times more than the increase in temperature resulting from the effect of insolation on the resistor (Δt_{ins} =18 °C).

4 Conclusion

Our study showed that even in the most severe emergency operating modes, heating in the lower resistor did not exceed the critical value (taken as around 150 °C, according to electrical insulation material characteristics). Maximum temperatures were determined in the central part of a voltage transformer. Its thermal conditions were influenced more by simulated voltage than by insolation.

We expect the resulting experimental values to be used in the development of algorithms for thermal self-diagnostics of digital transformers, and also algorithms for temperature correction of the transformed secondary voltage. The results should be taken into account in the development and verification of mathematical models of heat exchange between a transformer and its environment.

This research was conducted with financial support from the Ministry of Education and Science of the Russian Federation at Ivanovo State Power Engineering University within the framework of a federal target programme, "Research and Development in Priority Areas of Developing the Russian Scientific and Technological Complex in 2014-2020" concerning "A Multifunctional System based on Digital Current and Voltage Transformers for a Digital Substation" (Agreement No. 14.577.21.0276 on Granting Subsidies, dated 26 September 2017, Unique Identifier for Applied Scientific Research (Projects) RFMEFI57717X0276).

References

1. R. Hunt, M.L. Giordano, *Thermal overload protection of power transformers – operating theory and practical experience* (Georgia Tech, Atlanta, 2005)
2. L. L. Grigsby, *The Electric Power Engineering Handbook* (CRC Press, Boca Raton, 2001)
3. Z. Liang-dong, S. Hui, Power System Protection and Control, **4** (2010)
4. Ju. P. Aksenov, A. V. Golubev, V. I. Zavidey, Elektro, **1** (2006)
5. A. A. Yablokov, N. N. Smirnov, V. V. Tyutikov, V. A. Gorbunov, MATEC Web Conf. **141** (2017)
6. M. A. Ulchiekov, E. E. Bulba, MATEC Web Conf. **92** (2017)

Heat-mass transfer efficiency within the cooling towers with jet-film contact devices

Ilnur N. Madyshev[1], *Oksana* S. Dmitrieva[2], and *Andrey* V. Dmitriev[2,*]

[1]Kazan National Research Technological University, 420016 Kazan, Russia
[2]Kazan State Power Engineering University, 420066 Kazan, Russia

Abstract. Jet-film contact devices with special design elements for intensification of heat-mass transfer processes are proposed. An experimental apparatus has been created in order to study the interaction of water and air flows within these devices. It has been revealed that an increase in the ratio of mass flow rate of liquid and gas phases leads to a decrease in heat efficiency coefficient. Within the area of high values of specific loads, an increase in the local minimum of mass transfer efficiency is observed.

1 Introduction

Circulation water supply systems are one of the most important elements of technological complex at industrial enterprises and CHP plants. At the same time, efficiency of cooling the circulating water within the cooling towers depends largely on even distribution of the contact phases in the volume of nozzle [1].

Studies of the hydrodynamic situation within the current designs of contact devices reveal uneven distribution of liquid and gas phases in the cross section of the nozzle area, which prevents the intensification of heat-mass transfer processes and reduces the depth of cooling the circulating water.

Searching for the ways to intensify the heat-mass transfer within gas-liquid systems led to the development of fundamentally new apparatuses with vertical grids, drop-film sprinklers, and regular nozzles of increased throughput capacity [2-6]. According to the analysis of the most promising designs of contact devices, developed in recent years, each new design solution gives a slight gain in efficiency. The designs are mostly becoming more and more complex [7, 8].

2 Description of jet-film contact device

The jet-film contact devices [9], allowing self-distribution of the flowing liquid phase over the cross section of the apparatus, are developed. One of the modifications of contact device is shown in Fig. 1. The proposed device consists of parallel square drain cups *1* with vertical walls, required for maintaining the liquid level inside of them. Vertical corrugated

* Corresponding author: ieremiada@gmail.com

and perforated baffles *2* are the supports for drain cups *1* and have slotted holes for the installation of appropriate drain cups *1*. The drain cups *1* are open from the top end, but recurved leaves *3* in the form of circular segments are executed in the bottom of cups. These segments are required for the liquid distribution over the surface of vertical corrugated and perforated baffles *2*.

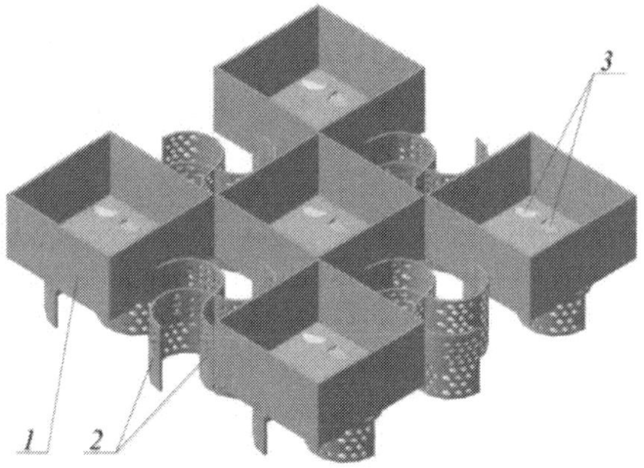

Fig. 1. Jet-film contact device: *1* – drain cup; *2* – corrugated and perforated baffles; *3* – recurved leaves.

The jet-film contact device, developed by the authors, operates as follows. Liquid, through the plurality of recurved leaves *3*, executed in the bottom of drain cups *1*, is dispersed in the form of jets to the below located vertical corrugated and perforated baffles *2*. When liquid jets move along the surface of vertical baffles *2*, the liquid is distributed with formation of a stable film flow. Thus, the coming down liquid film interacts with an ascending gas flow. After that the formed film, colliding with liquid surface inside of drain cups *1*, collapses. When distances between the drain cups, arranged at the same level, are equal to width of a drain cup, it provides stable, uniform gas flow, which leads to decrease in hydraulic resistance of the proposed jet-film contact device. Thus, organization of original interaction between gas and liquid allows stimulating the heat-mass transfer processes both in liquid and in gas phases at relatively simple design of apparatus.

3 Analysis of results of experimental research

Experimental research has been carried out in order to determine the efficiency of heat-mass transfer within the contact stage of jet-film device with corrugated and perforated baffles. Holes with a diameter of 5mm were executed on each baffle and located on tops of equilateral triangles with a step of 7.5 mm between the holes. Average air flow rate for the full cross-section of empty apparatus was varied in the range of 1.06-3.07 m/s; irrigation density – 11.0-44.85 $m^3/m^2 \cdot h$, temperature of the cooling air – 21.2-21.5°C; hot water temperature – 44.2-50.9°C. The relative air humidity during the experiment was 70%. In order to measure the temperature and relative air humidity at the inlet and outlet of contact stage, thermal hygrometer Testo 605i was used. Water temperature at the inlet and outlet of contact stage was measured by a two-channel regulator OVEN. Liquid flow rate was measured by variable-area flowmeter; the air flow velocity was measured by thermal anemometer Testo 405i with a relative error of not more than 5%.

The degree of improving the heat-mass transfer processes within the evaporative water cooler is usually characterized by the value of temperature coefficient (heat efficiency coefficient) [10]:

$$\eta_L = \frac{t_{L0} - t_{Lk}}{t_{L0} - t_{Lp}}, \tag{1}$$

where t_{L0} – water temperature at the inlet of contact stage, °C; t_{Lk} – water temperature at the outlet of contact stage, °C; t_{Lp} – equilibrium water temperature, i.e. dew-point temperature (theoretical limit of water cooling), °C.

It is known from early studies [11-13] that proportion of heat transfer by evaporation within evaporative water coolers is very high, thus, the heat balance equation with sufficient accuracy (without taking into account heat conductivity, convection and vapor condensation) can be recorded as follows:

$$Q = L_m c_L \left(t_{L0} - t_{Lk}\right), \tag{2}$$

where L_m – mass water flow rate, kg/s; c_L – mass specific heat of water, J/(kg·K).

The material balance (moisture balance) is determined by the equality between the amount of evaporated liquid and the increase in the moisture content of the air:

$$\Delta L_m = G_m \left(x_k - x_0\right), \tag{3}$$

where G_m – mass air flow rate, kg/s; x_0, x_k – moisture content of saturated air at the inlet and outlet of contact stage, kg/kg.

On the other hand, the amount of evaporated liquid can be determined by the equation:

$$\Delta L_m = \frac{Q}{r}, \tag{4}$$

where r – evaporation specific heat, J/kg.

Then the moisture content of saturated air at the outlet of contact stage can be determined as follows:

$$x_k = x_0 + \frac{\Delta L_m}{G_m}, \tag{5}$$

Thus, mass transfer efficiency coefficient of contact stage can be determined by the change in the moisture content of saturated air, assuming that the entire heat flow is consumed for the water evaporation:

$$E = \frac{x_k - x_0}{x_p - x_0}, \tag{6}$$

where x_p – equilibrium moisture content of saturated air, kg/kg.

The scheme in Fig. 2a shows that an increase in the ratio of mass flow rate of liquid and gas phases leads to a decrease in heat efficiency coefficient. When the irrigation densities are relatively low, the heat efficiency coefficient of contact stage can reach 37.5%. Within the area of high specific loads Lm/Gm, an increase in the local minimum of mass transfer efficiency is observed (Fig. 2b). For example, an increase in the irrigation density from 11.0 up to 44.85 $m^3/m^2 \cdot h$ leads to an increase in the average efficiency from 31 up to 60%. Further increase in mass transfer efficiency coefficient is due to the fact that the rate of increase in the amount of evaporated liquid prevails over the rate of increase in mass air flow rate.

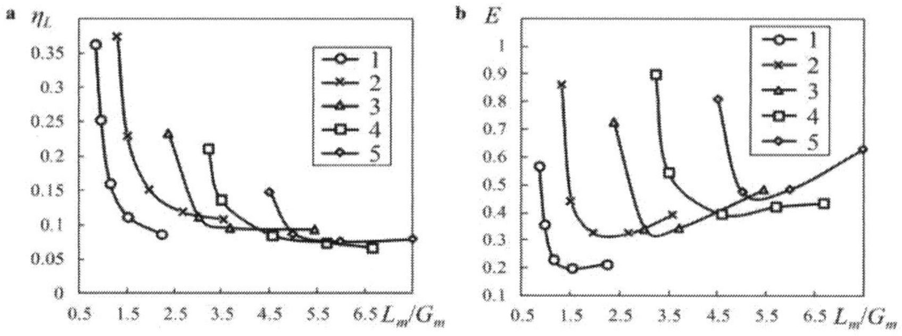

Fig. 2. Dependency of the change in heat (a) and mass (b) transfer efficiency coefficient of contact stage on ratio of mass flow rate of liquid and gas phases at different irrigation densities q, $m^3/m^2 \cdot h$: 1 – 11.0; 2 – 17.43; 3 – 24.3; 4 – 36.7; 5 – 44.85.

4 Conclusion

The conducted experimental research shows that the developed jet-film contact devices have a sufficiently high efficiency of heat-mass transfer, but, at the same time, modes, corresponding to a relatively low hydraulic resistance, should be chosen when conducting the processes. Thus, it can be stated that the use of the proposed contact devices at industrial enterprises and in the energy sector will ensure a decrease in the temperature of the cooled water within the cooling towers at low operating costs.

The research was conducted with funding from the grant of RF President No. MK-4522.2018.8.

References

1. A. G. Laptev, V. A. Danilov, I. V. Vishnyakova, Therm. Eng. **51**, 661 (2004)
2. E. V. Boev, S. P. Ivanov, V. G. Afanasenko, E. A. Nikolaev, Chem. Pet. Eng. **45**, 454 (2009)
3. J. Lee, Nucl. Eng. Des. **326**, 65 (2018)
4. S. P. Fisenko, A. I. Petruchik, A. D. Solodukhin, Int. J. Heat Mass Transfer **45**, 4683 (2002)
5. O. S. Dmitrieva, A. V. Dmitriev, A. N. Nikolaev, Chem. Pet. Eng. **50**, 169 (2014)
6. V. G. Afanasenko, F. Sh. Khafizov, N. F. Khafizov, S. P. Ivanov, E.V. Boev, Chem. Pet. Eng. **43**, 653 (2007)
7. O. S. Dmitrieva, A. V. Dmitriev, I. N. Madyshev, A. N. Nikolaev, Chem. Pet. Eng. **53**, 130 (2017)
8. C. Zheng, X. Chen, L. Zhu, J. Shi, Energy **150**, 653 (2018)
9. A. V. Dmitriev, O. S. Dmitrieva, I. N. Madyshev, A. N. Nikolaev, L. V. Kruglov, Patent RU No. 171022 (2017)
10. F. Kreith, R.F. Boehm, *Heat and Mass Transfer* (CRC Press LLC, Boca Raton, 1999)
11. G. V. Kuznetsov, D. V. Feoktistov, E. G. Orlova, S. Y. Misyura, V. S. Morozov, A.G. Islamova, Int. J. Heat Mass Transfer **126**, 161 (2018)
12. Y. Li, X. You, Q. Qiu, J. Li, Energy Convers. Manage. **52**, 53 (2011)
13. G. V. Kuznetsov, D. V. Feoktistov, E. G. Orlova, J. Eng. Phys. Thermophys. **89**, 317 (2016)

Study of heat transfer processes in the flowing part of hypersonic air-ramjet engine

Eugene Maslov[1,2*], *Irina* Zharova[1], *Valery* Faraponov[1], *Eugene* Kozlov[1], and *Vladislav* Matskevich[1]

[1]National Research Tomsk State University, 634050 Tomsk, Russia
[2]National Research Tomsk Polytechnic University, 634050 Tomsk, Russia

Abstract. The technique and results of the experimental-theoretical study of gas dynamics, heat transfer and the structure of gas flow in the flowing channel of a model hypersonic air-ramjet engine are presented for Mach numbers M = (5; 6).

1 Introduction

At hypersonic flight speeds of rockets at low altitudes, it is supposed to use solid fuel air-ramjet engine. Wherein, the air from the atmosphere, which enters through the air intake into the flowing channel of the engine with velocities M > 5 [1], is used as an oxidizer in the organization of combustion. The processes of combustion and burning-out (possibly, uneven) of the surface of the solid fuel element are determined by gas-dynamic and thermal parameters of flow in the flowing part of the engine. Thus, for design of engine facility, it is necessary to have of experimental studies data from that allow one to predict these parameters with a high degree of veracity for concrete designs [2]. It is especially important to solve the problem of assessing the thermal state of the flowing channel in the initial period after the switching-on of an installation. The presence of predictive estimates of the thermal state of the tract at the design stage of a specific design is necessary for a well-founded choice of the ignition method and the provision sustainable combustion of solid fuel. It is advisable to conduct preliminary studies with laboratory models, since full-scale stand experiments are costly.

The purpose of the study was to obtain objective information about the thermal state of the walls of the flowing channel of a model hypersonic air-ramjet engine at Mach numbers M = (5, 6).

2 Technique of the experimental study

An experimental study was carried out on a pulsed aerodynamic installation with heating of gas (up to $T_e \approx 700$ K) [3–5]. The aerodynamic installation provides the speed of the inflowing stream in the range of Mach numbers M = (2 ÷ 7). Steel axisymmetric profiled

* Corresponding author: maslov_eugene@mail.ru

nozzles [4] were used to create a hypersonic flow for Mach numbers M = (5, 6). The experiments were conducted on an axisymmetric model of an air-ramjet engine (Fig. 1, a).

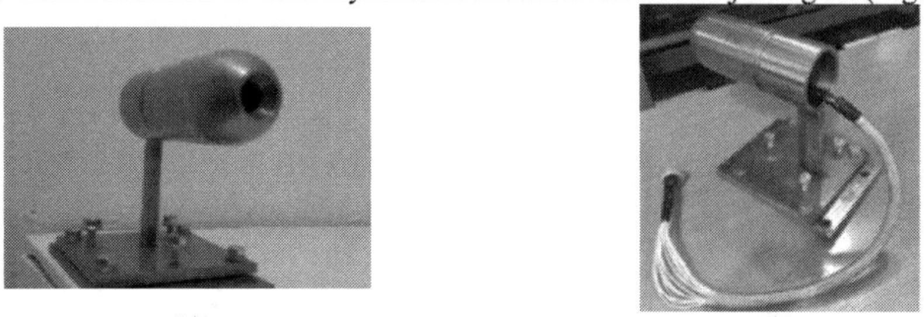

(a) (b)

Fig. 1. General view of model (a), and model with the installed thermo-probe (b).

The measurement of the gas temperature in the flowing channel was carried out by a thermocouple method [6]. A thermo-probe was manufactured for measure of the temperature along the wall of the model flowing channel. The thermos-probe represents a set of fluoroplastic rings of the equal diameter. Eight copper rings with a chromel-copel thermocouple were attached to the inner surface of each ring. The view of the axisymmetric model with the installed thermo-probe is shown in Fig. 1, b. The scheme (Fig. 2, a) and photographs of the thermo-probe in assembled form (Fig. 2, b) and the temperature sensor (Fig. 2, c) are shown.

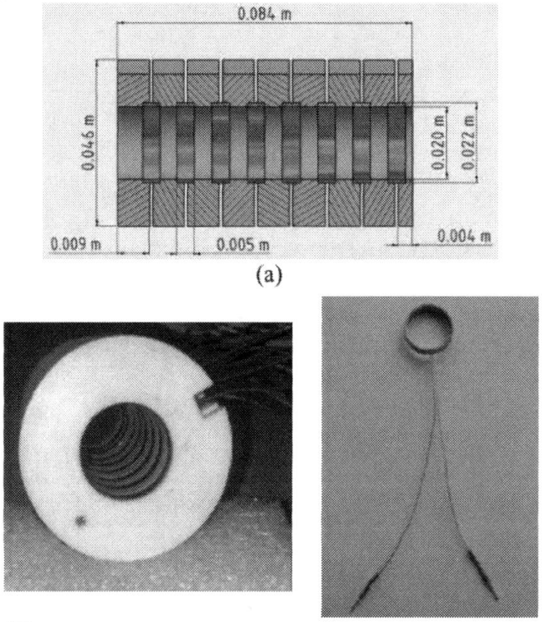

(a)

(b) (c)

Fig. 2. Scheme of the thermo-probe (a), the thermo-probe fully assembled (b) and the temperature sensor (c).

The temperature in the prechamber and in the exit section of the model was measured with the help of special temperature sensors during process of the experiments. The view of placed in the working part of the aerodynamic installation the axisymmetric model in complex with temperature sensors is shown in Fig. 3.

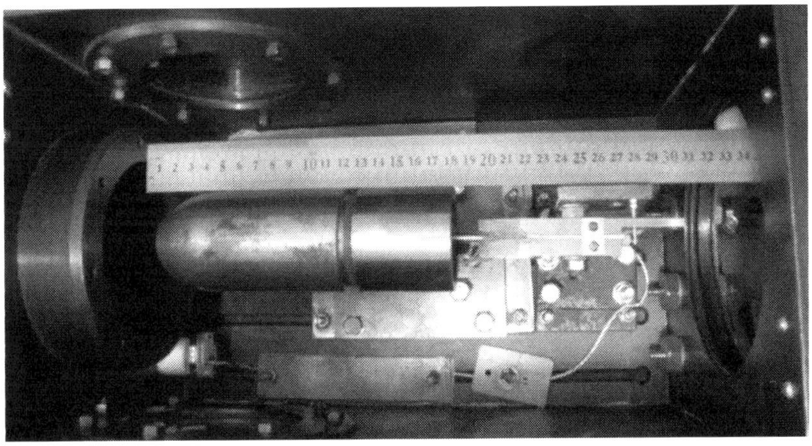

Fig. 3. Axisymmetric model located in the working part of the aerodynamic installation in complex with temperature sensors.

3 Results of the experimental study

Series of thermogasdynamic experiences were carried out for each Mach number M = (5, 6). The measured values of the static temperature in the prechamber and the stagnation temperature in the exit section of the model are shown in Fig. 4, 5.

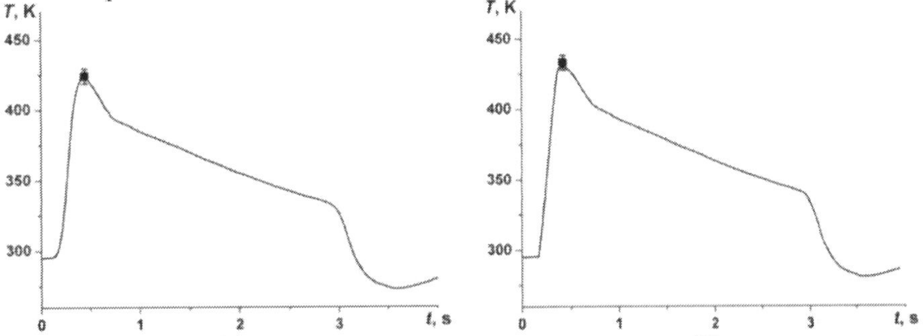

(a) (b)

Fig. 4. Dependence of the static temperature in the prechamber on time at M = 5 (a) and M = 6 (b).

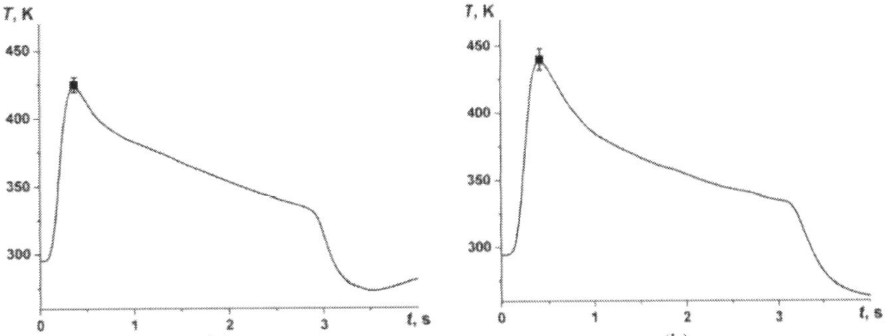

(a) (b)

Fig. 5. Dependence of the stagnation temperature in the exit section of the axisymmetric model on time for M = 5 (a) and M = 6 (b).

The temperature measurement data for flowing around an axisymmetric model are given in Fig. 6, a, 7, a, M = (5, 6). The data were obtained in one of the tests in a series of ten experiments. These are conditions for conducting a series of experiments: the heater temperature is 675 K, the work duration of the aerodynamic installation is 3 s. The maximum temperature values averaged over a series of experiments at the corresponding Mach numbers are given in Fig. 6, b, 7, b (x is the distance from the initial section of the model to the hot junction along the thermo-probe).

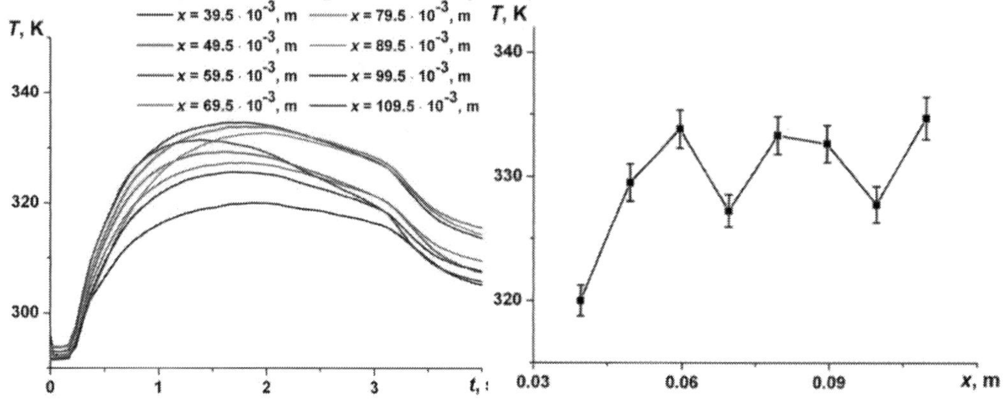

(a) (b)

Fig. 6. Typical temperature distribution along the surface of the flowing channel (a) and the maximum temperature values averaged over a series of experiments (b), M = 5.

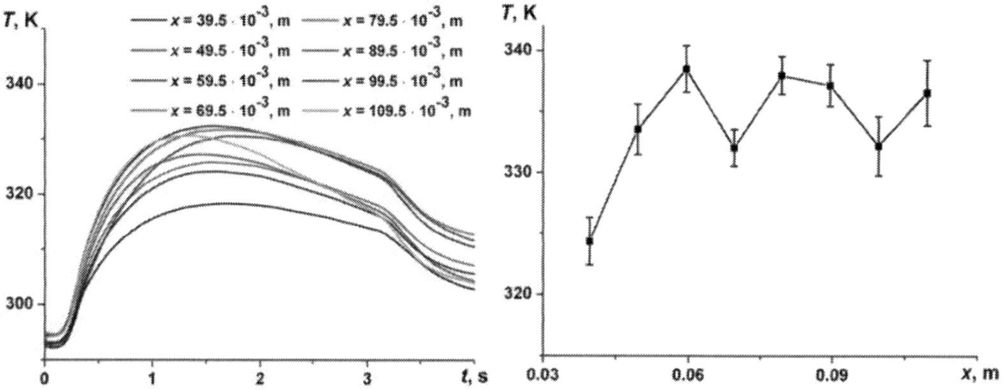

(a) (b)

Fig. 7. Typical temperature distribution along the surface of the flowing channel (a) and the maximum temperature values averaged over a series of experiments (b), M = 6.

4 Discussion of the results

Veracity of the data obtained with the help of the developed technique of experimental research is confirmed by their comparison with the results of calculations. Numerical analysis for M = (5; 6) was carried out on the basis of a mathematical model [7]. Numerically obtained distributions of Mach number and the temperature along the axis of symmetry of the flowing channel of the model for M = 5 are shown in Fig. 8. When $x = 0.042$ m (x is the distance from the initial model section to fixed point), the value of the Mach number was obtained experimentally (the point in Fig. 8, a) [5]. Comparison of the numerical and experimentally obtained Mach number values showed that the relative error is less than 10 %. The nature of the temperature change along the axis of the flowing channel (Fig. 8, b) correlates with the temperature measurement data along the thermo-probe surface and adequately describes the physical process [4].

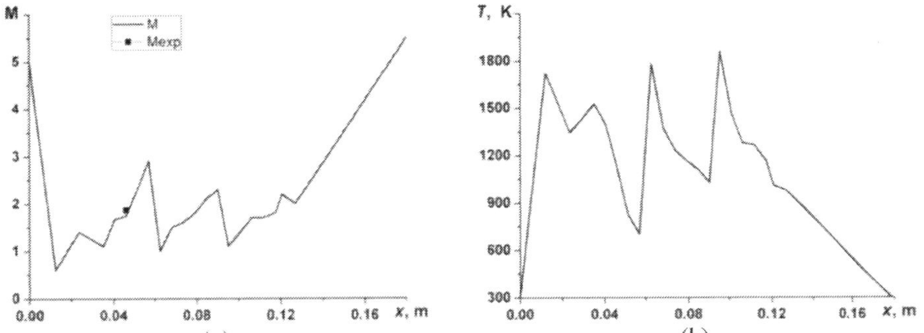

(a) (b)

Fig. 8. Distributions of the Mach number (a) and the temperature (b) along the axis of symmetry of the flowing channel of the axisymmetric model, M = 5.

The correlation of oscillatory regimes by Mach number, and temperature on the symmetry axis is noted. The Mach number decreases in the zones of compression shock, and the temperature increases. The Mach number increases in the decompression zones, and the temperature decreases.

5 Conclusion

The thermo-probe is developed, designed and implemented. The thermo-probe is equipped by temperature sensors with chromel-copel thermocouples. The working range of the measuring temperatures is $T = (200 \div 600)$ K.

The experimentally-calculating technique of studying gas dynamics, heat transfer and the structure of gas flow in the flowing channel of hypersonic air-ramjet engine are proposed.

Objective data about the heat transfer characteristics in the flowing part of the axisymmetric model of the air-ramjet engine were obtained for Mach numbers M = (5, 6) on the basis of the developed technique.

This research was supported by grant (No. 8.2.09.2018) from "The Tomsk State University competitiveness improvement program".

References

1. L. S. Yanovski, *Integral ramjet solid propellant air-jet engines. Fundamentals of theory and calculation: Reference book* (Moscow, Akademkniga, 2006) [in Russian]
2. B. V. Orlov, G. Yu. Mazing, A. L. Reidel, *Basics of design of rocket-ramjet engines for unmanned aerial vehicles* (Moscow, Mashinostroenie, 1967) [in Russian]
3. V. I. Zvegintsev, *Gasdynamic installations of short-term action: Reference book.* (Novosibirsk, Parallel, 2014) [in Russian]
4. E. Maslov, I. Zharova, V. Klochikhin, MATEC Web Conf. **23** (2015)
5. E. Maslov, I. Zharova, Ye. Kozlov, V. Faraponov, N. Savkina, N. Zolotorev, V. Matskevich, MATEC Web Conf. **115** (2017)
6. E. A. Maslov, I. K. Zharova, A. S. Zhukov, *Materials of the All-Russian Conference "XXXII Siberian Thermophysical Seminar": Reference book* (Institute of Thermophysics SB RAS, Novosibirsk, 2015) [in Russian]
7. V. A. Arkhipov, A. G. Egorov, S. V. Ivanin, E. A. Maslov, O. V. Matvienko, Combust. Explos. Shock Waves **46**, 1 (2010)

Influence of the water level in the work area on the hydrate formation process

Anton V. Meleshkin[1,*], and *Dmitriy* S. Elistratov[2]

[1]Novosibirsk State University, 630090 Novosibirsk, Russia

Abstract. This article describes a fundamentally new method for obtaining gas hydrates, based on self-organization in a closed working section of the cyclic boiling-condensation process of the gas-hydrate generator. A special feature of this method is the high rate of hydrate formation at low energy costs and, as a consequence, the expected efficiency of new technologies built on its basis over analogues. A series of experiments was performed, which shows the effect of the water level on the work site on the hydrate formation process.

1 Introduction

As early as the beginning of the 20th century, technological proposals for the storage and transportation of natural gas in the hydrated state appeared. This method draws the attention of experts for a long time because of the property of gas hydrates to concentrate large volumes of gas at relatively low pressures. Without a pipeline, a promising way of transporting gas is to convert it into a gas hydrate (solid) state and transport it at atmospheric pressure and low temperature (-10 ... -20 °C) [1]. The properties of gas hydrates, the main conditions and the features of their formation are described in [2-4] and some other works; the types of their crystallization, the mechanisms of formation and decomposition of gas hydrates are presented [5-11].

A method for producing gas hydrates based on shock-wave action on gas-liquid media was proposed, the intensification of the hydrate formation process is caused by the imposition of several factors such as an increase in the contact surface of hydrate-forming gas water due to the destruction of the bubbles by the shock wave, the transfer of the system far into the region of formation hydrates and the acceleration of bubbles with respect to water [12-13]. We continue to look for effective ways to intensify the formation of hydrates with the help of mechanical and thermal effects on gas-liquid media [14-17].

2 Experimental setup

The working site, which is a sealed vessel with a height of 740 mm and a section of 150X150 mm (Fig.1). Two walls of the vessel are made of an optically transparent material and allow observing the processes taking place inside the working area. Maintenance of the

* Corresponding author: mav@itp.nsc.ru

required temperature inside the installation is carried out using the built-in cooling system (water jacket). The lower cover of the installation does not have thermal insulation and is heated from the external environment. The temperature was measured by the DTS204-PT100 sensor, and the pressure measurement by the PD-100 sensor. The installation allows to carry out research in a wide range of temperatures (from -10 to 60 ° C) and pressure (from 0.1 to 10MPa).

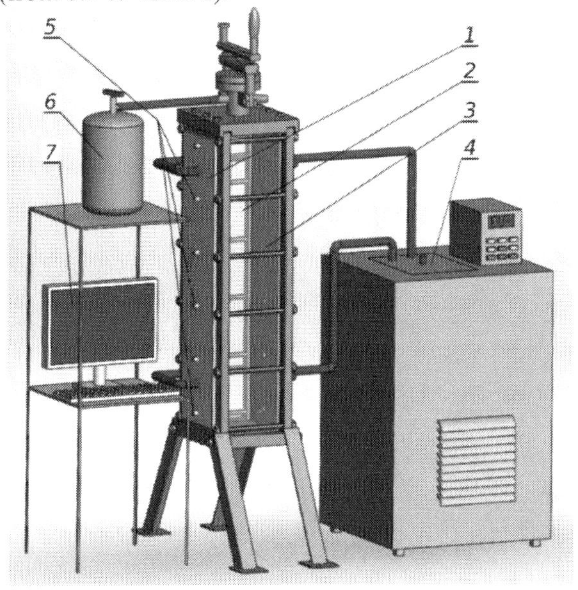

Fig. 1. The schematics of the unit: 1 – Side walls; 2 – Working area; 3 – temperature sensor; 4 – cryostat; 5 – pressure sensors; 6 – gas tank; 7 – PC.

Fig. 2. Gas-hydrate gained in the experiment.

3 Method and result

The essence of the experiment was as follows. The reaction vessel was filled with water, the volume of which varied from 1.5 to 6 liters. The water was cooled to a temperature of 7 °C, then a gaseous freon R-134A, weighing 0.3 kg, was fed into the chamber. Since the walls of the vessel had a negative temperature, Freon quickly liquefied and sank to the bottom. After the gas supply to the working area had ceased, the pressure equalized and the boiling began on the lower lid of the vessel. Due to intense boiling, a large number of bubbles of freon vapor formed. Rising to the surface of the vapor in the bubbles was cooled and conditions were created for the formation of Freon Hydrate. On the walls of the bubbles began to build up a hydrate film. Having reached the free surface of the liquid, the bubbles collapsed, leaving behind them hydrated flakes. Pairs of freon, not converted to hydrate, condense on cooled walls and descended to the bottom of the vessel where they were again included in the boiling process. From the description it can be seen that the process is cyclical and continues until all the liquid Freon, boiling off, goes to the hydrate (Fig. 2). The intensity of the process depends on both the removal of heat from the side walls and the supply of heat to the lower part of the installation, and from the distance traveled by the bubbles during boiling for each cycle. On fig. 3 shows the graphs of the growth of the gas hydrate "hat" from the distance traveled by a bubble from the bottom of the installation to the surface of the water.

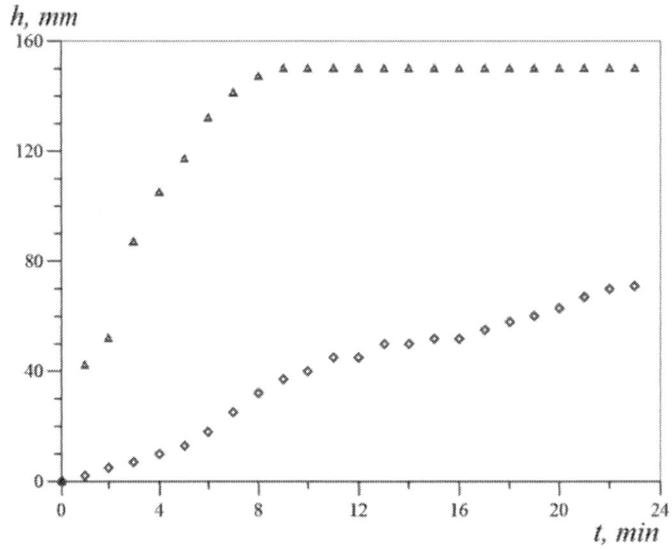

Fig. 3. Chart of growth of hydrate mass R-134A in the vessel depending on the level of water: *Triangles* – Height of water column is 50 mm, *Circle* - Height of water column is 230 mm.

4 Conclusion

A new method for obtaining gas hydrates is developed, based on self-organization of the cyclic boiling-condensation process of the gas-hydrate-forming agent in a closed water volume. This method is characterized by good energy efficiency. The influence of the height of the water column on the formation of hydrates is shown.

This work was supported by the Russian Science Foundation, project No. 18-19-00124.

References

1. J. Gudmundsson, M. Mork, O. Graff, Proc. 4th Intern. Conf. gas hydrates (2002)
2. Y. F. Makogon, *Hydrates of Hydrocarbons* (PennWell Books, Texas, 1997)
3. V. S. Yakushev, V. A. Istomin, Physics and Chemistry of Ice, 136 (1993)
4. E. D. Sloan, C. A. Koh, *Clathrate hydrates of natural gases* (CRC Press, Boca Raton, 2008)
5. V. Sh. Shagapov, N. G. Musakaev, M. K. Khasanov, Int. J. Heat Mass Transfer **84**, 1030 (2015)
6. V. Sh. Shagapov, M. K. Khasanov, N. G. Musakaev, Ngoc Hai Duong, Int. J. Heat Mass Transfer **107**, 347 (2017)
7. N. G. Musakaev, M. K. Khasanov, S. L. Borodin, Int. J. Heat Mass Transfer **118**, 455 (2018)
8. M. Z. Faizullin, A. V. Vinogradov, V. P. Koverda, Chem. Eng. Sci. **130**, 135 (2015)
9. S. Y. Misyura, Sci. Rep. **6**, 30324 (2016)
10. V. E. Nakoryakov, S. Y. Misura, S. L. Elistratov, A. Yu. Manakov, A. A. Sizikov, J. Eng. Thermophys. **22**, 169 (2013)
11. V. E. Nakoryakov, S. Y. Misura, J. Eng. Thermophys. **25**, 24 (2016)

12. V.E. Dontsov, A.A. Chernov, Int. J. Heat Mass Transfer **52**, 4919 (2009)
13. A.A. Chernov, V.E. Dontsov, Int. J. Heat Mass Transfer **54**, 4307 (2011)
14. A.A. Chernov, A.A. Pil'nik, D.S. Elistratov, I.V. Mezentsev, A.V. Meleshkin, M.V. Bartashevich, M.G. Vlasenko, Sci. Rep. **7**, 40809 (2017)
15. A.A. Chernov, D.S. Elistratov, I.V. Mezentsev, A.V. Meleshkin, A.A. Pil'nik, Int. J. Heat Mass Transfer **108**, 1320 (2017)
16. V.E. Nakoryakov, A.N. Tsoi, I.V. Mezentsev, A.V. Meleshkin, J. Eng. Thermophys. **23** (2014)
17. V.E. Nakoryakov, A.N. Tsoi, I.V. Mezentsev, A.V. Meleshkin, T and A. **21**, 279 (2014)

Modelling of thermal and stress-strain state of transformable space structures from hybrid composite materials

Konstanitin Mikhaylovskiy[1,*], and *Pavel* Prosuntsov[1]

[1]Bauman Moscow State Technical University, 105005 Moscow, Russia

Abstract. The strikes of the meteoroids represent a serious danger for transformable space structures. Multilayer plates and shells made of hybrid composite materials can be effectively used for protection and thermal control of such structures. The results of the analysis of the thermal regime and shock resistance of a multilayer structure flying in a circular orbit 400 km high are presented.

1 Introduction

Prospective exploration of the near-Earth space and the Solar System is associated with large-scale orbital and planetary structures with advanced shells made from hybrid composite materials. The basic concept of the TransHab (Transit Habitat) inflatable space modules was developed in the 1990s by NASA [1-4]. They proposed the following combination of layers: the atomic oxygen protective layer, the thermal protection system, the micrometeoroid/ orbital debris shield, the seal interface, the restraint layer, the triply redundant bladder system, and the inner layer. In 2016, the Bigelow Expandable Activity Module (BEAM) was successfully docked to the International Space Station (ISS) for full-scale tests [2]. In Russia, RSC Energia has created a prototype of an inhabited module with a multilayer transformable shell [5, 6]. Expandable space structures from composites can also be used in small spacecraft designed to capture and deorbit space debris [7, 8].

One of the key problems in creating expandable structures with multilayer elements is the selection and justification of the materials and the layer thicknesses with regard to the operating conditions in space, including impacts of meteoroids and space debris.

2 Statement of the task and the method modelling

The object of investigation is a space debris catcher (**Fig. 1**) and a shell of a prospective ISS module. It is assumed that the catcher flaps and the module shell have identical multilayer structure. Layer 1 (top) comprised the 10 mm thick screen vacuum thermal insulation (EVTI-I type) coated with the metallized arimid fabric (product identification number 5397-92). The screen vacuum shield is composed of several 20-micron polyimide film

* Corresponding author: kosmiv@yandex.ru

layers separated by a fiberglass scrim. Layer 2, a part of the meteoroid protective system, was a 15-layer fabric stack. The fabric (product identification number 8601) has the following properties: 190 g/m^2 linear density, plain weave, 15 warp threads by 15 weft threads, 58 tex. Layer 3 was the 40 mm screen vacuum shield. Layer 4 was a 20-layer arimid metallized fabric stack, with 4 mm total thickness.

The thermal state modeling did not take into account the possibility of the screen vacuum shield compacting under load, the presence of elastic shell deployment elements, which may act as "thermal bridges", as well as the power and sealing shell layers with low thermal resistance due to their small thickness (0.5 to 1.0 mm). Both layers of the screen vacuum shield and both layers of the meteoroid protection were considered as a continuous material without internal microstructure. The thermal physical and physical mechanical characteristics of the materials were assumed to be temperature-independent.

For the stress-strain modeling, the following assumptions were introduced:

- interaction on the shell boundary layers, as well as with the spacecraft elements is carried out with the help of surface-to-surface contact finite elements, with friction between layers accounted for; inner degrees of freedom of the shell layers are disregarded;
- the layers of the screen vacuum shield and the arimid metallized fabric (8601) were regarded as 3D orthotropic materials. Large deformations and displacements in the materials, including quasi-plastic ones, were also taken into account. In addition, the heating and heat propagation effects under high-velocity impact were evaluated. For stress analysis, the maximum stress failure criterion was used (temperature-dependent). It was taken into consideration that the arimid fabric stack consists of 0.22 mm thick layers with 2.4% critical stress level;

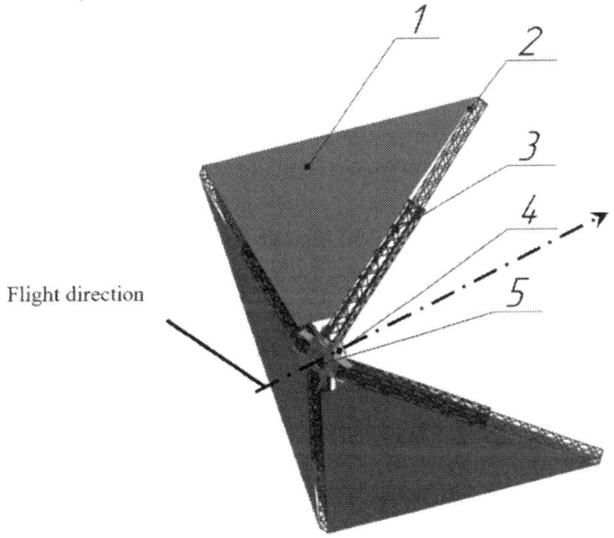

Fig. 1. Space debris catcher, one of the flaps not shown: 1 – debris catcher flap; 2, 3 – webbed load-bearing element; 4 – base; 5 – mount.

3 Results and discussion

The temperature state of an expendable shell was modelled for a 400-km orbit and 51.6° inclination. The temperature variation dynamics on the outer and back surfaces of the multilayer shell was determined for the maximum value of direct solar radiation flux. It was established that the temperature on the external surface of the screen vacuum shield varied within a wide range: from – 98°C to + 84°C. At the same time, the temperature of the meteoroid protection for the variously oriented parts of the trap stabilized almost completely after 30 days of orbital flight, and was at the level acceptable for the arimid fiber (+ 11°C to – 17°C) (Fig. 2). This result confirms that the 10 mm thickness is sufficient for the outer layer of the screen vacuum shield.

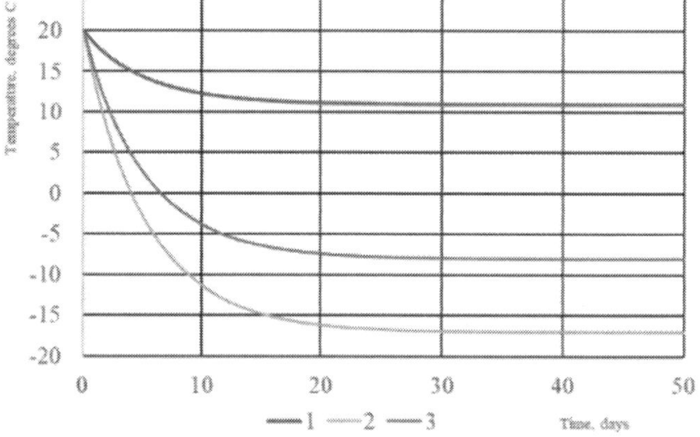

Fig. 2. Change in the temperature of the outer layer of the meteoroid protection for elements oriented towards the Earth (1), away from the Earth (2) and in the flight direction (3), as a function of time elapsed after launch, °C.

To assess the stability of the multilayer flap wall/ shell, the consequences were analyzed for a high-speed impact at 90° and at 45 ° with an aluminum or ice debris particle, 0.5 to 10 mm in diameter at 4000 m/s – 7000 m/s. As seen in Fig. 3, at the moment of penetration, local temperature rises abruptly to about 500 K, which points to the significance of the thermal stability of the materials used.

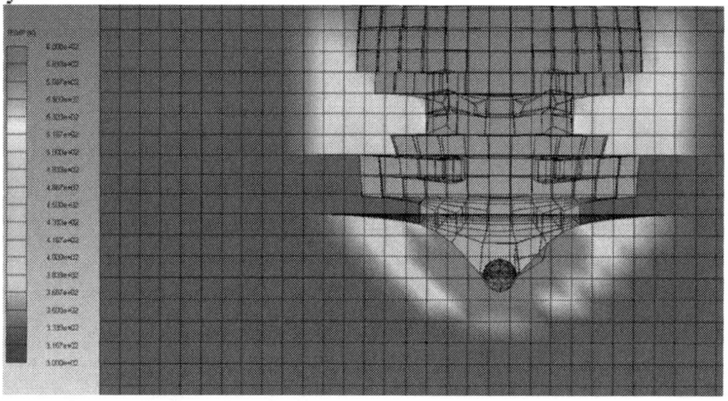

Fig. 3. The temperature distribution in an element of the multilayer structure at a high-velocity impact (5000 m/s) of an aluminum particle, 1 mm in diameter (cross-section shows the impact center), K.

Parametric modeling established that for a high-speed impact of a 1 mm aluminum particle at 5000 m/s at 90° to the surface, the instantaneous compaction, penetration and local compression of the first screen vacuum layers stack results in a noticeable extrusion of the destroyed material, as well as the propagation of the stress wave in the material. Then the particle penetrates the arimid layers and perforates them, forming numerous layer separation areas. Some of the patricle's kinetic energy converts into the elastic stretching and tearing of the arimid layers (threads). The residual kinetic energy of the particle and the accelerated part of the arimid layers is sufficient for the total deformation and failure of the second stack of the screen vacuum shield layers. The particle is partially destroyed in the second stack of arimid layers, with the fragments penetrating the stack. The characteristic size of the failure zone for the first arimid stack is 2.26 mm.

4 Conclusion

The proposed multilayer structure design can provide protection against ice debris particles up to 2 mm in diameter at up to 5000 m/s impact velocity and against aluminum debris particles up to 1 mm in diameter at up to 4.500 m/s impact velocity.

In our country there is a fairly large practical experience in the design, production and testing of large-size space structures, allowing to take into account the features of the thermal regime in the implementation of the project of the transformable space debris catcher [9-14].

Certain results of this research were obtained with funding from Ministry of Education and Science of Russian Federation within the incentive 1.2 of the federal targeted programme "R&D in priority development areas of Russian science and technology sector in 2014-2020" (Agreement for grant N 14.574.21.0146 from September 26th, 2017, the unique identifier for the project RFMEFI57417X0146).

References

1. K. J. Kennedy, J. Raboin, G. Spexarth, G. Valle, *Inflatable Structures Technology Handbook. Chapter 21; Inflatable Habitats* JSC-CN-6300 (NASA, Johnson Space Center, 2000)
2. BEAM Fact Sheet: Demonstrating Technologies For Deep Space Habitation. URL: https://www.nasa.gov/sites/default/files/atoms/files/2016-march-beam-factsheet-508.pdf (Accessed 01.03.2018)
3. R. T. Joyce, L. D. Carpenter, J.-M. Chang, D. L. Akin, *47-th Int. Conf. on Environmental Systems* (2017)
4. J. Hinkle, R. Timmers, A. Dixit, J. Lin, J. Watson, *50-th AIAA/ASME/ASCE/AHS/ASC Structures, Structural Dynamics, and Materials Conference* (2009)
5. I. I. Khamits, I. M. Filippov, L. S. Burylov, Kosmicheskaya tekhnika i tekhnologii **2**, 23 (2016) [in Russian]
6. N. A. Goldenko, Ye. P. Buslov, V. A. Fel'dshteyn, *Proceed. of Mechanics and Mathematical Modelling in Engineering Conf. ded. to the 100th anniv. of V. I. Feodos'yev* (2016) [in Russian]
7. B. Lyndon, *Handbook for Designing MMOD Protection* (NASA, Texas, 2009)
8. W. Ley, K. Wittmann, W. Hallmann, *Handbook of Space Technology* (John Wiley & Sons, Manhattan, 2009)
9. S. V. Reznik *Materials and coatings in extreme conditions. View in the Future* (Bauman MSTU Publ. House, Moscow, 2002)
10. V.A. Zavoruev, L.A. Kudryavin, V. I. Khalimanovich, Technical textiles **16** (2007)
11. L. V. Denisova, D.Yu. Kalinin, S. V. Reznik, Herald of the Bauman MSTU. Ser. Natures'. Science, **1** (82) (2011)
12. S. Reznik, Proc. *1-st Int. Workshop on Advanced Composite Materials and Technologies for Aerospace Applications* (Wrexham, UK, 2011)
13. S. V. Reznik, O. V. Denisov, P. V. Prosuntsov, V. P. Timoshenko, and A. V. Shulyakovskii, Polymer Sci. **6**, 3 (2013)
14. S. V. Reznik, P. V. Prosuntsov, K. V. Mikhailovsky, I. R. Shafikova, IOP Conf. **153** (2016) 012001 doi:10.1088/1757-899X/153/1/012001

Nonisothermal desorption of droplets of LiBr salt solution on a heated wall

S. Y. Misyura [1,*], and *V.* S. Morozov[1]

[1]Kutateladze Institute of Thermophysics, 630090 Novosibirsk, Russia

Abstract. Nonisothermal desorption and crystallization of droplets of aqueous salts solution of H_2O/LiBr were studied experimentally. A droplet was placed on a horizontal heated wall. The initial concentration of salt C_0 varied within 0-53 %. The wall temperature T_w varied within 80 - 235 °C and ambient air pressure was 1 bar. It was shown that the desorption time t_1 decreases sharply with increasing T_w and at $T_w < 150$-160 °C. The desorption time t_1 is the time until the onset of crystallization. The desorption time t_1 is quasi-constant at $T_w > 160$ °C and increases with growth in C_0.

1 Introduction

The droplet evaporation was investigated in a considerable number of works [1-6]. The evaporation of drops is widely used in modern technologies [1, 3]. Gas-droplet flows are used in technical apparatuses [6]. When burning gas hydrate of methane a vapor-droplets flow is formed and a combustion temperature is significantly lowered [7]. The significant decrease in the combustion temperature sharply reduces the efficiency of the combustion technology of this natural raw material. The desorption rate and the heat transfer coefficient depend on wall properties and wettability [3]. The behavior of the desorption depends on a solution convection and a degree of turbulence of a gas flow [8, 9]. It is important to investigate the desorption process of aqueous salt solutions for a wide range of technical tasks. The behavior of salt solutions differs from the behavior of one-component liquid [10-14]. The nucleate boiling of an salt solution droplet can be divided into several characteristic regimes [14]. The following typical stages of heat and mass transfer can be distinguished depending on wall temperature T_w: 1) evaporation below boiling temperature T_b; 2) nucleate boiling; 3) transitional stage; 4) boiling at Leidenfrost temperature T_L. Heat and mass transfer in the salt solution was investigated in [15]. Crystallization during desorption of a salt solution was considered in [16-18]. The onset of crystallization in a droplet is realized at a droplet edges and in a solution layer at a distance from the side heated wall [16, 18]. The crystallization kinetics depends on the heat and mass transfer and a degree of a solution supercooling [17]. Aqueous salt solutions can be divided into two kinds depending on the desorption behavior. The structural wall enhances heat transfer and increases the rate of the droplet evaporation [19].

* Corresponding author: misura@itp.nsc.ru

2 Experimental data

The experiments were carried out at relative air humidity of 40 %, ambient air temperature was 22 °C and ambient air pressure was 1 bar. The droplet of aqueous salt solution of LiBr was placed on the horizontal heated wall. The values of initial mass concentration of solutions C_0 were determined by the standard densimeters. Fig.1 represents the behavior of the time of crystallization onset t_1 for different wall temperature T_w and C_0. With increasing the wall temperature for $T_w < 150$ °C, the heat transfer coefficient increases, and the evaporation time t_1 of the droplet decreases. The desorption time t_1 is the time until the onset of crystallization. The desorption time t_1 is quasi-constant at $T_w > 160$ °C and increases with growth in C_0. It is probable that the constant value of t_1 is due to the attainment of T_s (T_s is the surface temperature of the drop) and the free convection velocity in the liquid of the maximum values (the values of these parameters reach a maximum and then practically do not change). At the boiling crisis of a one-component liquid the growth of the wall temperature leads to an increase in t_1. However, for a salt solution, the behavior of t_1 is different. With increasing wall temperature (at the boiling crisis), the time t_1 tends to a constant value.

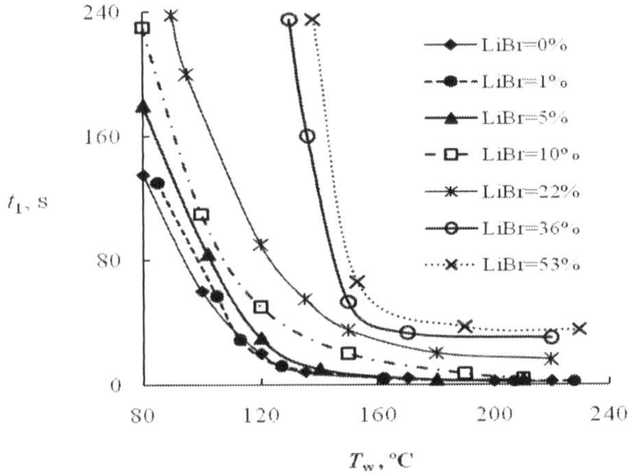

Fig. 1 The time of crystallization onset t_1 ($V_0 = 50$ µl).

There is a stable vapor film under a water droplet at the Leidenfrost temperature T_L. For the droplet of salt solution, this stable vapor film is absent. There are three time modes of the droplet evaporation (for $T_w > 130$-160 °C): 1) In the initial time period, the drop jumps and periodically touches the wall. Touching is accompanied by a spray of liquid. 2) The wall cools down after a while and the drop begins to evaporate like a sessile drop (without detachment from the wall). 3) The evaporation and heat transfer are realized in the presence of a crystalline crust on the droplet surface.

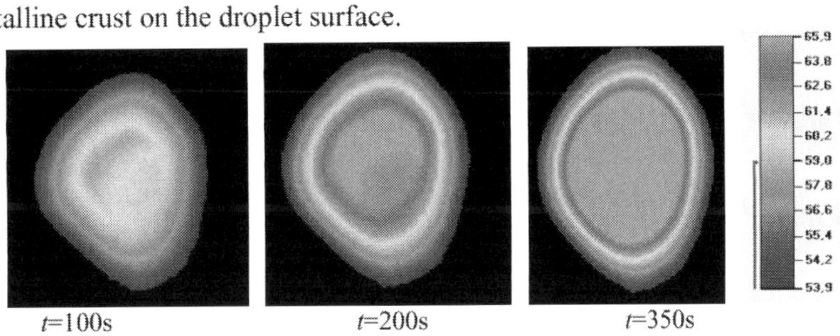

Fig. 2. Droplet thermal imaging (T_w=70 °C; C_0 = 22%).

Thermal imaging of the droplet of LiBr salt solution is shown in Fig. 2. With increasing salt concentration, the temperature T_s on the droplet free surface becomes more uniform due to a suppression of liquid convection. A sharp increase in the concentration of salt leads to a multiple increase in the viscosity of the solution and to the suppression of convection inside the drop.

3 Conclusions

Experimental studies of droplet of water and aqueous salt solution of H_2O/LiBr have been carried out.

Experimental data have shown that the behavior of aqueous solutions and water is qualitatively and quantitatively different. The desorption time t_1 decreases sharply with increasing T_w for T_w < 150-160 °C. The desorption time t_1 is quasi-constant at T_w > 160 °C and increases with growth in C_0.

The obtained experimental dependences for t_1 can be used for the calculation of the desorber cycles of a thermal absorption pump.

The study of the evaporation of salt solutions layers was funded by FASO Russia.

References

1. G.V. Kuznetsov, Int. J. Heat Mass Transfer **126,** 161 (2018)
2. S.Y. Misyura, Exp. Therm. Fluid Sci. **84,** 190 (2017)
3. G.V. Kuznetsov, J. Eng. Thermophys. **89,** 317 (2016)
4. S.Y. Misyura, Chem. Eng. Research Design **129,** 306 (2018)
5. G.V. Kuznetsov, Int. J. Heat Mass Transfer **96,** 20 (2016)
6. P.A. Strizhak, J. Hazardous Materials **338,** 148 (2017)
7. S.Y. Misyura, Sci. Rep. **6,** 30324 (2016)
8. V.P. Lebedev, Fluid dynamics **28,** 624 (1993)
9. S.Y. Misyura, Sci. Rep. **7,** 14759 (2017)
10. S.Y. Misyura, Int. J. Therm. Sci. **124,** 76 (2018)
11. S.Y. Misyura, Int. J. Heat Mass Transfer **116,** 667 (2018)
12. V.E. Nakoryakov, J. Eng. Thermophys. **20,** 338 (2011)
13. S.Y. Misyura, Chem. Eng. Research Design, **126,** 153 (2017)
14. V.E. Nakoryakov, J. Eng. Thermophys. **25,** 24 (2016)
15. V.E. Nakoryakov, Int. J. Heat Mass Transfer **55,** 6514 (2012)
16. S.Y. Misyura, Cryst Growth Des. **18,** 1327 (2018)
17. A.A. Chernov, Int. J. Heat Mass Transfer **108,** 1320 (2017)
18. S.Y. Misyura, Appl. Therm. Eng. **139,** 203 (2018)
19. S.Y. Misyura, Int. Comm. Heat Mass Transfer **96,** 7 (2018)

Evaporation of layers of salt solutions

S. Y. Misyura [1,*], and *V.* S. Morozov[1]

[1]Kutateladze Institute of Thermophysics, 630090 Novosibirsk, Russia

Abstract. Nonisothermal evaporation of layers of water and aqueous salts solutions of $H_2O/LiBr$, $H_2O/CaCl_2$ and $H_2O/LiCl$ was studied experimentally. The liquid layer was placed on a horizontal heated wall. The initial concentration of salt C_0 was 10 %. The wall temperature T_w = 75 °C and ambient air pressure was 1 bar. It was shown that the heat flux q increases for water for the final evaporation stage and falls for salt solutions due to the increase in salt concentration C and due to a significant drop in the equilibrium partial pressure of water vapor.

1 Introduction

The evaporation was investigated in a considerable number of works [1-6]. The droplet and liquid evaporation is widely used in modern technologies [1, 3, 6]. Gas-droplet flow is used in technical apparatuses [6]. When burning methane hydrate vapor-gas-droplets flows are formed and a combustion temperature is significantly lowered [7]. The behavior of a layer evaporation depends on wall properties and wettability [3]. The evaporation rate depends on free convection in the solution and a degree of turbulence of a gas flow [8, 9]. It is important to investigate the evaporation process of aqueous salt solutions for a wide range of technical tasks. The behavior of salt solution layer differs from the behavior of one-component liquid [10-14]. The nucleate boiling in a salt solution can be divided into several characteristic modes [9, 10]. Heat and mass transfer in aqueous salt solutions was investigated in [15]. Crystallization on liquid surface during evaporation of a salt solution was considered in [16]. The crystallization of a salt solution depends on the heat and mass transfer [16-18]. Aqueous salt solutions can be divided into two types depending on the behavior of the evaporation rate [19]. Measurements of the velocity field in gas-droplet flow using the optical method PIV were presented in [20].

2 Experimental data

Fig. 1 shows the main elements of the experimental setup. The experiments were carried out at ambient air temperature 20-21 °C, relative air humidity of 30-35 %, and ambient air pressure 1 bar. The aqueous salt solution layers of LiBr, $CaCl_2$, LiCl and water were placed on the horizontal heated wall. The values of initial mass concentrations C_0 of a solution were determined by the densimeters. The interfacial temperature of liquid T_s was determined with the help of thermal imager (NEC R500). Thermocouples for wall

* Corresponding author: misura@itp.nsc.ru

temperature measurement T_w were located near the wall surface (T_w = const = 75 °C). A change in the solution mass was recorded by means of the balance (5). The initial thickness of the layer in all experiments was 3-3.1 mm.

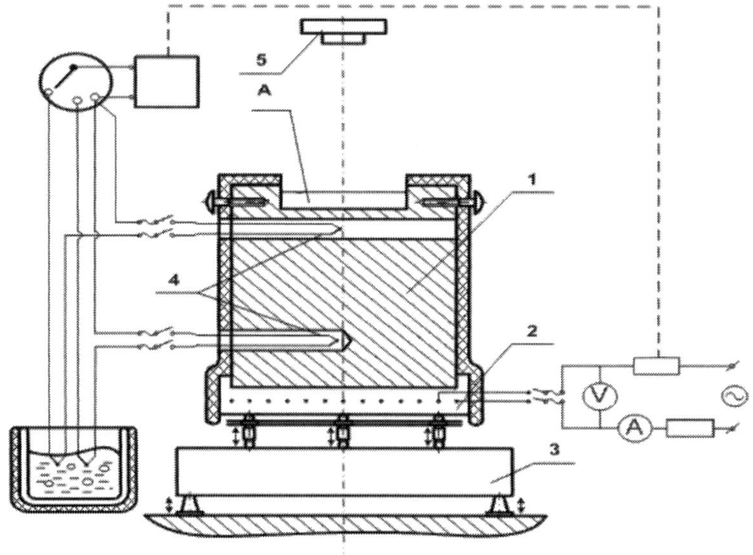

Fig. 1. Experimental setup: 1 – titanium cylinder, 2 – heater, 3 – balance, 4 – thermocouples, 5 – film and video facilities, IR imager.

The evaporation rate of the layer j was determined as $j = \Delta m/\Delta t$ (m is the liquid mass and t is the time). The maximum error of j was 11-13 % and the heat flux q was 13-16 %. Changes in the heat flux q over time are presented in Fig. 2.

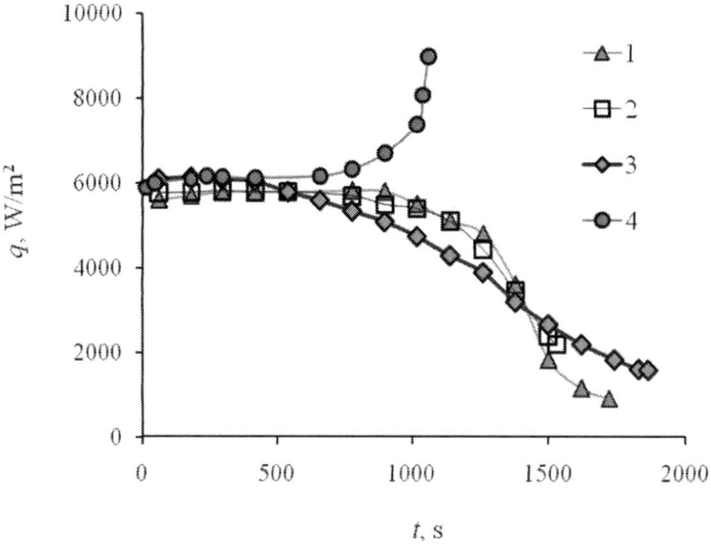

Fig. 2. Changes in the heat flux q over time: 1 – H_2O/LiBr; 2 – $H_2O/CaCl_2$; 3 – H_2O/LiCl; 4 – water (T_w = 75 °C; C_0=10 %; the initial layer thickness is 3 mm).

At the initial stage, the heat flux q is approximately constant for both water and salt solutions. After 500 seconds, the heat flux increases for water and falls for salt solutions due to the increase in salt concentration. The lowest heat flux corresponds to the LiBr salt concentration C = 65 % near the crystallization point. In accordance with the equilibrium curves, an increase in the concentration of salt in the salt solution leads to a drop in the equilibrium partial pressure of water vapor (p_s for the water layer was 0.3-0.35 bar and for the aqueous solution of LiBr was 0.15-0.17 bar at t = 1000-1100 s).

3 Conclusions

Experimental studies of liquid layers of water and aqueous salt solutions of LiBr, LiCl and $CaCl_2$ have been carried out.

Experimental data have shown that the behavior of aqueous solutions and water is qualitatively and quantitatively different. For the salts solutions the heat flux q decreases many times with increasing salt concentration, and for the water layer, the heat flux q significantly increases for the final stage of evaporation.

The study of the evaporation of salt solutions layers was funded by FASO Russia.

References

1. G.V. Kuznetsov, Int. J. Heat Mass Transfer **126,** 161 (2018)
2. S.Y. Misyura, Exp. Therm. Fluid Sci. **84,** 190 (2017)
3. G.V. Kuznetsov, J. Eng. Thermophys. **89,** 317 (2016)
4. S.Y. Misyura, Chem. Eng. Research Design **129,** 306 (2018)
5. G.V. Kuznetsov, Int. J. Heat Mass Transfer **96,** 20 (2016)
6. P.A. Strizhak, J. Hazardous Materials **338,** 148 (2017)
7. S.Y. Misyura, Sci. Rep. **6,** 30324 (2016)
8. V.P. Lebedev, Fluid dynamics **28,** 624 (1993)
9. S.Y. Misyura, Sci. Rep. **7,** 14759 (2017)
10. S.Y. Misyura, Int. J. Therm. Sci. **124,** 76 (2018)
11. S.Y. Misyura, Int. J. Heat Mass Transfer **116,** 667 (2018)
12. V.E. Nakoryakov, J. Eng. Thermophys. **20,** 338 (2011)
13. S.Y. Misyura, Chem. Eng. Research Design **126,** 153 (2017)
14. V.E. Nakoryakov, J. Eng. Thermophys. **25,** 24 (2016)
15. V.E. Nakoryakov, Int. J. Heat Mass Transfer, **55,** 6514 (2012)
16. S.Y. Misyura, Cryst Growth Des. **18,** 1327 (2018)
17. A.A. Chernov, Int. J. Heat Mass Transfer **108,** 1320 (2017)
18. S.Y. Misyura, Appl. Therm. Eng. **139,** 203 (2018)
19. S.Y. Misyura, Int. J. Heat Mass Transfer **125,** 610 (2018)
20. G.V. Kuznetsov, Appl. Therm. Eng. **131,** 340 (2018)

Influence of water-methanol solution additives on hydrocarbon fuel combustion in burner

Aleksandr S. Naumkin[1,*], *Boris* V. Borisov[1], and *Aleksandr* G. Nigay[1]

[1]National Research Tomsk Polytechnic University, 634050 Tomsk, Russia

Abstract. Mathematical model of burning disposal a water-methanol solution by the gas fuel is presented. Two-dimensional model in the ANSYS Fluent Software is made. Thermal distribution of the flare and distribution of methane concentration are derived. We analysed the results of mathematical modelling.

1 Introduction

Ecological aspects of human life are becoming more important every year. Disposal of waste without harm to the environment is costly; therefore, the economic efficiency in choosing methods of utilization is tFhe important factor for the any industry [1].

In this paper, we consider the disposal of the water-methanol solution. This solution is a waste of the technological cycle of extraction and preparation for transportation by pipeline transport of natural gas. Most of the methanol is separated in the distillation column, but the residue contains a small volume of methanol in water, which is harmful to the environment. Currently, disposal of water-methanol solution is produced by incineration. About 60 commercial companies in Russia use the GFU-5.

2 Research method and data

The processes accompanying disposal by the example of burning a water-methanol solution is simulated. This is the basis for developing a methodology for calculating the considerate process. A water-methanol solution is the following volumetric composition: H2O-99.67%, CH3OH-0.33 and the fuel is a natural gas of the following volumetric composition: CH4-98.9%; C2H6-0.12%; C3H8-0.01%; C4H10-0.01%; CO2-0.06%; N2-0.9%.

LLC "TyumenNIIgiprogaz" has developed GFU-5 which shown in Fig. 1 [2].

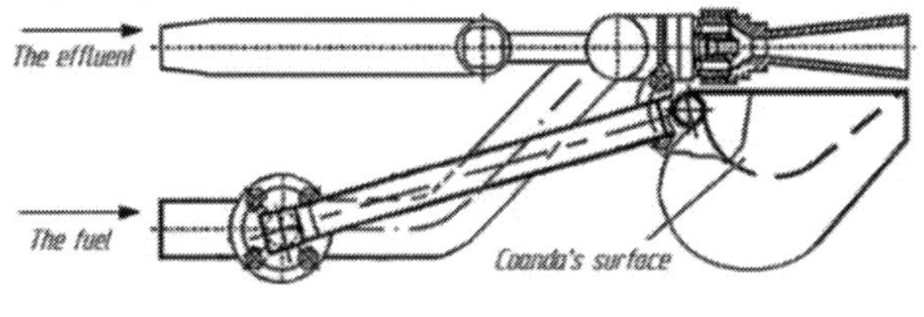

Fig. 1. Flare facility of GFU-5.

* Corresponding author: alexnaumserg93@mail.ru

Fuel gas consumption for atomization of industrial effluents is 0.1478 m^3/s. Flow rate of combustible gas to Coanda's surface (Fig. 1) is 0.0739 m^3/s. Consumption of water water-methanol is 0.278 kg/s.

The dimensions of the droplets formed at a given ratio of the gas and liquid phases were 55 μm.

The stationary model of the flame propagation in a two-dimensional setting is consider. Part of the fuel, aimed at spraying industrial effluents, moves inside the cylindrical pipeline. The industrial effluents comes from a coaxially located pipeline, smaller diameter, into the fuel flow through the nozzle.

The stationary model is solved by the establishment method. The process of mixing different phases is described by a system of differential equations. The system contains the equations of mass, momentum and energy conservation [3]:

$$\begin{cases} \dfrac{\partial\rho}{\partial t} + \nabla(\rho\vec{v}) = S_m \\ \dfrac{\partial}{\partial t}(\rho\vec{v}) + \nabla(\rho\vec{v}\vec{v}) = -\nabla p + \nabla(\bar{\bar{\tau}}) + \rho\vec{g} + \vec{F} \\ \dfrac{\partial}{\partial t}(\rho e) + \nabla[\vec{v}(\rho e + p)] = -\nabla\left(\sum_i h_i J_i\right) + S_h \end{cases} \tag{1}$$

where $\bar{\bar{\tau}} = \mu[(\nabla\vec{v} + \nabla\vec{v}^T) - 2/3\nabla\vec{v}I]$ – viscous stress tensor; I – unit tensor; μ – dynamic viscosity, Pa·s; t – time, s; ρ – density, kg/m^3; $\vec{v}$ – velocity vector, m/s; J – diffusion current density , kg/(m^2·s); F – external force, N; g – gravitational acceleration, m/s^2; p – static pressure, Pa; $e = h - p/\rho + v^2/2$ – specific total energy, J/kg; $h = \sum_i h_i J_i$ – specific total enthalpy, J/kg; $h_i = \int_{T_{ref}}^{T} c_{p,i} dT$ – specific total enthalpy of component i, J/kg; T_{ref} =298.15 – reference temperature, K; $c_{p,i}$ – heat capacity at constant pressure of component i, J/(K·kg); $S_m = \sum_i^N (Y_i S_i)$ – source of mass, kg/(m^3·s); $S_h = -\sum_i h_i^o R_i / M_j$ – the volumetric heat sources of the chemical reactions, J/(m^3·s); R_i – the net source of chemical species i due to reaction, mol/(m^3·s); h_i^0– standard state enthalpy, J/mol; M_i – mole volume of components i, m^3/mol.

Part of the gaseous fuel that passes through the nozzle forms a gas-droplet mixture with a water-methanol solution. Another part of the gaseous fuel flowing around the Coanda's surface comes to the periphery of the torch, where there is a mutual diffusion of the two streams and the environmental air to form a fuel mixture.

We model the diffusion process of a multicomponent mixture in accordance with the Maxwell-Stefan equations. Diffusion pressure is low [4]:

$$\sum_{\substack{j=1 \\ j\neq i}}^{N} \frac{X_i X_j}{D_{ij}}(\vec{V_j} - \vec{V_i}) = d_i - \frac{\nabla T}{T}\sum_{\substack{j=1 \\ j\neq i}}^{N} \frac{X_i X_j}{D_{ij}}\left(\frac{D_{T,j}}{\rho_j} - \frac{D_{T,i}}{\rho_i}\right), \tag{2}$$

where X_i – mole fraction *of* component i; V_i – diffusion velocity *of* component i, m^2/s; D_{ij} – binary mass diffusion coefficient, m^2/s; D_T – thermal diffusion coefficient, m^2/s; T – temperature, K; d_i - particle diameter, m.

The Maxwell diffusion coefficient is considered equal to the binary diffusion coefficient for an ideal gas. The barodiffusion effect, excluding the gravitational effect, is written by the following expression:

$$\vec{J_i} = \rho_i \vec{V_i}, \tag{3}$$

$$d_i = \nabla X_i, \tag{4}$$

$$\sum_{\substack{j=1 \\ j\neq i}}^{N} \frac{X_i X_j}{D_{ij}}\left(\frac{\vec{J_j}}{\rho_j} - \frac{\vec{J_i}}{\rho_i}\right) = \nabla X_i - \frac{\nabla T}{T}\sum_{\substack{j=1 \\ j\neq i}}^{N} \frac{X_i X_j}{D_{ij}}\left(\frac{D_{T,j}}{\rho_j} - \frac{D_{T,i}}{\rho_i}\right) \tag{5}$$

In order to find the diffusion flux of a chemical species use Fick's law:

$$\vec{J_i} = -\sum_{j=1}^{N-1}\left(\rho D_{i,m} + \frac{\mu_t}{Sc_t}\right)\nabla Y_i - D_{T,i}\frac{\nabla T}{T}. \tag{6}$$

where Y_i – the mass fraction of species i; Sc_t – turbulent Schmidt number; $\mu_t = \rho C_\mu k^2/\epsilon$ – eddy viscosity, Pa·s; C_μ =0.09 – constant [5]; ϵ – dissipation rate, m²/s³; k – eddy kinetic energy, m²/s²; $D_{i,m} = (1 - X_i)/\sum_{ij\neq i}(X_i/D_{ij})$ – the mass diffusion coefficient for species i, m²/s.

The equation of the diffusion flux vector of the mass component through the generalized diffusion coefficients A and B is consider. The following expression determin the diffusion coefficient of the component i in j:

$$D_{ij} = [D] = [A]^{-1}[B]. \tag{7}$$

In the matrix of generalized diffusion coefficients according to the Fick's law, the coefficients A and B are determined by the following expression, according to [6]:

$$A_{ii} = -\left(\frac{X_i}{D_{iN}}\frac{M_w}{M_{w,N}} + \sum_{\substack{i=1 \\ j\neq 1}}^{N}\frac{X_i}{D_{ij}}\frac{M_w}{M_{w,i}}\right), \tag{8}$$

$$A_{ij} = X_i\left(\frac{1}{D_{ij}}\frac{M_w}{M_{w,j}} - \frac{1}{D_{iN}}\frac{M_w}{M_{w,N}}\right), \tag{9}$$

$$B_{ii} = -\left(X_i\frac{M_w}{M_{w,j}} + (1 - X_i)\frac{M_w}{M_{w,i}}\right), \tag{10}$$

$$B_{ij} = X_i\left(\frac{M_w}{M_{w,j}} - \frac{M_w}{M_{w,N}}\right), \tag{11}$$

where M_w – the molecular weight, kg/kmol.

The coefficients of thermal diffusion is determined in accordance to [7] by the following expression:

$$D_{T,i} = -2.59\times 10^{-7}T^{0.659}\left[\frac{M_{w,i}^{0.511}X_i}{\sum_{i=1}^{N}M_{w,i}^{0.511}X_i} - Y_i\right]\cdot\left[\frac{\sum_{i=1}^{N}M_{w,i}^{0.511}X_i}{\sum_{i=1}^{N}M_{w,i}^{0.489}X_i}\right], \tag{12}$$

The turbulence effects are determined in according by standard $k - \epsilon$ model. This model are includes the transport of turbulent kinetic energy and the dissipation rate of turbulent kinetic energy equations [8]:

$$\frac{\partial}{\partial t}(\rho k) + \frac{\partial}{\partial x_j}(\rho k u_i) = \frac{\partial}{\partial x_j}\left[\left(\mu + \frac{\mu_t}{\sigma_k}\right)\frac{\partial k}{\partial x_j}\right] + G_k - \rho\epsilon, \tag{13}$$

$$\frac{\partial}{\partial t}(\rho\epsilon)+\frac{\partial}{\partial x_j}(\rho\epsilon u_i)=\frac{\partial}{\partial x_j}\left[\left(\mu+\frac{\mu_t}{\sigma_\epsilon}\right)\frac{\partial\epsilon}{\partial x_j}\right]+C_{1\epsilon}\frac{\epsilon}{k}C_k-C_{2\epsilon}\rho\frac{\epsilon^2}{k}, \quad (14)$$

where $G_k=-\rho\overline{u_i'u_j'}\partial u_j/\partial x_j$ - the production of turbulence kinetic energy, kg/(m·s^3); k – turbulence kinetic energy, m^2/s^2; ϵ – turbulent dissipation rate, m^2/s^3; $C_{1\epsilon}$ =1.44 – empirical constant; $C_{2\epsilon}$ =1.92– empirical constant; $C_\mu=0.09$ – empirical constant; $\sigma_k=1.0$ – empirical constant; $\sigma_\epsilon=1.3$ – empirical constant [5].

The chemical reactions are simulated in according by the mass conservation law of components i.

$$\frac{\partial\rho Y_i}{\partial t}+\nabla(\rho\vec{v}Y_i)=-\nabla\vec{J_i}+R_i+S_i\,, \quad (15)$$

where R_i – the net rate of production of species i by chemical reaction, kg/(s·m^3); S_i – rate of creation by addition from the dispersed phase, kg/(s·m^3); Y_i – the mass fraction of species i.

The combustion model from a single gross reaction for each fuel component is considered. The reaction rate is calculated in according by the vortex dissipation model. Turbulent mixing affects the rate of chemical reactions. The lower value from the speed of the diffusion process and the rate of chemical reactions in according by the Arrhenius law is taken to determine the reaction rate [10].

$$R_{i,r}=v'_{i,r}M_{w,i}A\rho\frac{\epsilon}{k}min_R\left(\frac{Y_R}{v'_{R,r}M_{w,R}}\right), \quad (16)$$

$$R_{i,r}=v'_{i,r}M_{w,i}AB\rho\frac{\epsilon}{k}\frac{\sum_P Y_R}{\sum_j^N v''_{R,r}M_{w,j}}, \quad (17)$$

where A = 4.0 – empirical constant; B = 0.5 – empirical constant; Y_P the mass fraction of product P; Y_R – the mass fraction of reagent R; N – number of reactions; $v'_{i,r}$ – stoichiometric coefficient of reagent i in reaction r; $v''_{i,r}$ – stoichiometric coefficient of product i in reaction r; $M_{w,i}$ – the molecular weight of species i, kg/mol.

The chemical substance i formation rate is determined as the sum of the its formation rates according by Arrhenius law for each reaction:

$$R_i=M_{w,i}\sum_{r=1}^{N_R}R_{i,r}\,, \quad (18)$$

The fuel reaction heat is determined in according by the known calorific value:

$$\Delta H=\sum_{r=1}^{N}h_i^0(v''_{i,r}-v'_{i,r})\,, \quad (19)$$

The heat transfer equation is solved in order to determine the thermal state:

$$\frac{\partial T}{\partial t}=a\nabla^2T+\frac{S_h}{c_p\cdot\rho}, \quad (20)$$

where $a=\lambda/(c_p\cdot\rho)$ – coefficient of thermal diffusivity, m^2/s.

The specific heat capacity of each component does not depend on temperature. The specific heat capacity of the mixture is determined as the sum of the mass fractions of each component:

$$c_p = \sum_i Y_i c_{p,i} \, , \tag{21}$$

The used boundary condition for the boundary G3 (Fig. 2) mean, that gas fuel flow enters through the pipeline dy80 with defined mass flow and the velocity profile according by the turbulent regime:$\partial \upsilon / \partial n$=const.

The boundary condition for the boundary G4 describes admission the flow of the water-methanol solution with defined mass flow rate:$\partial v / \partial n = const.$

The boundary condition for the boundary G2 describes admission the flow of air with temperature equal to 0°C.

The soil profile including the hill limits the burning area of the torch. This is described in according the boundary conditions G5. The lower boundary describes a wall, with the condition of adhesion, according by fallowing equations: $\partial T / \partial n = 0$; $\partial c / \partial n = 0$; $\partial v / \partial n = 0$.

The boundary condition for the boundary G1 is described by soft boundary conditions.

3 Results and discussion

These differential equations are solved in the software module ANSYS Fluent in a two-dimensional flat formulation. The results of the mathematical experiments are presented as a distribution of temperature and component composition (Fig. 2).

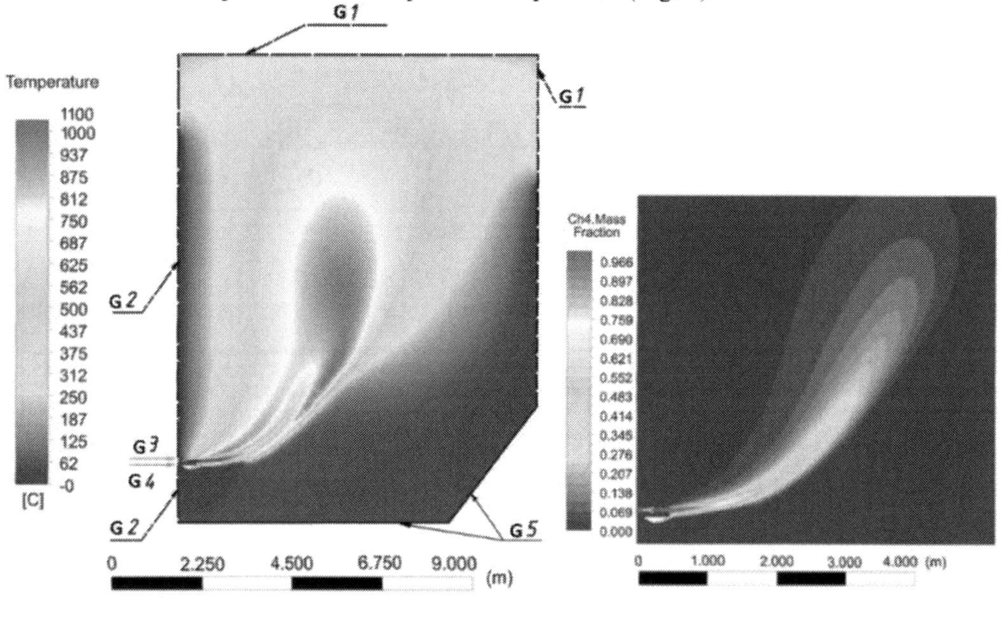

Fig. 2. Thermal distribution of the flare and distribution of methane concentration.

The facility is shown in Fig. 3 with flare geometric parameters when combusted defined mass flow the gas fuel and the water-methanol solution.

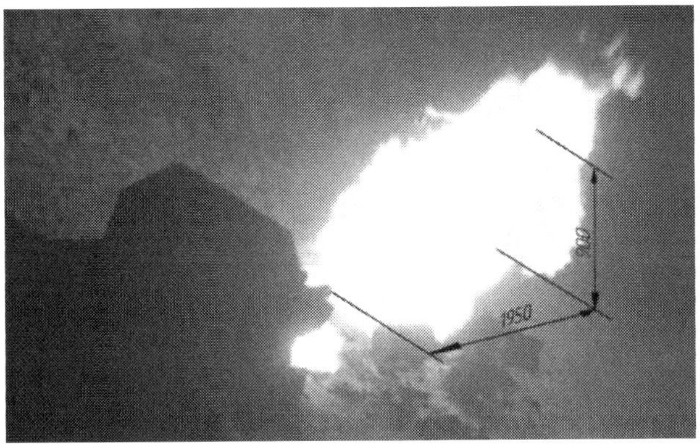

Fig. 3. The facility with flare geometric parameters.

The results of mathematical experiments satisfactorily correspond to the real dimensions of the facility flame. The formation of carcinogenic substances does not occur. The average temperature at the periphery of spraying the solution is about 1000 °C. The mathematical model allows us to determine the heat flux in the necessary direction.

The mathematical model may be the basis for further study of waste disposal in burner facilities.

References

1. M. Vasilevsky, A. Razva, A. Alibek, B. Roman, MATEC Web Conf. **100**, 01054 (2016)
2. N. A. Sozonov, Exposition Oil Gas **4**, 90, (2014)
3. G. K. Batchelor, *An Introduction to Fluid Dynamics* (Cambridge Univ. Press, Cambridge, 1967)
4. H. J. Merk, Appl. Sci. Res. **8**, 73, (1958)
5. B. E. Launder, D. B. Spalding. *Lectures in Mathematical Models of Turbulence* (Academic Press, London, 1972)
6. R. Taylor, R. Krishna, *Multicomponent Mass Transfer* (John Wiley and Sons, New York, 1993)
7. K. Y. Kuo, *Principles of Combustion* (John Wiley and Sons, New York, 1986)
8. T. H. Shih, W. W. Liou, A. Shabbir, Z. Yang, J. Zhu, Computers Fluids **24**, 227 (1995)
9. B. F. Magnussen, B. H. Hjertager, *On mathematical models of turbulent combustion with special emphasis on soot formation and combustion.In 16th Symp. on Combustion* (The Combustion Institute, Pittsburgh, 1976)

Improving the quality of coal is one of the factors in intensification of combustion processes in coal-based energy

Yuriy Patrakov[1], *Svetlana* Semenova[1], and *Anna* Usanina[1,2*]

[1]The Federal Research Center of Coal and Coal Chemistry of SB RAS, 650000 Kemerovo, Russia
[2]Tomsk State University, 634050 Tomsk, Russia

Abstract. For flotation enrichment of coals, spent engine oils modified with ozone were used. Preliminary ozonation contributes to the increase in the share of low-boiling components in the compositions of spent oils, also it reduces the viscosity of reagents and improves their flotation properties when coals are enriched. By flotation of coal, using ozonized spent oils, the yield of concentrate significantly increases, ash content is reduced, mineralization of waste flotation is increased, and as a result, the selectivity of the process is increased as well.

1 Introduction

The high ash content of ordinary energy coals significantly reduces the caloric content of fuel, increases the cost of energy produced, and also causes difficulties in the operation of boiler equipment. Therefore, one of the strategies in the coal industry development is the widespread introduction of enrichment technologies for energy coals in order to obtain high-calorie fuel.

The development of highly mechanized methods of production entails the formation of large quantities of fine coal with particle size < 0.1 mm. Nowadays, the only effective way of enriching the thin sludge is froth flotation, which is based on the difference of physico-chemical properties (wettability) in the surface and in the particles of coal and rock.

To improve flotation properties of coal particles, surfactants is flotation reagents are used, which are divided into collectors and blowing agents. Petrochemical products and wastes are mainly used in the flotation enrichment of coal. Non-polar reagents are employed as collectors: diesel fuel, kerosene, gas oil, airplane fuel, and mineral oil. Heteropolar reagents are employed as frothing agents, for example, waste of recycling technical alcohols [1]. For the solution of ecological problems, (as flotation reagents) it is offered to use the spent motor oils which are formed in huge quantities, including operation of motor transport of the mining enterprises.

To obtain reagents with both frothing and collection properties, the traditional non-polar flotation reagents are modified by chemical or physicochemical means. For more effective

* Corresponding author: usaninaanna@mail.ru

flotation, the reagent must have a polar heteroatom (O, S, N) or a functional group (OH, C=O, N–H, S–H, etc.) [2].

Ozonization is an effective method of oxidative modification. Some advantages of ozone over the traditional oxidants are used in chemistry and chemical technology (mineral acids, hydrogen peroxide, potassium permanganate, molecular, oxygen, etc.). They are as follows: low reactivity; selectivity for specific types of bonds and heteroatoms; mild oxidation conditions (with no need for catalysts, high temperatures or high pressure); and no byproducts to remove, with a corresponding reduction in production costs [3].

The aim of this work is to investigate the influence of ozonization on the composition of the spent motor oil so as to improve its flotation properties in coal enrichment.

2 Experimental setup and study technique

As the object of the study spent mineral oil was used, which were collected from coal mines in Kuzbass, where heavy trucks are operated. The characteristics of the spent motor oil are: density (at 20 °C) 874 kg/m^3; kinematic viscosity (at 50 °C) 41.2 sSt; moisture content 1.5 %; ash content 1.2 %; mechanical impurities 1.0 %.

A bubbling reactor was used for the spent motor oil ozonization in condition of room temperature and atmospheric pressure. The ozone concentrations at the reactor input and output were determined by UV gas analysis (on the basis of the UV absorption of ozone at a wavelength of 254 nm).

A laboratory flotation machine (volume 1 L) was employed. The experimental results were assessed in terms of the yield of concentrate and the yield of wastes, the ash content of the concentrate and the wastes, and the selectivity of the process.

3 Experimental results

The spent oils have been tested for flotation of coals of different grades, different stages of metamorphism and petrographic composition, which can potentially be used as energy coals (Table 1). The table shows the parameters of coal quality: A^d – ash, V^{daf} – yield of volatile substances, y – thickness of the plastic layer, R_o – reflection index of vitrinite, Vt – vitrinite, Sv – semivitrinite, I – inertinite, L – liptinite.

Table 1. Parameters of coal samples.

Sample code	Coal	A^d, %	V^{daf}, %	y, mm	R_o, %	Maceral composition, %			
						Vt	*Sv*	*I*	*L*
1	Sibir'coal	20.4	38.1	17	0.81	85	2	11	2
2	Krasnobrodskaya coal	8.5	26.5	10	1.07	47	10	42	1
3	Tomusinskaya coal	15.6	23.3	13	1.25	50	19	30	1
4	Anzherskaya coal	16.5	19.3	10	1.32	49	5	45	1

The success of flotation largely depends on the molecular structure of coals and the presence of functional oxygen-containing groups on the surface of coal particles [4]. For example, low-metamorphosed coals have increased amounts of hydrophilic carboxyl and hydroxyl groups. This has a negative impact on the efficiency of flotation and requires increased costs of hydrophobizing reagents. Coals of medium and high stages of metamorphism consist mainly of hydrophobic hydrocarbon aliphatic-aromatic fragments which are similar to oil-based reagents [5]. Characteristics of the investigated coals are shown in Table 2.

Table 2. Chemical composition of coal samples.

Sample code	Content, % (*daf*)					Atomic ratio		Functional composition, mg-equ/g		
	C	H	N	*S*	O	H/C	O/C	COOH	OH	C=O
1	83.5	5.9	2.3	0.3	8.0	0.85	0.07	0.17	0.32	0.89
2	86.8	4.9	2.2	0.2	5.7	0.68	0.05	0.16	0.25	0.71
3	88.9	4.2	2.3	0.3	4.3	0.57	0.04	0.10	0.18	0.74
4	89.6	3.9	2.1	0.3	4.1	0.52	0.03	0.08	0.15	0.55

The chemical composition of the waste oil is substantially transformed depending on the amount of ozone absorbed. With the increase in the degree of ozonation in the oil content, oxygen content increases as well while carbon, hydrogen and sulfur decreases. The acid number of the oil increases, the viscosity decreases. The optimal degree of ozonation was 11 g of ozone per kg of product (Table 3).

Table 3. Chemical composition and properties of initial and ozonized oil samples.

Oil sample	Content, %					Atomic ratio		Change in viscosity, %
	C	H	N	S	O	H/C	O/C	
Without ozonization	85.4	13.3	0.2	0.3	0.8	1.9	0.01	-
With 1-h ozonization (11 g/kg)	84.4	12.6	0.1	0.2	2.3	1.8	0.02	-10.5
With 1-h ozonization (22 g/kg)	83.4	11.8	0.1	0.2	4.5	1.7	0.04	- 4.6

The IR spectroscopy method can be used to identify changes in the fragmented composition of hydrocarbons. In the IR spectra of the ozonized samples, the absorption intensity of aromatic CH bonds (3040 cm^{-1}) declines; and the proportion of methyl, methylene (2920, 1380, 720 cm^{-1}), and alicyclic (970 cm^{-1}) fragments increases. We note increased intensity of the absorption bands for oxygen-bearing groups: O–H bonds of alcohols and phenols (3400 cm^{-1}), C=O bonds of lactones and anhydrides (1770 cm^{-1}), C = O bonds of aliphatic (1730 cm^{-1}) and aromatic (1710 cm^{-1}) acids, aromatic ketones (1650 cm^{-1}, C–O bonds of esters, furans, and lactones (1260 cm^{-1}), S = O bonds of sulfoxides (1300, 1150, 1050 cm^{-1}) (fig. 1).

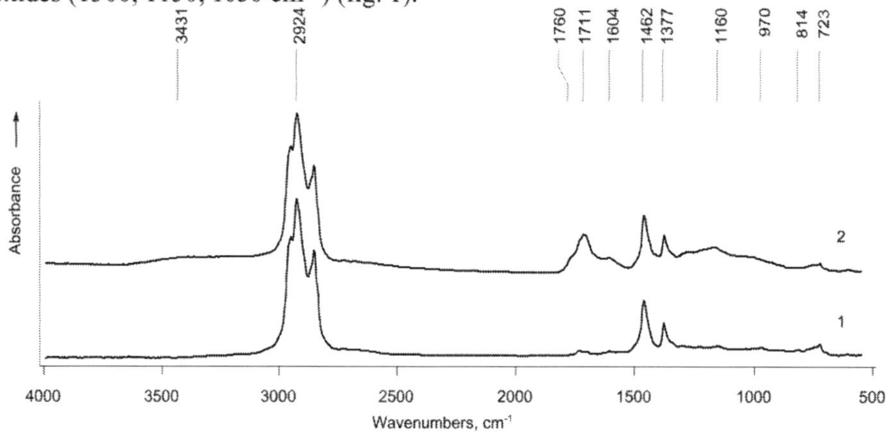

Fig 1. IR spectra of initial (*1*) and ozonized (*2*) oil samples.

In the flotation of the coal samples using unmodified spent motor oil, the effectiveness of the process increases with the increase in the metamorphic development of the coal and decrease in its oxygen content. In other words, the reagent with non-polar properties has the best adsorption properties with respect to the hydrocarbon fragments of coal with greater metamorphic development (Fig. 2).

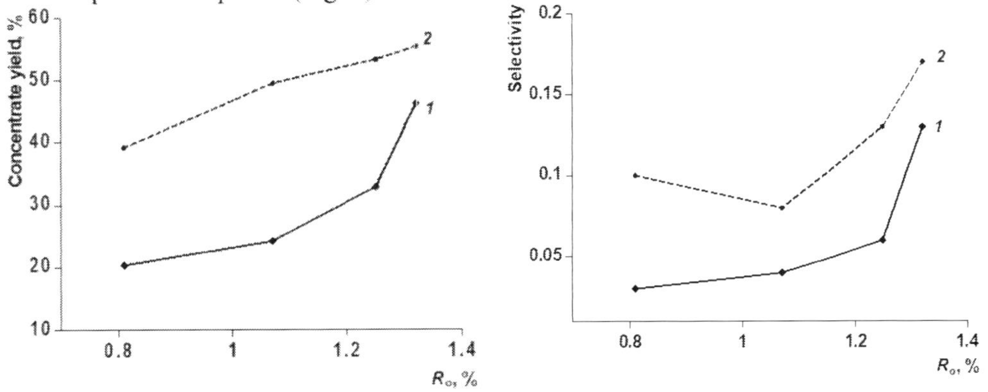

Fig. 2. Results of flotation of coals in different metamorphism stage using initial (1) and ozonized (2) spent oils.

Using ozonized spent oils as flotation agent significantly improves the performance of flotation. Thus bigger effect is observed for low - and average metamorphosed coals (to R_o = 1.1 %). The use of ozonized oil significantly changes the flotation outcomes: the concentrate yield increases by a factor between 1.2 (for coal with R_0 = 1.32 %) and 1.9 (for coal with R_0 = 0.81 %), with the decrease in ash content of the concentrate and th increase in ash content of the waste. That corresponds to the increase in the selectivity by factors of 1.1 and 3.3 respectively.

Increased flotation activity of ozonized oil, apparently, contributes to the reduction of viscosity, as well as the appearance of the oil as oxygen-containing and light non-polar compounds. These hydrocarbons can be more easily fixed in the form of a film on the surface of coal particles [3]. At the same time, the positive flotation effect is achieved without the use of additional reagents - modifiers and foaming agents. It can be assumed that the use of analytical processing allows to obtain reagents of complex action with collecting and foaming properties on the basis of spent oils.

The greater flotational activity of the ozonized oil with respect to coal from early metamorphic stages (with R_0 < 1.1%) may evidently be attributed to greater hydration of the coal particles, because of the presence of hydrophilic carboxyl and hydroxyl groups at their surface. Those groups form hydrogen bonds with the polar groups of the ozonized hydrocarbons in the coal's organic mass. That promotes the adsorption of the reagent.

Coal concentrates obtained by flotation using ozonized oils not only have reduced ash content, but are also enriched with petrographic components of the vitrinite group. Vitrinite has a positive effect on the technological quality of coal: the yield of volatile substances, the heat of combustion and others. The content of mineralized suppressive components of the inertinite group with flotation enrichment of coals of complex petrographic composition decreases.

4 Conclusion

Involvement of spent oils in the technology of coal enrichment helps to expand the directions of their utilization and to solve important environmental problems.

The use of ozonolytic modification of spent oils contributes to the production of complex reagents with collecting and foaming properties. High results in the enrichment of coals of complex composition are achieved without additional use of modifying and emulsifying additives.

The higher flotation activity of ozonized oils is manifested for low-metamorphosed coals, which associated with the presence of an increased number of hydrophilic oxygen-containing groups on the surface of coal particles.

Improving flotation properties of ozonized spent oils can be called as education in their composition heteropolar compounds, and light non-polar components, which enhance hydrophobicity of the coal particles.

References

1. A. A. Abramov, *Flotation: collecting reagents, in Collection of Research Works vol. 7.* (Mountain book, Moscow, 2012) [in Russian]
2. H. Khoshdast, A. Sam, Open Miner. Process. J. **4**, 25 (2011)
3. Yu. F. Patrakov, S. A. Semenova, M. S. Klein, T. E. Vahonina, Coal Chem. **4**, 22 (2017) [in Russian]
4. S. Dey, Fuel Process. Technol. **94**, 151 (2012)
5. J. Soni, S. K. Tripathy, P. K. Banerjee, Int. J. Miner. Process. **134**, 97 (2015)

Optimization of construction of a shell for launcher in the water medium in the supercavitation regime

Vladimir Arkhipov[1], *Nikolay* Zolotorev[1], and *Ksenia* Perfilieva[1,*]

[1]National Research Tomsk State University, 634050 Tomsk, Russia

Abstract. New scheme of a shell, providing increase in range and stability of the movement in the water medium due to realization of the supercavitation regime on all trajectory of its movement, has been considered. The results of the analysis in the efficiency of the movement of a shell in the water environment in the supercavitation mode are presented.

1 Introduction

When the shell moves in the water medium, the resistance force increases sharply, since the water density is three orders of magnitude higher than the air density [1]. To increase the firing range and ensure sufficient speed of the shell at the end point of the trajectory, it is necessary to increase the initial speed of the V_0 shell, reduce the area of the mid-beam section of the S_M and reduce the resistance coefficient of the C_x. Increasing the muzzle velocity V_0 is inefficient because it increases the strength of the resistance. As the caliber of the shell decreases, its mass and, consequently, the initial kinetic energy decreases, which leads to a decrease in the firing range. A significant reduction in the resistance coefficient can be obtained when the solid moves in the super-cavitation mode [2]. In this case, a cavity filled with water vapor is formed around the solid with a special head cavitator.

In this paper, the scheme of the shell for firing in an aqueous medium was considered, which allows to increase the range of motion of the target element in the aqueous medium and increase the stability of its motion.

2 The scheme of the shell for launcher in the water

To increase the range and stability of movement in the water environment, a new scheme of the shell [3] is proposed (Fig. 1). Due to the motion of the shell in the barrel of the weapon, ignition of the end surface of the retarder 16 occurs. The decelerator 16, prevents the ignition of the solid fuel charge 2 by the combustion products of the propellant charge. After the shell leaves the barrel of the gun with an initial velocity V_0, its movement takes place in the water medium in the super-cavitation mode. After a certain period of time t_*, the velocity of the shell is reduced to a critical value V_*, at which the super-cavitation mode

* Corresponding author: k.g.perfiljeva@yandex.ru

is not implemented. At the time $t_{ign} = t_*$ the retarder 16 completely burns, and the combustion products of the retarder initiates the pyrotechnic igniter 6 which is fixed on it. The products of combustion of the igniter 6 are received in the through axial channel 7 of the solid propellant charge 2 and burns it. The combustion products of the solid fuel charge 2 enter through the through axial channel 7 into the communicating through channel 10 of the ballistic tip 3 and push the cylindrical plug 14 out of the contraction of the 11 through channel 10. In this case, the combustion products of the solid fuel charge 2 flow into the aqueous medium through the Central nozzle 5 and through the narrowing of the 14 channel 10. Part of the combustion products through the radial channels 15 enter the annular gap 9 and flow into the water medium through the annular nozzle 13 at an angle β to the longitudinal axis of the shell in the direction of its stern.

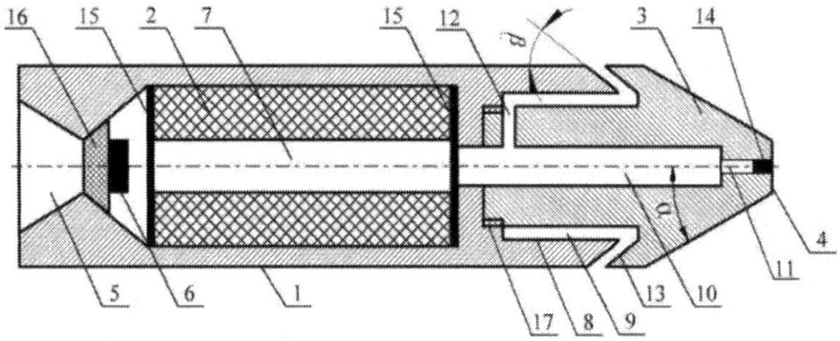

Fig. 1. Scheme of the shell to fire in the water environment.

Thus, the additional injection of the combustion products of the solid fuel charge 2 through the narrowing 11 and the annular nozzle 13 ensures the preservation of the super-pressure mode at the velocities of the shell $V < V_*$. In this case, the flow of combustion products through the Central nozzle 5 and the annular nozzle 13 creates an additional reactive force, which increases the speed and range of motion of the shell.

3 Analysis of the effectiveness of the shell motion in the water environment in the supercavitation mode

Consider the movement of the shell caliber 0.03 m in the water environment at a depth of H = 20 m with an initial (muzzle) velocity V_0 = 300 m/s.

The hydrostatic pressure at this depth is:

$$p(H) = 10^5 (1 + 0.1H) . \tag{1}$$

The choice of the critical value of the shell velocity V_* at which the solid fuel charge is ignited is determined by the following factors. The main criterion for the super-cavitation motion of a shell in an aqueous medium is the number of cavitation [4]:

$$\sigma = \frac{2[p(H) - p_0]}{\rho V^2} , \tag{2}$$

where p_0 = 2336.8 Pa is the pressure of saturated water vapor during bubble cavitation.

The super-cavitation mode is realized at values $\sigma < \sigma_* = 0.06$ [4]. From the formula (2) the ratio for the velocity of the shell follows, which provides its motion in the super-cavitation mode:

$$V \geq V_* = \sqrt{\frac{2[p(H)-p_0]}{\rho \cdot \sigma_*}}. \tag{3}$$

Thus, the value depends on the depth of immersion H, where the shell moves [5]. Map modes of movement of the shell in the coordinates (H, V) is shown in Fig. 2. The line that defines the boundary of the supercavitation region corresponds to the value $\sigma = \sigma_* = 0.06$.

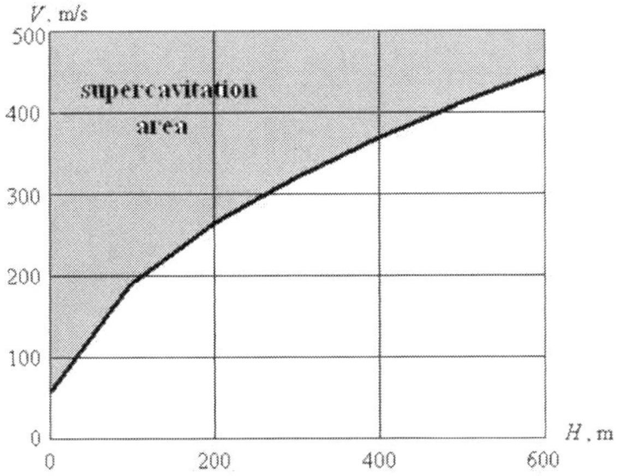

Fig. 2. Map modes of movement of the shell in the coordinates (H, V).

Choose a solid fuel charge with a diameter D = 0.026 m and a length l = 0.2 m. Set the charge channel in the form of a star, then the combustion area will correspond to the area of the cylindrical channel with a diameter d = 0.013 m.

As a solid fuel, we choose a mixture fuel TPQ-3011A, the characteristics of which are given in Table 1.

Table 1. Characteristics of fuel mixture TP-Q-3011A [6].

Parameter	P_f, **kg/м³**	u_1, **m/s**	**v**	k	R, **J/(Kg·K)**	T_p, **K**
Value	1630	$0.69 \cdot 10^3$	0.32	1.26	378	1260

The main characteristics of the solid-fuel engine, obtained by the results of the intra-ballistic calculation of the SRM, carried out by the method [6] are given in Table 2.

Table 2. Characteristics of the solid-fuel engine of the shell.

Parameter	P, **N**	G, **kg/s**	u, **m/s**	t_c, **s**
Narrowing of the nozzle	17.4	0.02	870	2.145
Nozzle assembly	23.5	0.01	2349	

Consider the characteristics of the motion of models in an aqueous medium with an initial velocity V_0 = 300 m/s., V_* = 100 m/s. The calculations of the inert shell (M1) and the shell with rdtt (M2). The main parameters of the models are shown in Table 3.

Table 3. Model parameters.

Model	m, **kg**	ρ, **kg/m³**	D_k, **m**	D_n, **m**	C_x
M1	1	10^3	$5 \cdot 10^{-3}$	$30 \cdot 10^{-3}$	0.82
M2					0.46

When the model moves along the trajectory, two sections are realized. In the first section the model moves with the working SRM, in the second section the model moves as an inert body.

The classical equation of motion of a body of constant mass m in a continuous medium has the form:

$$m\frac{dV}{dt} = -C_x S_M \frac{\rho V^2}{2}, \quad (6)$$

where m – the mass of the model; C_x – dimensionless coefficient of resistance; S_M – the area of the mid-body section.

The main criterion of super-cavitation motion of the shell in the water medium is the number of cavitation (2). Evaluation [7-8] show that when the values of the cavitation $\sigma < \sigma_* = 0.06$ super-cavitation model experiences less resistance than the same model with the continuous flow. At high speeds and low depths of immersion $\sigma << 1$, and therefore $C_x = 0.82$ the equation of motion (6) has the form:

$$\frac{dV}{dt} = -kV^2, \quad (7)$$

where $k = C_x \frac{\rho S_k}{2m} = \text{const}$.

Integrating equation (7) (with $u = u_0$ for $t = 0$), we obtain a formula for the dependence of the velocity of the inert model on time:

$$V(t) = \frac{V_0}{1 + kV_0 t} \quad (8)$$

When the model moves on the active part of the trajectory, the equation of motion of the model (6) changes taking into account the thrust of the rocket engine and has the following form

$$m\frac{dV}{dt} = P - C_x S_k \frac{\rho V^2}{2}, \quad (9)$$

where P is the thrust force.

The equation of motion (9) is presented as

$$\frac{du}{dt} = -ku^2 + a, \quad (10)$$

where $a = \frac{P}{m} = const$.

Integrating equation (10), we obtain an algebraic formula for calculating the speed of the model with the working SRM from time

$$u(t) = A\frac{(C + e^{-B \cdot t})}{(C - e^{-B \cdot t})}, \quad (11)$$

The distance traveled by the models over time is determined by the equation

$$S = u \cdot t \tag{12}$$

The ignition delay time t_{ign} is chosen to be equal to the time t_*, during which the shell speed will decrease to the value V_*

$$V_* = \frac{V_0}{1 + kV_0 t_*}, \tag{13}$$

From (13) the ratio follows to determine the ignition delay time of the solid fuel charge which is determined by the ratio

$$t_{ign} = t_* = \frac{1}{k}\left(\frac{V_0}{V_*} - 1\right) = \frac{2m_0}{C_x \rho V_0 S_k}\left(\frac{V_0}{V_*} - 1\right), \tag{11}$$

The results of parametric calculations of the motion of models in the water environment are given in Table 4.

Table 4. The results of parametric calculations of the motion of models in the water environment.

Model	t_*, s	V_*, m/s	L_*, m	L_f, m
M1	0.82	100	131.08	136
M2				305

The calculated dependences of the speed of the models on time are shown in Fig. 3, and the distances covered by the models are shown in Fig. 4, where the inert shell (M1), missiles with SRM (M2).

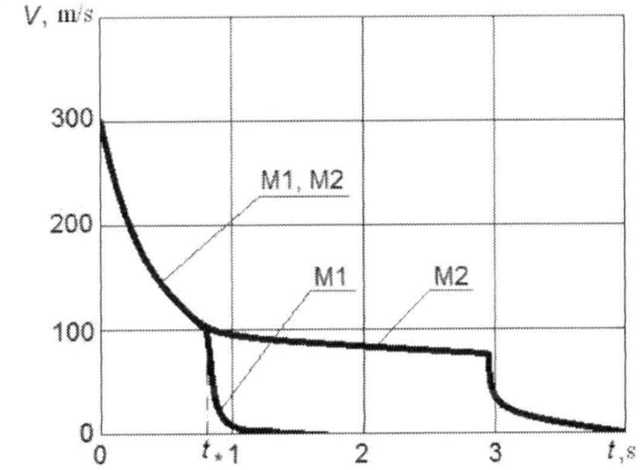

Fig. 3. The estimated velocity models M1, M2.

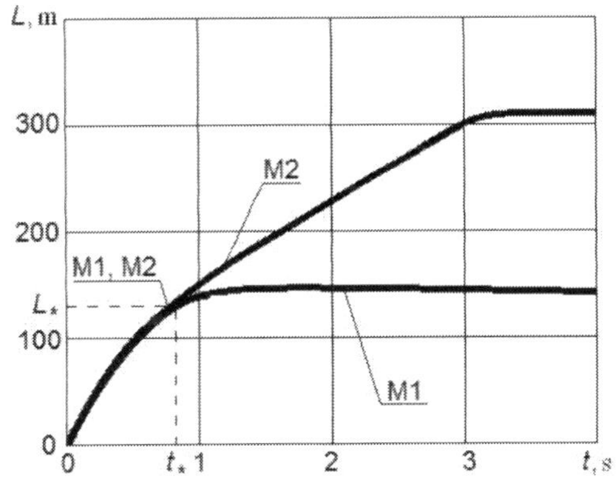

Fig. 4. The distance covered by the models M1, M2.

4 Conclusion

- A modified scheme of the shell for firing in the aquatic environment, was presented, which allows to increase the range of motion in throwing the item in the aquatic environment and increase the sustainability of its movement due to the provision of the mode of supercavitation in all its trajectory.
- The analysis of the dependence of the shell velocity on the depth of immersion was carried out and the boundary of the supercavitation area was determined.
- Certain parameters of solid fuel engine were determined, ensuring the preservation of the regime of supercavitation at speeds of projectile motion $V < V_*$.
- External ballistic series of calculations of the motion of inert shell (M1) and shell with solid propellant motors (M2) in the aquatic environment were determined.
- It is determined that for the new shell scheme when moving in the water environment in the super-pressure mode, the speed and the distance traveled far exceed the corresponding values when moving the model without a cavitator.

This research was supported by "The Tomsk State University competitiveness improvement programe"

References

1. M. E. Deich, *Technical gas dynamics* (Moscow, Gosenergoizdat, 1961) [in Russian]
2. Yu. D. Vlasenko, Applied hydro mechanics **2**, 26 (2000) [in Russian]
3. V. A. Arkhipov, A. I. Konovalenko, K. G. Perfilieva, Shell for launcher in the water medium. The application on patent RU 2017135869, (Int.Cl.: F42B 5/02, 2017) [in Russian]
4. Yu. N. Savchenko, A. N. Swarovski, Appl. fluid mech. **11**, 69 (2009) [in Russian]
5. G. Thomas, Torpedo body form and gas layer control. Patent US No. 3205846, (Int.Cl.: F42B 19/12, 1965)
6. A. A. Shishkov, S. D. Panin, B. V. Rumyantsev, *Working processes in rocket engines solid fuels* (Moscow, Machine Building, 1988) [in Russian]
7. Yu. N. Savchenko, V. N. Semenenko, S. I. Putilin, Appl. fluid mech. **1**, 79 (1999) [in Russian]
8. Yu. N. Savchenko, G. Yu. Savchenko, Appl. fluid mech. **6**, 78 (2004) [in Russian]

Analytical solutions of the segregation problem

Andrey A. Pil'nik[1,*], and *Andrey A.* Chernov[1]

[1]Kutateladze Institute of Thermophysics, 630090 Novosibirsk, Russia

Abstract. New analytical solutions of the segregation problem were found. The segregation is caused by the dissolved gas displacement by a moving crystallization front. The displacement is caused by the low solubility of the volatiles in solid, so the dissolved gas is withdrawn from the crystallizing melt towards the surrounding melt. Two different modes of crystallization were considered. The first one is the case of constant crystal growth rate which takes place in the case of intensive heat removal or when only initial stage of crystallization process is considered. The second mode is the "normal growth law" which is a self-similar solution of the crystal growth problem. In this case the crystal growth rate is inversely proportional to a square root of time. In contrast to the already existing solutions here the solidification shrinkage is taken into account.

1 Introduction

The crystallization processes are very common in the sphere of modern materials production. Often, solidified material contains pores and other imperfections. One mechanism of pore formation is the liquid rupture due to the solidification shrinkage [1]. Another mechanism of pore formation is the segregation during solidification. The melts often contain significant amount of dissolved volatiles and other impurities. Combined with the cooling regime these imperfections determine the final morphology of the sample [2,3]. Although, different methods of degassing and purification are available for the most cases it is not realistic to use them from the economical point of view. Most technological processes do not require the use of highly pure or homogeneous materials and depending on the tolerances even a pore formation to a limited extent can be acceptable. Of course some technological processes require the material completely without pores or at least impose strict constrains over the number of pores and other types of imperfections.

For the both cases it is highly important to be able to predict the conditions for pore formation caused by the segregation process during solidification. The pore formation depends on the dissolved gas concentration field during the whole process. So, the analytical solutions obtained in the article can be of great help for initial estimation of the cooling rate needed to achieve the desired level of porosity.

* Corresponding author: pilnik@itp.nsc.ru

2 Problem statement

We will consider the problem of segregation of dissolved gas by the moving plane crystallization front. The problem can be stated as the following diffusion problem.

$$\frac{\partial C}{\partial t}+v_l\frac{\partial C}{\partial x}=D\frac{\partial^2 C}{\partial x^2} \tag{1}$$

where t is the time; x is the coordinate with the origin at the substrate surface; C is the mass concentration of the volatile dissolved in melt; D is the diffusion coefficient of the volatile dissolved in melt.

The equation (1) differs from the basic form of diffusion equation by containing the additional term caused by solidification shrinkage. Melt velocity v_l can be found from the continuity equation. For the case of incompressible liquid $v_l=-(\bar{\rho}-1)(dX/dt)$, where $\bar{\rho}=\rho_s/\rho_l$ is the density ratio; dX/dt is the crystal growth rate; X is the crystallization front coordinate ($X=0$ at $t=0$).

In the general case crystallization goes according to the following scenario [4,5]. During the first stage of the process (nonequilibrium crystallization) the supercooling of the melt near crystallization front is equal to the initial one, so the crystal growth is determined only by the kinetics of phase transition. Because of the latent heat of phase transition release the supercooling near the phase transition surface starts to decrease and asymptotically goes to zero. This stage can be labeled "equilibrium crystallization". Thermodynamic equilibrium establishes at the phase boundary (so called Stefan problem). In this case crystal growth is determined only by the heat transfer at the phase boundary and the crystal growth rate becomes inversely proportional to the square root of time.

At the inception of the process the dissolved gas concentration equals to C_0. We will study the case when $C_s<C_0<C_l$, where C_s and C_l are solubility in solid and solubility in liquid respectively.

The boundary conditions are derived from the conservation of the concentration at large distance from the crystallization front and the mass balance at the phase boundary:

$$(C)_{x\to\infty}=C_0; \tag{2}$$

$$-D\left(\frac{\partial C}{\partial x}\right)_{x=X}=\bar{\rho}\frac{dX}{dt}\left[(C)_{x=X}-C_s\right]. \tag{3}$$

Because of the fact that the range of coordinate x changes in time it is problematic to use it in finding analytical solutions. Therefore, we will use coordinate characterized by the immobility of crystallization front: $\hat{x}=x-X$.

It is common while solving the heat and mass transfer problems to seek for the solution in dimensionless form, so we will be using following dimensionless variables: $\bar{C}=\dfrac{C-C_0}{C_0-C_s}$; $\tau=t/t_0$; $\bar{x}=x/x_0$; $\bar{X}=X/x_0$; $d\bar{X}/d\tau=v_0^{-1}dX/dt$; $\delta=\bar{x}-\bar{X}$, where $t_0=D/v_0^2$ and $x_0=D/v_0$.

In the general form the formulated boundary value problem can only be solved numerically. But in certain cases of known crystal growth laws analytical solutions of segregation problem can be found.

3 Analytical solutions of the problem

3.1 The case of crystal growth rate inversely proportional to the square root of time

Let us consider the case of "normal growth law". When the supercooling at crystallization front asymptotically goes to zero (equilibrium crystallization) and the crystal growth rate becomes inversely proportional to the square root of time [5]: $dX/dt = \beta_T\sqrt{a/t}$; $X(t) = 2\beta_T\sqrt{at}$, where a is the thermal diffusivity of the melt; $\beta_T = \beta_T(Ku)$ is the dimensionless coefficient (a function of Kutateladze number $Ku = L/(c\Delta T_0)$); L is the specific heat of fusion; c is the heat capacity; ΔT_0 is the initial melt supercooling. In dimensionless form: $d\bar{X}/d\tau = \beta_0/\sqrt{\tau}$; $\bar{X} = 2\beta_0\sqrt{\tau}$, where $\beta_0 = \beta_T\sqrt{Le}$; $Le = D/a$ is the Lewis number.

Using dimensionless variables τ, δ it becomes obvious that the solution of the problem becomes self-similar. Such solution but for the case of zero solidification shrinkage was obtained in [2]. Using variable $\eta = \delta/\sqrt{\tau}$ the diffusion equation can be reduced to the ordinary differential equation:

$$\frac{d^2\bar{C}}{d\eta^2} + \left(\eta/2 + \beta_0^*\right)\frac{d\bar{C}}{d\eta} = 0 \tag{4}$$

with boundary conditions

$$\left(\frac{d\bar{C}}{d\eta}\right)_{\eta=0} = -\beta_0^*\left(\bar{C}+1\right);\left(\bar{C}\right)_{\eta\to\infty} = 0 \tag{5}$$

where $\beta_0^* = \bar{\rho}\beta_0$ is the only parameter depending on shrinkage appearing in equations describing the process. Solving this equation we can obtain:

$$\bar{C}(\eta) = \frac{\sqrt{\pi}\beta_0^* erfc\left(\eta/2 + \beta_0^*\right)}{\exp\left(-\beta_0^{*2}\right) - \sqrt{\pi}\beta_0^* erfc(\beta_0^*)} \tag{6}$$

3.2 The case of constant crystal growth rate

Let us consider another case when the melt supercooling at the crystallization front stays the same during the crystallization process. Crystal growth rate depends only on the supercooling, so it also remains constant: $dX/dt = v_0 = const$; $X = v_0 t$ (in dimensionless form: $d\bar{X}/d\tau = 1$; $\bar{X} = \tau$).

If solidification shrinkage is not taken into account the most obvious variables for the problem are τ and δ (so that the phase transition surface is immobile) as was done in [2]. Using these variables the problem can be stated in the following form:

$$\frac{\partial\bar{C}}{\partial\tau} - \bar{\rho}\frac{d\bar{X}}{d\tau}\frac{\partial\bar{C}}{\partial\delta} = \frac{\partial^2\bar{C}}{\partial\delta^2} \tag{7}$$

$$\left(\bar{C}^{*}\right)_{\tau=0}=0;\left(\frac{\partial \bar{C}}{\partial \delta}\right)_{\delta=0}=-\bar{\rho}\frac{d\bar{X}}{d\tau}\left[\left(\bar{C}\right)_{\delta=0}+1\right];\left(\bar{C}\right)_{\delta\to\infty}=0 \qquad (8)$$

The use of the variables $\tilde{\tau}=\bar{\rho}^{2}\tau$, $\tilde{\delta}=\bar{\rho}\delta$ allows us to bring the equations to the form they had in the case of zero shrinkage. So, this substitution allows to take the shrinkage into account without any additional complications.

By making a substitution $\bar{C}^{*}=\bar{C}\exp\left(\tilde{\tau}/4+\tilde{\delta}/2\right)$ the convective term can be excluded from the diffusion equation (7). The boundary value problem transforms into the following form:

$$\frac{\partial \bar{C}^{*}}{\partial \tilde{\tau}}=\frac{\partial^{2}\bar{C}^{*}}{\partial \tilde{\delta}^{2}}, \qquad (9)$$

$$\left(\bar{C}^{*}\right)_{\tilde{\tau}=0}=0;\left(\frac{\partial \bar{C}^{*}}{\partial \tilde{\delta}}\right)_{\tilde{\delta}=0}=-\bar{C}^{*}/2-\exp\left(\tilde{\tau}/4\right);\left(\bar{C}^{*}\right)_{\tilde{\delta}\to\infty}=0. \qquad (8)$$

The exact solution of the problem can be found using the Laplace transformation. The solution has the following form:

$$\begin{aligned}\bar{C}(\tau,\delta)&=\bar{\rho}\sqrt{\frac{\tau}{\pi}}\exp\left(-\frac{\delta^{2}}{4\tau}-\frac{\bar{\rho}\delta}{2}-\frac{\bar{\rho}^{2}\tau}{4}\right)+\\&+\frac{1}{2}\left(1-\bar{\rho}\delta+\bar{\rho}^{2}\tau\right)\exp\left(-\bar{\rho}\delta\right)erfc\left(\frac{\delta}{2\sqrt{\tau}}-\frac{\bar{\rho}\sqrt{\tau}}{2}\right)-\frac{1}{2}erfc\left(\frac{\delta}{2\sqrt{\tau}}+\frac{\bar{\rho}\sqrt{\tau}}{2}\right).\end{aligned} \qquad (9)$$

4 Conclusions

The problem of dissolved gas segregation during solidification was formulated. The case of the plane crystallization front was studied. Solidification shrinkage was taken into account. Exact solutions of the problem were found for two different cases of crystal growth law: the constant crystal growth rate (non-equilibrium crystallization) and the case when crystal growth rate is inversely proportional to the square root of time (equilibrium crystallization). The obtained solutions allow us to understand the dissolved gas concentration field dynamics. In the case of equilibrium crystallization the solution is a self-similar one. It was shown that the obtained solutions are on the same complexity level as the solutions for the case of zero solidification shrinkage. Therefore, there is no justification for omitting the shrinkage from the equations as it was done traditionally.

The study was funded by FASO Russia.

References

1. A. A. Chernov, A. A. Pil'nik, Int. J. Heat Mass Transfer **55** (2012)
2. A. I. Fedorchenko, A. A. Chernov, Int. J. Heat Mass Transfer **46** (2003)
3. A .A. Chernov, A. A. Pil'nik, D. R. Islamov, J. Cryst. Growth **450** (2016)
4. A. A. Chernov, J. Eng. Thermophys. **18** (2009)
5. A. A. Chernov, A. A. Pil'nik, JETP Lett. **102** (2015)

Mathematical modeling of heating of a dimetallic plate by a high-energy concentrated radiation flux

Mikhail Purin[1,*], and *Ekaterina* Zakharova[1]

[1]National Research Tomsk Polytechnic University, 634050 Tomsk, Russia

Abstract. Mathematical modeling of heat transfer processes proceeding together under conditions of intense phase transformations (melting and evaporation of metals) under laser action on a dimetallic target has been carried out. Based on the results of numerical simulation, it is established that the laser power exerts a significant influence on the dynamics of the melting process. It is shown that evaporation of metal in the boundary layer of the plate can also have a significant effect on the heat transfer characteristics. It is established that metal vapors forming a gas mixture in the near-wall region can absorb laser radiation and cause the phenomenon of optical resonance with increasing their critical concentration, in other words, when the metal melts, optical breakdown is possible.

1 Introduction

Laser treatment of metals in industry is becoming increasingly important due to its advantages such as high processing accuracy, the absence of mechanical damage on the processed products, and as a result, a lower consumption of metal for tolerances and allowances of blanks, as well as the ability to work with complex geometric shapes [1,2]. It should be noted that industrial laser metal processing plants are less harmful for the environment, since they do not generate waste industrial oils that are used to cool and lubricate moving parts of mechanical metalworking machines.

In connection with the growing interest in laser processing of metal, studies of the processes that occur during laser processing become topical, however, field experiments are quite complex and expensive. In these cases, the most suitable is the apparatus of mathematical modeling, which makes it possible to carry out studies of the processes under consideration with high accuracy and efficiency without any material costs.

The melting process is accompanied by a change in the structure of the material with a change in the thermophysical characteristics: heat capacity, density, thermal conductivity. It should be noted that under the conditions of laser action, intense heating of the metal takes place in a limited region. Accordingly, in the plate structure it is possible to form an essentially nonlinear and spatially-distributed fusion front, the shape of which does not yield to polynomial approximation.

* Corresponding author: mvp21@tpu.ru

A significant contribution to the solution of the multidimensional formulation of Stefan problem was made by V.I. Mazhukin [3-5], A.A. Samarskii [6,7], and L.I. Rubinshtein [8]. However, the methods of [3-8] are based on the explicit separation of the fusion front, and this requires the use of a very complex mathematical apparatus based on a combination of the experimental results of the melting process and the apparatus of differential topology that takes into account the dynamic changes in the calculated difference grids.

The aim of the work is mathematical modeling of the processes of heat and mass transfer proceeding together under conditions of intense phase transformations (melting and evaporation of a metal) under the action of a laser.

2 Physical statement of the problem

At the initial instant of time (t = 0), a laser beam is incident on the bounded region of the dimetallic target, as a result, the metal of both layers begins to heat up. When the phase transition temperatures are reached, the process of metal melting is initiated, resulting in a heterogeneous structure consisting of liquid and solid phases of metals with different thermophysical characteristics. With continued heating, the process of evaporation of the metal into the gas region is initiated. The metal vapor forms a vapor-gas mixture in the external environment, which has a significant effect on the radiation transfer characteristics.

When solving the problem, it was assumed that the thermophysical characteristics of metals depend on the position of the fusion front. The thermophysical characteristics of metals and gas, the heat of melting and evaporation of metals are given in Table 1. The laser beam is directed perpendicular to the target surface.

Table 1. Basic physical characteristics in the system "Dimetallic plate-gas".

Thermophysical characteristics of metals							
	Solid phase			Fluid phase			
	c, J/(kg·K)	λ, W/(m·K)	ρ, kg/m³	c, J/(kg·K)	λ, W/(m·K)	ρ, kg/m³	
Aluminum	951.3	240	2712	1090	90.7	2368	[9]
Plumbum	127.5	32.9	11340	148	15.12	10592	[9]
Thermophysical characteristics of gas							
	c, J/(kg·K)		λ, W/(m·K)		ρ, kg/m³		
Argon	519		0.0394		1.661		[10]
Heat of fusion, heat of vaporization and melting point of metals							
	Q_{melt}, kJ/kg		Q_{evp}, kJ/kg		T_{melt},K		
Aluminum	393		10900		933.3		[9]
Plumbum	24.3		860		600.4		[9]

3 Mathematical formulation of the problem

The mathematical model of the process under consideration includes the energy equations for metals in dimensionless form, taking into account the dependence of thermophysical characteristics on temperature:

The energy equation for the upper plate:

$$\frac{C_1(\theta)P_1(\theta)}{Fo_1}\frac{\partial\theta}{\partial t} = \frac{\partial}{\partial X}\left(\Lambda_1\frac{\partial\theta}{\partial X}\right) + \frac{\partial}{\partial Y}\left(\Lambda_1\frac{\partial\theta}{\partial Y}\right) + Pom_1, Pom_1 = \frac{2\cdot Q_1\cdot W_1\cdot\delta\cdot l^2}{h\cdot\lambda_0\cdot T_0}, \quad (1)$$

The energy equation for the lower plate:

$$\frac{C_2(\theta)P_2(\theta)}{Fo_2}\frac{\partial\theta}{\partial t}=\frac{\partial}{\partial X}\left(\Lambda_2\frac{\partial\theta}{\partial X}\right)+\frac{\partial}{\partial Y}\left(\Lambda_2\frac{\partial\theta}{\partial Y}\right)+Pom_2, Pom_2=\frac{2\cdot Q_2\cdot W_2\cdot\delta\cdot l^2}{h\cdot\lambda_0\cdot T_0}, \quad (2)$$

here $C = c / c_0$, $P = \rho / \rho_0$, $\Lambda = \lambda / \lambda_0$, $\theta = T / T_0$ are the dimensionless heat capacity, density, thermal conductivity, and temperature, respectively; c_0, ρ_0, λ_0, T_0 are the dimensionless parameters $X = x / l$, $Y = y / l$ are dimensionless coordinates; Pom_1, Pom_2 - Pomerantsev criteria for the melting process for the upper and lower plates, respectively; Fo_1, Fo_2 - Fourier numbers for the upper and lower plates, respectively; W_1, W_2 - melting rates of plates, kg / ($m^2 \cdot s$); Q_1, Q_2 - heat of melting of metals, kJ / kg; δ is the Dirac delta function; h, l - linear dimensions of plates, m.

The energy equation for the described process in dimensionless quantities for a gas:

$$\frac{1}{Fo_3}\frac{\partial\theta}{\partial t}=\frac{\partial^2\theta}{\partial X^2}+\frac{\partial^2\theta}{\partial Y^2}+PomI\cdot divE_{isl}, PomI=\frac{I_0\cdot l}{\lambda\cdot T}, \quad (3)$$

here I_0 is the intensity of the laser radiation, W/m^2; $E_{isl} = I/I_0$ is the dimensionless radiation intensity.

The diffusion equation in the gas region:

$$\frac{1}{Fo_d}\frac{\partial C}{\partial t}=\left(\frac{\partial^2 C}{\partial X^2}+\frac{\partial^2 C}{\partial Y^2}\right), Fo_d=\frac{t\cdot D_{12}}{l^2}, \quad (4)$$

here C is the mass concentration of metal vapor in the gas, kg/kg; Fo_d is the Fourier diffusion number.

The diffusion coefficient was determined according to [11]:

$$D_{12}=2{,}66\cdot 10^{-2}\cdot\frac{\sqrt{T^3\cdot(M_1+M_2)/(2\cdot M_1\cdot M_2)}}{P\cdot\sigma_{12}^2\cdot\Omega^{(1,1)*}(T^*)}, \frac{m^2}{s}, \quad (5)$$

here σ_{12} is the parameter of the function expressing the potential interaction energy of the molecules of the first and second components; $\Omega^{(1,1)*}(T^*)$ is the reduced interaction integral of the molecules of the first and second components; M_1, M_2 are molar masses, kg / kmol; P is the pressure, Pa.

The reduced interaction integral of the molecules of the first and second components [11]:

$$Q_{12}^{(1,1)}\cdot\left(\frac{k\cdot T}{J_2}\right)^m\cdot\alpha_2^{-\frac{2}{3}}=a\cdot\exp\left(b\cdot\frac{J_2}{J_1}\right), \quad (6)$$

here J_1, J_2 is the ionization energy of the metal atom and the gas molecule, eV; α is the gas polarizability coefficient; $k = 0.863\cdot 10^{-4}$ eV / K is the Boltzmann constant; m is the temperature dependence of $D_{12}\sim T^n$; a, b are coefficients that depend on the values of J_1.

The radiation transfer equation [12]:

$$\Omega\cdot grad\, I+\chi\cdot I=\frac{\chi\cdot\sigma\cdot T^4}{\pi}, \quad (7)$$

here Ω is the direction of the ray; I - radiation intensity, W / m^2; χ is the absorption coefficient of the medium, m^{-1}; σ is the Stefan-Boltzmann constant, $W\cdot m^{-2}\cdot K^{-4}$.

The physical constants for equations (5-7) are given in [8,10,13]. The system of equations (1-7) was solved under the following boundary conditions (Fig. 1):

$$t=0;\ 0<x<l;\ 0<y<H_1;\ T(x,y)=T_s;$$

$$t=0;\ 0<x<l;\ H_1<y<H_2;\ T(x,y)=T_s;$$

$$t=0;\ 0<x<l;\ H_2<y<H_{gas};\ T(x,y)=T_{gas}$$

$$\left.\frac{\partial\theta}{\partial X}\right|_{x=0}^{0<y<H_{gas}} = \left.\frac{\partial\theta}{\partial X}\right|_{x=l}^{0<y<H_{gas}} = \left.\frac{\partial\theta}{\partial Y}\right|_{y=0}^{0<x<l} = \left.\frac{\partial\theta}{\partial Y}\right|_{y=H_g}^{0<x<l} = 0, t>0;$$

$$\Lambda_2(T)\left.\frac{\partial T}{\partial Y}\right|_{y=H_2}^{0<x<x_1} = \Lambda_{gas}\left.\frac{\partial T}{\partial Y}\right|_{y=H_2}^{0<x<x_1}, t>0;$$

$$\Lambda_2(T)\left.\frac{\partial\theta}{\partial Y}\right|_{y=H_2}^{x_1<x<x_2} = \Lambda_{gas}\left.\frac{\partial\theta}{\partial Y}\right|_{y=H_2}^{x_1<x<x_2} - (Ki_{EV}\text{-}\ Ki_L)\cdot\Psi(X_L;Y_L), t>0,$$

here $Ki_{EV} = \frac{W_{EV}\cdot Q_{EV}}{T\cdot\lambda}$ and $Ki_L = \frac{I_0\cdot l\cdot(1-k_{ref})}{T\cdot\lambda}$ – Kirpichev criteria; k_{ref} – reflection coefficient.

$$\Lambda_2(T)\left.\frac{\partial\theta}{\partial Y}\right|_{y=H_2}^{x_2<x<l} = \Lambda_{gas}\left.\frac{\partial\theta}{\partial Y}\right|_{y=H_2}^{x_2<x<l}, t>0;$$

$$\Lambda_1(T)\left.\frac{\partial\theta}{\partial Y}\right|_{y=H_2}^{0<x<l} = \Lambda_2(T)\left.\frac{\partial\theta}{\partial Y}\right|_{y=H_2}^{0<x<l}, t>0.$$

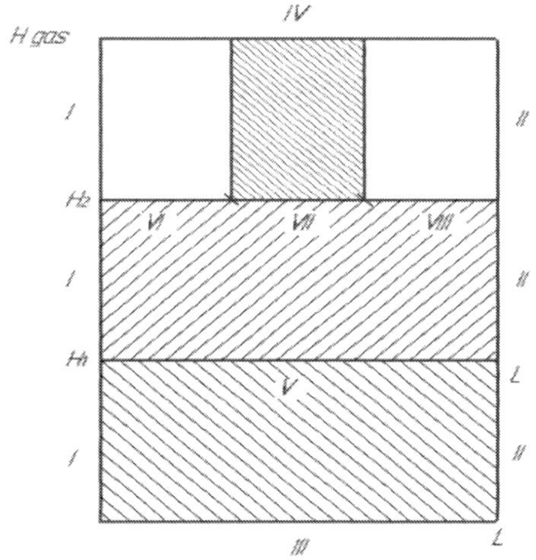

Fig. 1. Scope of the problem.

The problem was solved by the method of finite differences using the implicit four-point difference template with the use of discontinuous coefficients [13, 14]. The nonstationary multidimensional Stefan problem was solved by the method of implicit separation of the fusion front [14].

4 Results

When solving the problem, a dimetallic plate was considered, the upper layer of which was made of aluminum, and the lower one was made of lead. The choice of these materials is due to their extensive use in industry.

Fig. 2 (a, b) shows the temperature and concentration of metal vapors as a function of the intensity of the incident radiation.

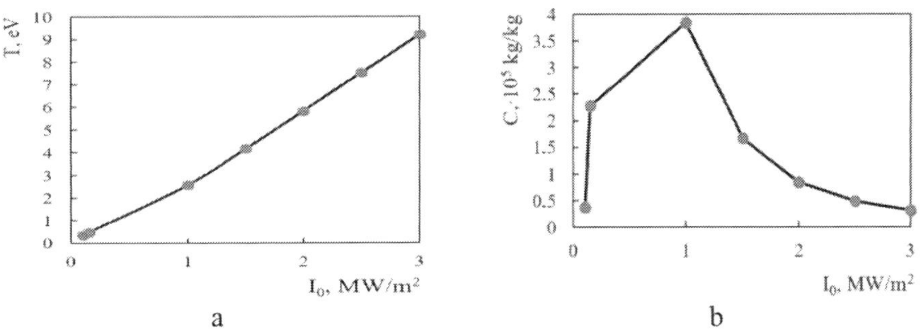

Fig. 2. Dependence of temperature (a) and metal vapor concentration (b) in the gas on the intensity of the incident radiation at the upper plate-gas interface in the middle of the upper plate.

With increasing intensity of the incident radiation, the concentration of metal vapor in the gas region increases to a critical value, and then begins to decrease. This can be explained by the intensive outflow of metal vapors from the region under consideration, owing to a greatly increased diffusion coefficient. According to (5), the diffusion coefficient depends nonlinearly on the temperature. With a continuous increase in the radiation intensity, starting from a certain instant of time, the diffusion coefficient becomes so large that an intensive outflow of metal vapors occurs and the concentration decreases.

Fig. 3 shows the temperature field in the dimetallic plate (a) and in the dimetallic plate under optical breakdown (b).

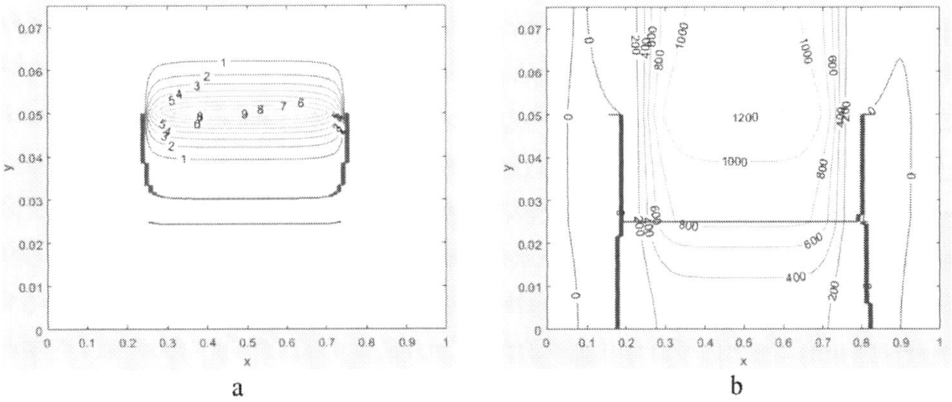

Fig. 3. Temperature field (eV) (a) and fusion fronts (b) in a dimetallic plate.

According to the results of mathematical modeling, it is established that with sufficiently intense laser action (> 900 MW/m^2), the effect of optical breakdown in the near-wall gas region is possible. This is due to the fact that metal vapors, forming a gas mixture in the near-wall region, can substantially absorb laser radiation, as a result, an intense, even exponential temperature increase is possible. The latter leads to an increase in the optical thickness of the wall gas layer, and also forms zones with intense self-heating in the gas region as a result of the phenomenon of reradiation. In other words, in some cases of metal melting, there are risks of destruction of the metal preform and damage to the laser.

As can be seen from Fig. 3,a, melting of the lower plate can begin even before the moment of contact with it of the melting front of the upper plate.

5 Conclusion

It is established that the laser power exerts a significant influence on the dynamics of the melting process. It is shown that the evaporation of metal into the near-surface gas layer can also have a significant effect on the heat transfer characteristics. It is established that metal vapors, forming a gas mixture in the near-wall region, can substantially absorb laser radiation and cause the phenomenon of optical resonance.

In conclusion, we can say that in order to ensure a fairly stable process of melting by a laser beam, it is necessary to organize a constant flow of inert gas in the melting and welding region, which ensures the entrainment of metal vapors from the melting region.

References

1. G. A. Abilsiitov, V. S. Golubev, V. G. Gontar, *Technological lasers: directory* (Mashinostroenie, Moscow, 1991)
2. A. G. Grigoriants, A. A. Sokolov, A. A. Grigor'yants *Laser cutting of metals.* (Higher education, Shk., 1988)
3. V. I. Mazhukin, A. V. Shapranov, *Mathematical modeling of the processes of heating and melting metals* (Preprints of IPM. MV Keldysh, Moscow, 2012)
4. V. I. Mazhukin, A. V. Shapranov, *Molecular-dynamic modeling of heating and melting of metals* (Preprints of IPM. MV Keldysh, Moscow, 2012)
5. O. N. Koroleva, V. I. Mazhukin, Zh. Vychisl. Math. And Math. Fiz. **46**, 887 (2006)
6. A. A. Samarsky, P. N. Vabishchevich, *Computational heat transfer* (Editorial URSS, Moscow, 2003)
7. A. A. Samarskii, B. D. Moiseenko, Zh. Vychisl. Math. And Math. Fiz. **5**, 816 (1965)
8. V. S. Chirkin, *Thermophysical properties of materials* (Fizmatgiz, Moscow, 1959)
9. V. V. Pomerancev. *Collection of problems on the theory of combustion: Textbook for high schools* (Energoatomizdat, Leningrad, 1983)
10. B. N. Chetverushkin, *Mathematical modeling of radiating gas problems* (Nauka, Moscow, 1985)
11. M. P. Vukalovich, V. A. Kirillin, S. A. Remizov, *Thermodynamic properties of gases* (Mashgiz, Moscow, 1953)
12. N. B. Vargaftik, *Reference book on thermophysical properties of gases and liquids* (Nauka, Moscow, 1972)
13. S. V. Syrodoy, N. Y. Gutareva, R. I. Taburchinov, MATEC Web Conf. **72**, 01109 (2016) DOI: 10.1051-matecconf-20167201109
14. S. V. Syrodoy, N. Y. Gutareva, K. A. Bugaeva, MATEC Web Conf. **72,** 01035 (2016) DOI: 10.1051/matecconf/20167201035

The modeling of the dust deposits on the surfaces of cyclone scrubber

Michail Vasilevsky[1], *Aleksandr* Razva[1,*], *Dmitri* Zaharov[1], and *Danil* Shabirov[1]

[1]National Research Tomsk Polytechnic University, 634050 Tomsk, Russia

Abstract. The phenomena at interaction of agglomerate of disperse material with the surface of the inertial device are considered. It is revealed that tension in the interactions between the agglomerate surface can reach values comparable to atmospheric pressure, which is the reason for the formation of solid deposits, what is the reason of formation of strong deposits which, in turn, are defined by the autohesion communications of particles forming agglomerate.

1 Introduction

Dust removals of gases is an integral part of technological processes in the production of power, chemical, metallurgy, construction, food and other products. One of the most common and reliable methods of cleaning industrial gases from suspended particles is mechanically treated in a dry apparatus. A counter flow cyclone dust collector is a key element in systems of gas purification.

In this apparatus, a vortex is organized in such a way that the disperse phase of the aerosol state at the receiver is converted to a dispersed body in which the particles touch each other and interact with each other. The process of dust separation in the cyclone consists of several stages: swirling flow, concentration and withdrawal of particles from the separation zone, forming a layer in the receiver, removing the collected dust from the receiver. Violation of any of stages leads to poor performance of the collector. One of the reasons for reducing the effectiveness of these devices is the formation of deposits of particles on the enclosing surfaces of the separation space and hanging dust volume dust collector. Entering the gas cleaning system particulate matter consists of aggregates of primary particles, aggregated particles, and is along with the carrier medium of the aerosol. In another state it is in the form of deposits or is a bulk body. Agglomerates are formed as a result of turbulent coagulation, when disrupting the flow of deposits from surfaces [1], when handling materials, with segregation of particles as a result of its transport to different devices [2, 3]. In the cyclone itself, migratory-turbulent and gradient near-wall agglomerations are significant.

2 Theoretical bases and design values

* Corresponding author: razva@mail.ru

We will consider the phenomena at interaction of agglomerate with a surface. At elastic collision of solid bodies no changes of macroscopic mechanical properties of the contacting materials. In this case, depending on the arising interaction force, the contact area is determined by the Hertz formula, which relates the Poisson coefficients and the elasticity moduli of interacting bodies [4]. In most studies on the impact of elastic bodies assume the lawful hypothesis of the static Hertz problem and accept further that the ratio between the contact force and the convergence remains at impact the same as under static loading. To determine the time duration of impact, solve the problem of the motion of the particle during its interaction with the barrier, and is the same for the duration of the process of deepening particles and rebound. Such interaction is characteristic of integral, rigid particles and the ricochet of particles and abrasive wear of surfaces is his consequence.

The agglomerates are composed of many particles, so are heavier compared to the mass of the particles. Depending on the ratios of the radial and tangential velocity of the agglomerate approaching it to the surface is at different angles (Fig.1). The angle of incidence determines the nature of its interaction with the surface. The sinter has a density several times lower than the density of the particles of which it consists. Therefore ricochet hardly possible. Depending on the angle of interaction and between the agglomerate will either be eroded, disperse and again become stronger, or Vice versa, will condense and form deposits. If sediments are solid, they are not destroyed by the flood. In Fig. 1 shows the trajectory of the agglomerate in the channel and conversion scheme of the agglomerate A (dotted line) in the compacted bundle B.

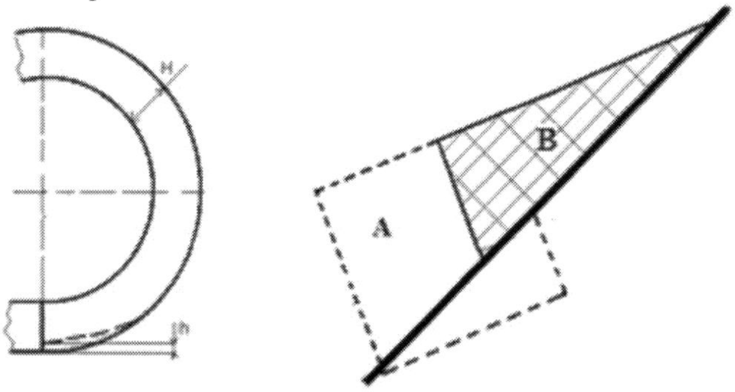

Fig. 1. The trajectory of the agglomerate in the channel and conversion scheme of the agglomerate A (dotted line) in the compacted bundle B.

Agglomerate consisting of small particles, under the influence of centrifugal force moves relative to the periphery of the gas and the interaction with the wall passes through the stages of compaction, deformation, aftereffect. The change of momentum equals the impulse of force generated in the formed contact surface. The interaction time and deformation is determined by the speed of approach of the agglomerate to the wall and the size of the area of "flattening". In addition, can be used the equation of rate of change of energy that binds the kinetic energy and the rate of change of energy of the peripheral forces applied to the deformable surface. All these data allow predicting a state and settings of the agglomerate, if you know the parameters of the layer of particles in the deposition.

The velocity of the particle relative to the gas in curved flow in the radial direction can be found from the dimensionless equations [5].

$$\frac{\overline{W_{\Phi}^{2}}}{\bar{\bar{r}}} = \Psi \ Re_{\delta} \ \frac{Re_{\delta}^{2}}{24StkRe_{\delta,w}}, \qquad (1)$$

$$Re_{\delta,w} = \frac{W_0 \delta}{v}, \tag{2}$$

$$Stk = \frac{\delta^2}{18v} \frac{\rho_\delta}{\rho} \frac{W_0}{R_c}, \tag{3}$$

$$StkRe_{\delta,w} = \frac{\rho_\delta}{\rho} \frac{W_0^2}{R_c} \frac{\delta^3}{18v^2}, \tag{4}$$

$$Re_\delta = \frac{\Delta U \delta}{v}, \tag{5}$$

where ρ, ρ_δ– the density of gas and particles, δ – particle diameter, v – kinematic viscosity of gas, ΔU – particle velocity relative to the gas in the radial direction, W_0 – gas inlet velocity, $\bar{W}_\phi = \frac{W_\phi}{W_0}$ – dimensionless peripheral velocity of a particle, R_c– the base value of the radius of the cyclone. In [5] it was shown that in the range $Re_\delta = 1-100$ the equation (1) can be approximated dependence

$$Re_\delta = (StkRe_{\delta,w} \frac{\bar{W}^2}{\bar{R}})^{0.75}, \tag{6}$$

(1), but in the range $Re_\delta = 500-100000$ equation (1) we approximate dependence

$$Re_\delta = 1.7\left(Re_{\delta,w}^2 \frac{\rho_\delta}{\rho} \frac{\delta}{R}\right)^{0.5}. \tag{7}$$

Let $W_0 = 20(\text{m/s})$, $\frac{\rho_\delta}{\rho}\frac{1}{18} = 27$, $R_c = \frac{R_{out} + R_{inner}}{2} = \frac{0.4+0.32}{2} = 0.36(\text{m})$, $v = 1.5\times10^{-5}(\text{m}^2/\text{s})$,

$$Stk = \frac{\rho_\delta}{\rho} \frac{1}{18} \frac{\delta^2}{v} \frac{W_0}{R_c} = 10^{-4}\delta^2, \tag{8}$$

$$Re_{\delta,w} = \frac{w_0^\delta}{v} = 1.33\delta, \tag{9}$$

$$Re_{\delta,w} = StkRe_{\delta,\omega}{}^{0.75} = 12.38\times10^4\delta^{2.25}, \tag{10}$$

where δ – particle diameter in microns. $\delta_* = (10^4/1.33)^{1/3}$. Value δ_* corresponds to the value $Re_\delta = 1(\text{m/s})$.

$$\Delta U = Re_\delta \frac{v}{\delta} = Stk^{0.75} Re_{\delta,w}^{0.75} \frac{W_0}{Re_{\delta,w}} = W_0 \frac{Stk^{0.75}}{Re_{\delta,w}^{0.25}} = 186\times10^{-4}\delta^{1.25}, \tag{11}$$

where δ – diameter of agglomerate particle substitute in microns. And on (7):

$$\Delta U = Re_{\delta}\frac{\nu}{\delta} = 1.7\left(Re^{2}_{\delta,w}\frac{\rho_{\delta}}{\rho}\frac{\delta}{R}\right)^{0.5} = 1.7W_{0}\left(\frac{\rho_{\delta}}{\rho}\frac{\delta}{R_{c}}\right)^{0.5} = 1.249\sqrt{\delta}\,. \tag{12}$$

Agglomerations near the wall pass through a section with a decreasing flow velocity, and at the wall itself the flow velocity is zero. The circumferential velocity of the gas at a distance from the wall $W(y) = \sqrt{\frac{\tau_w}{\rho}}(9.5 + 1.32\ln\eta)$, $\eta = \frac{y}{\nu}\sqrt{\frac{\tau_w}{p}}$ [6], $\tau_w = 0.5C_f\rho W_{\Delta}^2/2$, $C_f \approx 0.006$, W_{Δ} – the value of the circumferential velocity at the boundary of the wall zone, $W_{\Delta} = 0.95W_0$. Let us assume that on the half of the wall zone of intensive fall of speed $\Delta \approx 0.03R$ the agglomerate is not affected by the entrainment force of the flow, and this zone is agglomerated by inertia. We estimate the value of the final rate of approach of the agglomerate to the wall. We denote by $x=R-y$, the speed of the approaching near-wall zone ΔU_{Δ}, falling speed on a wall in the normal direction ΔU_{wa} the equation of a decelerated motion of agglomerate in the radial direction has an appearance $m\frac{d\Delta U}{dt} = \xi\frac{\pi\delta_a^2}{4}\frac{\Delta U^2}{2}\rho$, $m = \frac{\pi\delta_a^3}{6}\rho_a$, $\Delta U = \frac{dx}{dt}$, $\frac{d\Delta U}{dt} = \frac{d\Delta U^2}{2dx}$. After integration, we will receive the following expression

$$\frac{\Delta U_{wa}}{\Delta U_{\Delta}} = \sqrt{\exp\left(-\frac{3}{2}\xi\frac{\rho}{\rho_a}\frac{\Delta}{2\delta_a}\right)}\,. \tag{13}$$

We will consider the simplest process of consolidation of agglomerate at his normal contact interaction with a barrier. Problems of the process of deformation and destruction of a rigid-plastic body (striker) when struck against a rigid obstacle are given in [7]. The difference between the processes in the interaction of a dispersed body with an obstacle from the process of interaction of a rigid-plastic body lies in the variability of its density. Experiments show that, at impact rates above a certain value, the agglomerate breaks down to form fragments and adherent fragments; at lower velocities, the agglomerate reduces its length, compacts and adheres to the surface. The agglomerate in the form of a cylinder produces a flat-ended impact on the surface of a motionless absolutely rigid barrier. The impact velocity V_0 is directed along the cylinder axis perpendicular to the obstacle plane. The impact process is accompanied by the compression of a dispersed medium with a small change in the cross-sectional area. This can occur in the case of the interaction of particles inside the cylinder, in which each contact is inelastic, and the particles are attracted to each other by autohesive forces. This is typical for particles smaller than 10 (μm). The initial length of the cylinder is L1, the length in the compressed state is L2 (Fig. 2). Upon impact, the dispersed medium from a volume with a length L1-L2 passes into a volume with a length L2. Suppose that before the impact and after impact, the density of the medium in the volumes is distributed uniformly. The condition of preservation of mass of the disperse environment in position of cylinders 1 and 2 gives the following of ratios:

$$\rho_1 L_1 = \rho_2 L_2\,. \tag{14}$$

The equation of pulses in the interaction of the agglomerate with the wall has the form $\rho_1 L_1\frac{dV}{dt} = \sigma_s$, $dt = \frac{dl}{V_0}$, where t – is a variable time, l – variable length, V - variable speed,

σ_s– contact voltage on the wall. The last equation can be written in the difference form $\rho_1 L_1 \frac{dV_0^2}{\Delta l} = \sigma_s$. Separating the factors and taking into account that $\Delta l = L_1 - L_2$, using equality (14), we obtain the relation between density, velocity, and voltage.

$$\sigma_s = V_0^2 \frac{\rho_1}{l \frac{\rho_1}{\rho_2}} . \tag{15}$$

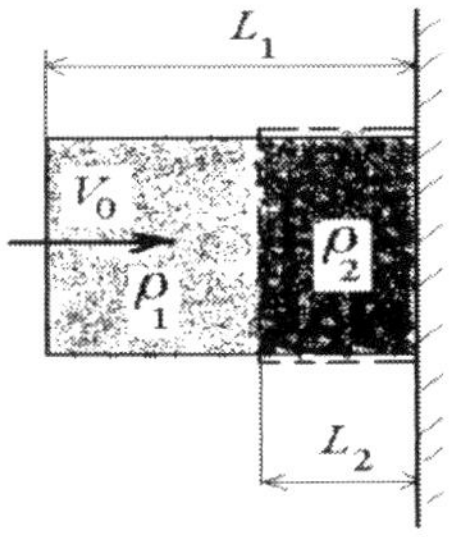

Fig. 2. Scheme of interaction of a cylindrical agglomerate with a surface at right angles.

The density of agglomerate defines its strength parameters, which, in turn, are defined by autohesive internal forces. Therefore, the assumption of a constant density of the dispersed medium in the agglomerate leads to the assumption of a constant stress, which determines the strength during the interaction period. It follows from equation (3) that the contact voltage, which under the above assumptions is also the voltage in the agglomerate at the end of the contact period, is determined by the density and velocity of the agglomerate prior to impact and the density after impact. However, it is clear that the voltage inside the agglomerate should decrease to the free end, so the length-averaged voltage should be less than the voltage at the end. We represent the stress distribution along the length in a volume with the length L_2 in the form

$$\sigma = \sigma_s (1 - \frac{l}{L_2})^k , \tag{16}$$

with $\bar{\sigma} = \frac{1}{L_2} \int_0^{L_2} \sigma_s \left(1 - \frac{l}{L_2}\right)^k dl = \frac{\sigma_s}{k+1}$, whence $k = \frac{\sigma_s}{\bar{\sigma}} - 1 = \frac{\rho_1}{\rho_2} \frac{2}{\left(1 - \frac{\rho_1}{\rho_2}\right)^2} - 1$.

Table 1. Results of calculating the speed of agglomerates in the radial direction.

δ (μm)	50	100	200	400	800
Re_δ	8.2	39.2	186	607	1690
ΔU_Δ (m/s)	2.4	5.8	13.9	23	32
$\frac{\Delta U_{wa}}{\Delta U_\Delta}$	0.31	0.48	0.69	0.83	0.91
ρ_2/ρ_1=0.7					
σ_s (kPa)	0.553	7.75	92	364	847
ρ_2/ρ_1=0.4					
σ_s (kPa)	0.276	3.87	46	182	423

After the termination of influence of all loadings (the after-effect period) agglomerate in a new state appears in balance. The agglomerate is attracted by the aerodynamic forces from the side of the flow, the gravitational force, and on the other, attracts the wall force due to the adhesion and autohesion of the particles. In the new state, there is no pressure force at the free end, whereas at the contact end there are adhesion forces that depend on the "clamping" force during the interaction, the surface forces acting on the lateral surface are acted upon by autohesive forces. Formula (13) and formulas (15, 16) allow the calculation of the contact stress and evaluate the stress distribution of the agglomerate in the period of its interaction with the wall. The calculation is carried out for two values of the ratios of the densities $\rho_2/\rho_1=0.7$ and $\rho_2/\rho_1=0.4$ [8], $\rho_1=300$ (kg/m^3). In the formula (3) $\Delta U_{wa} \equiv V_0$

Calculation of stresses in the agglomerate itself (table 1) shows fast reduction of tension by the free end.

3 Conclusion

Thus, information on density of deposits can be also information on the speed of approach of agglomerate to a surface and about his weight. Estimates show that the stresses in the interactions of the agglomerate with the surface can reach values comparable to atmospheric pressure, which is the reason for the formation of strong deposits, which in turn are determined by the autohesive bonds of the particles forming the agglomerate.

References

1. A. D. Zimon, *Adhesion of dust and powders* (Chemistry, Moscow, 1976)
2. V. G. Levich, *Physicochemical hydrodynamics* (Fizmatgiz, Moscow, 1959)
3. E. Bulba, N. Ivanova, MATEC Web Conf. **141**, 01014 (2017)
4. E. Bulba, N. Ivanova, MATEC Web Conf. **110**, 01020 (2017)
5. M. V. Vasilevsky, E. G Zykov, *Raschet effektivnosti ochistki gaza v inertsial'nykh transportnykh sredstvakh* (TPU Publishing House, Tomsk, 2005) [In Russian]
6. A. N. Shtym, *Aerodynamics of cyclone-vortex chambers* (Far Eastern University, Vladivostok, 1985)
7. A. Y. Sagomonyan, *Dynamics of piercing the obstacle* (Moscow Univ. Publ., Moscow, 1988)
8. K. V. Nekrasova, A. S. Razva, E. G. Zykov, M. V. Vasilevsky, *Energetika: ekologiya, nadezhnost', bezopasnost': materialy dokladov trinadtsatoy Vserossiyskaya nauchno-tekhnicheskaya konferentsiya* (TPU Publishing, Tomsk, 2007) [In Russian]

Thermal regimes of space composite structures. Part I

Sergey Reznik[1,*]

[1]Bauman Moscow State Technical University, 105005 Moscow, Russia

Abstract. Composite materials have a unique combination of mechanical and thermal characteristics. This allows you to create lightweight, durable, heat-resistant design. In the article the common features for all space composite structures were highlighted. The article deals with the thermophysical problems arising in the design of space composite structures associated with strict requirements for their weight and functional efficiency. The feasibility of using combined heat transfer models to improve design accuracy has been shown.

1 Composition materials for aerospace applications

1.1 Brief history of composite materials

Composite materials (CM) consist of two or more components with a definite interface [1]. Industrial production of CM dates back to the late 1930s. The aviation industry has been one of the main consumers of CM ever since. Starting from the mid-1950s, CM technology has been developing in rocket and space technology parallel to and independent of the aviation industry. This "branching" was caused by fundamentally different operating requirements for the thermal regimes of aircraft, ballistic missiles and spacecraft. In rocket and space technology, the operating level of thermal loads, temperatures, heating / cooling rates and temperature gradients is fundamentally different from that of aviation. A reentry vehicle is typically subjected to a heat flux with $50 \cdot 10^5$ W/m^2 density its heat shield withstands temperature gradients exceeding 1000 degrees/mm. An orbiter is exposed to the temperature changes ranging from plus 150°C on the illuminated portion of the orbit to minus 150°C on the shadow portion all of these is still out-of-limit for the conventional aircraft [2]. It should be noted, however, that the development of hypersonic aircraft in the near future may partially alleviate this difference.

First-generation CM were based on a polymer matrix: impregnated timber, glass fiber reinforced plastics (GFRP) and asbestos-filled plastics. Impregnated timber, a combination of birch timber lamina and phenolic resin, was used in power structures of military aircraft, in air rudders of R-7 missiles [3, 4]. There were attempts to use it for liquid fuel tanks for of ballistic missiles [5]. GFRP comprising glass fabrics, tapes or yarns and epoxy or phenolic resin was used in radio-transparent fairings for airplanes and missiles, solid rocket motor bodies, and launch pod containers [6, 7]. Asbestos-reinforced plastic was used in the first

* Corresponding author: sreznik@bmstu.ru

ablative thermal protection for the warheads of ballistic missiles and space vehicles that helped overcome the "thermal barrier" [8, 9].

The second-generation of CM includes carbon fiber reinforced plastics (CFRP) and organic fiber reinforced plastics (OFRP). CFRP proved an excellent material for any power structure that required high strength, stiffness and linear dimensions stability: wing skins and body parts, antenna reflectors etc. [10, 11]. OFRP took time to be introduced into aerospace technology and were used for solid rocket motor bodies and meteoroid protection screens [12].

Third-generation CM are metal composites and carbon or ceramic matrix composites. Metal matrix composites with boron fibers reinforcement were employed in a number of load-bearing structures, such as trusses, frames and landing gear [13]. The nose cones and leading edges in the Space Shuttle and Buran were made of carbon-matrix and carbon fiber reinforced CM. Additional surface and in-depth oxidation protection guaranteed operation in 100 flight cycles: up to 1650°C maximum surface temperatures for up to 20 minutes per flight [14, 15]. Current research in ceramic matrix and carbon/ceramic fiber composites has shown that these materials can work at temperatures exceeding 2000° C in high-speed erosion and chemically active gas flows [16, 17].

1.2. General approaches to spacecraft thermal design

Spacecraft thermal design is a multistage, iterative process [18]. When choosing design solutions that will enable the specified thermal regime for a new spacecraft, engineers must account for the design requirements, operating conditions, available material and production resources. It is customary to consult the knowledge data bases containing information on previous research, on characteristics of structural, thermal protection and thermal insulating materials, coatings, liquid and gaseous media, and aerodynamic information etc. These databases can contain information on mathematical models, thermal physics, aerodynamics, radio physics, optics, problem solving methods and software. Apart from computational and theoretical data, they contain information on composite materials manufacturing technologies, laboratory testing methods of materials and coatings, ground and flight testing methods of structures, test equipment, instruments and sensors.

Verification of mathematical models is embedded into the design process. This is a procedure for structural and /or parametric identification based on inverse heat transfer problems using experimental data. These inverse problems are stated in the extreme form, where the unknown quantities are determined as a result of minimization of the residual functional between theoretical and experimental data for the similar spatial points of temperature measurements in the entire experiment. Reliable mathematical models enable selection of optimal design solutions.

2 Earth orbit cleanup

2.1 Transformable space debris capture device

Space debris is becoming a major problem for the practical activity in the outer space. There are a number of ways to clean up the near-Earth space from debris [19, 20]. One of the devices for cleaning the near-Earth orbits can be debris capture vehicles. To manufacture these spacecraft, transformable load-bearing composite elements can be used in combination with multilayer hybrid fabrics to withstand collision with space debris elements.

To date there are no recycling methods for the expired space structures, so the only option is to dispose of them by deorbiting into the atmosphere or transferring them into graveyard orbits away from main operational orbits. In any case, this will require special capture devices: nets, harpoons or catchers that will play an important though intermediate role in the cleanup process.

One of the catcher options can be a transformable tetrahedral pyramid with 6.3, 5.0 and 5.0 m sides and 6.3x6.3 m^2 base. The open base is directed funnel-like towards the flight, capturing the space debris elements (Fig. 1). Four telescopic rods (6 m total length, 200x100 mm^2 base cross-section) are used to deploy the catcher from the its stow position.

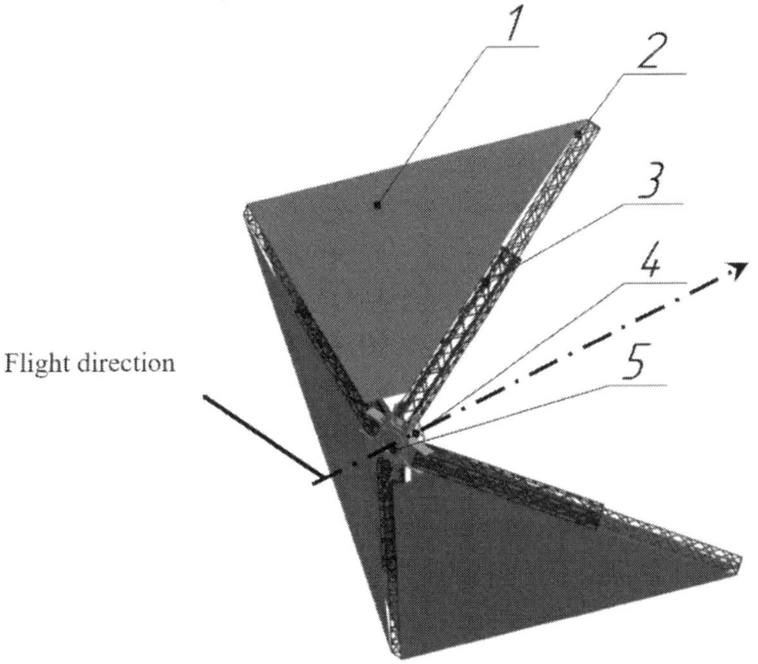

Fig. 1. Space debris catcher, one of the flaps not shown: 1 – debris catcher flap; 2, 3 – webbed load-bearing element; 4 – base; 5 – mount.

The walls of the catcher should be strong enough to withstand the impacts of space debris. The flaps are proposed to be 200 mm thick consisting of two separate screens. The top layer of each screen contains a layer of screen-vacuum thermal insulation (SVTI-I type), 20 mm thick, coated with metallized aramid fabric (product identification number 5397-92). The screen-vacuum thermal insulation consists of 20 μm thick polyimide film layers separated by a fiberglass scrim. The second layer of the screen serving as debris protection is a 4 mm thick fabric stack comprising 20 layers of CBM fabric (product identification number 8601), 190 g/m^2 linear density, plain weave, 15 warp threads by 15 weft threads, 58 tex. The back layer of the screen is another layer of screen-vacuum thermal insulation, 20 mm thick. The distance between the screens is 150 mm.

2.2 Modelling thermal state of a space debris catcher

When in the near-earth space, the catcher will be subjected to harmonical fluctuations in temperature. To understand the level of the walls' stability, the thermal regime of the trap must be determined with and without impact.

The aim of the work is to investigate the temperature state of multilayer fabric stacks when exposed to the heat radiation from the Sun and Earth.

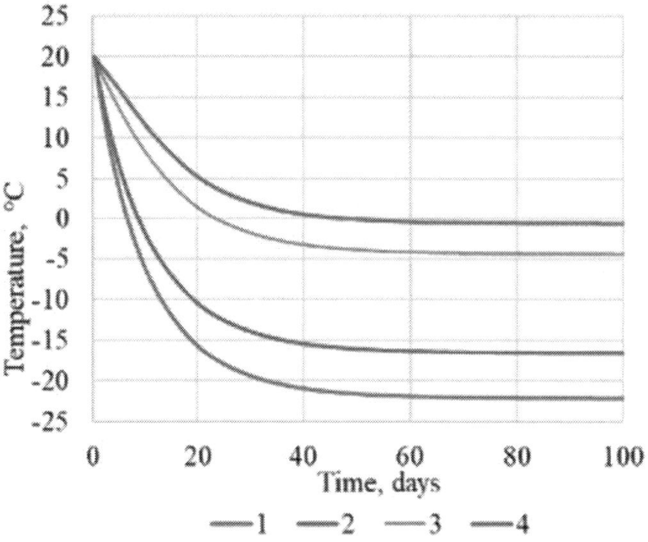

Fig. 2. Temperature of debris protection aramid fabric: 1 – first layer in the upper (relative to Earth) part of the catcher; 2 – second layer in the upper part; 3 – first fabric layer in the lower (relative to Earth) part of the catcher; 4 – second layer in the lower part.

The thermal regime modeling did account for the screen-vacuum insulation compacting under load or the elastic shell deployment elements potentially acting as "thermal bridges". Both the screen-vacuum insulation and debris protection layers were considered as a continuous material free from internal microstructure. The thermal physical characteristics of the materials were assumed to be independent of temperature.

With regard to the thermal loads, it was assumed that the direction of the catcher geostationary flight coincided with its symmetry axis, and the transverse axis was directed at the Earth. Under these conditions the catcher will be unevenly illuminated by the Sun, which will cause significant temperature changes along its surface. It was established that the outer surface temperature can vary from minus 150°C to plus 120°C, which is outside the recommended temperature range for the aramid fabric. However, the double-sided screen-vacuum thermal insulation makes it possible to reduce the temperature fluctuations to the acceptable values (Fig. 2).

The author would like to acknowledge A.A. Alekseev, A.V. Azarov, W.P.P. Fisher, M.L. German, P.S. Grinchuk, S.A. Gusev, O.V. Denisov, A.D. Novikov, K.V. Mikhailovsky, A.M. Mikhalev, L.Ya. Paderin, N.V. Pavlyukevich, I.R. Shafikova and M.S. Tret'yak for their scientific cooperation.
Separate parts of this work are fulfilled with financial support on contract from Russian Ministry of Education and Science on project on Agreement 14.574.21.0146 from 26 September 2017 (Unique identifier of the project RFMEFI57417X0146).

References

1. V.V. Vasil'ev, Yu.M. Tarnopolsky, *Composite materials* (Mashinostroenie, Moscow, 1990)
2. A.G. Bratuhin, M.A. Pogosyan, V.I. Surov, L.V. Tarasenko, *Structural and functional materials of modern aircraft construction* (MAI Publ. House, Columbia, 2007)
3. L.I. Anikhovskaya, Klei. Germetiki. Tekhnologii **5** (2005)
4. E.N. Kablov, G.M. Gunaev Alexei Tikhonovich, *Proc. Int.-Industry Conf. on Composites in Aerospace Material Science* (VIAM, Moscow, 2009)

5. R.N. Kanin, N.N. Tikhonova, V.G. Degtyar, *SDB-385 DB Mashinostroeniya, SRC "DB named after acad. V.P. Makeev"* (Voennii' Parad, Moscow, 2007) [In Russian]
6. P. Morgan, *Glass Reinforced Plastics* (Iliffe, London, 1954)
7. A.K. Burov, G.D. Andrievskaya, *Synthetic fibrous anisotropic structures* (AN USSR, Moscow, 1952)
8. T.A. Heppenheimer, *Facing the Heat Barrier: A History of Hypersonics* (NASA History Series, Washington, DC, 2007)
9. S. Reznik, *Proc. 1-st Int. Workshop on Advanced Composite Materials and Technologies for Aerospace Applications* (Wrexham, UK, 2011)
10. A.I. Meleshko, S.P. Polovnikov, *Carbon, carbon fiber, carbon composites* (Science Press, Moscow, 2007)
11. S.L. Bazhenov, A.A. Berlin, A.A. Kul'kov, V.G. Oshmyan, *Polymer composite materials* (Intellect, Dolgoprudnyi, 2010)
12. E.F. Kharchenko, *Extremely high strength reinforced organoplastic* (NTC Informtrchnika, Moscow, 1999)
13. Yu.A. Kurganova, A.G. Kolmakov, *Structural metal-matrix materials* (Bauman MSTU Publ. House, Moscow, 2015)
14. Yu.V. Sokolkin, A.M. Votinov, A.A. Tashkinov, A.M. Postnykh, A.A. Chekalkin, *The technology and design of carbon-carbon composites and constructions* (Science. Fizmatlit, Moscow, 1996)
15. M.Ya. Gofin, *Heat-resistant and thermal protection structures of reusable aerospace vehicles* (ZAO "TF MIR", Moscow, 2003)
16. S.M. Johnson, *Proc. High Temperature Ceram. Matrix Comp. Conf.* (Bayreuth, Germany, 2010)
17. S.P. Walker, K. Daryabeigi, J.A. Samareh, R. Wagner, W. Waters, *Proc. 20th AIAA International Space Planes and Hypersonic Systems and Technologies Conf.* (Glasgow, UK, 2015)
18. O.M. Alifanov, *Basics of identification and design of thermal processes and systems* (Logos, Moscow, 2001)
19. M. Shan, J. Gup, E. Gill, Progress in Aerospace Sciences **80** (2009)
20. J.N. Pelton, *New solutions for the space debris problem* (Springer International Publishing, New York, 2015)

Thermal regimes of space composite structures. Part II

Sergey Reznik[1,*]

[1]Bauman Moscow State Technical University, 105005 Moscow, Russia

Abstract. Composite materials have a unique combination of mechanical and thermal characteristics. This allows you to create lightweight, durable, heat-resistant design. In the article the common features for all space composite structures were highlighted. The article deals with the thermophysical problems arising in the design of space composite structures associated with strict requirements for their weight and functional efficiency. The feasibility of using combined heat transfer models to improve design accuracy has been shown.

1 Space antenna reflectors

1.1 Evolution of onboard space antennas

The first satellites were equipped with non-directional rod antennas, so to receive their weak signals large receiving ground antennas were required. Exploration of the solar system (first Venus and Mars, then Jupiter, Saturn, Uranus, Neptune and Mercury), as well as the development of the high-apogee elliptical orbits and geostationary orbit (GEO) called for devices to concentrate the radio waves – reflectors. As a rule, reflectors are open, gently sloped parabolic shells. Today reflectors can be fixed-shape, adaptive, shape-shifting within a predetermined range and transformable, with the antenna aperture diameter being much greater than that of the launch vehicle fairing (typically over 4 meters). Reflectors operate in L, Ka, Ku, V, and Q bands in 1-70 GHz range for the stationary and mobile "land - board of the satellite – land" radio communication, as well as the inter-satellite communication.

In space reflectors are unevenly heated by the heat radiation from the Sun and the Earth, which causes thermal deformations and shape deviation. To enhance the shape stability, a three-layer layout is most commonly used, where two thin composite shells are joined together by honeycombs of metal foil. Frames on the convex surface are used to increase the structural stiffness and to fix the reflector to the spacecraft. The areal density of the three-layer reflectors is 3.5-5.0 kg/m^2. A different layout, where ribs are positioned on the convex surface, enables stiff structures with a still lower areal density. Modern communication satellites are often equipped with several reflectors, so a lower areal density can noticeably decrease the structure weight. However, the choice of the ribs arrangement, shape, height and thickness requires further research.

* Corresponding author: sreznik@bmstu.ru

1.2 Modern approaches to the design of space antenna reflectors

Design requirements for the advanced space antenna reflector: accuracy of reflector surface (root-mean-square error less than Λ/16, Λ is wavelength and displacement 0.1 mm); strength and stability during spacecraft launch; high dynamic characteristics (high frequencies of structures, not less than 100 Hz); areal mass of the reflector assembly not greater than 2 kg/m^2; more than 15 years' operating cycle in space.

The aim of the work is to select the design layout of ultra-lightweight space antenna reflector from CFRP with high shape and size stability for advanced communication systems.

Why are CFRP best suited for antenna reflectors? Because CFRP have: high strength (unidirectional material); 1.3-2.1 GPa ultimate tensile strength; high stiffness (unidirectional material): 280-400 GPa tensile modulus; low temperature coefficient of linear expansion of $\pm 0.5 \cdot 10^{-6}$ K^{-1}; thermal conductivity, within $\lambda_{\perp}$=0,4-0,8 W/(m·K) range;λ_{II}=8-70 W/(m·K) (300 W/(m·K) for pitch-based carbon fibers); relatively small density (1500-1650 kg/m^3); moderate cost ($1000 per kg); variety of manufacturing methods (infusion, hand layup, winding, etc.).

Many modern communication satellites are located in the geostationary orbit. At the same time, the spacecraft is constantly above one very large section of the earth surface, since it moves with the same angular velocity as Earth. In the geostationary orbit, spacecraft and its antennas are constantly illuminated by the direct solar heat flux for most part of the year. Usually, calculations are based on the average value of this flux, the so-called solar constant of 1368 W/m^2. Twice a year, during the 45 days of the spring and autumn equinox, the spacecraft is in the Earth's shadow. The maximum time spent in the shadow is about 71 minutes, where the spacecraft is exposed to the Earth own radiation, about 5.5 W/m^2. The experience of estimating the thermal regimes in GEO shows that even such weak thermal effects should not be neglected, otherwise the spacecraft temperature values are calculated erroneously low, especially in the case of thin-walled structures prone to rapid cooling due to low heat capacity. Because of these operating conditions, the thermal regimes for space antenna reflectors are estimated both the illuminated and for the shadow areas of the orbit.

The design process includes the following steps: selecting one design layout; defining the thermal loads on the reflector in GEO; creating the geometric and finite element model of the reflector; FE modelling of the thermal state using "Siemens space system analysis"; selecting the method of joining the reflector to the spacecraft; FE modeling of the stress-strain state using "Siemens NX Nastran"; determining the maximum thermal deformation and displacement of the reflecting surface; analyzing the results and comparing them with other design layout options.

1.3 A hard-to-achieve combination of lightweight and stiffness

The authors and their colleagues have presented the design studies of reflector structures made of carbon fiber with various ribs arrangements on the convex surface of the parabolic shell [1-4]. This paper, by way of example, presents the mathematical modeling results of thermal regimes for different design variants of a reflector: isogrid, five-pointed star and six-pointed star ribbed reinforcement arrangement (Table 1, 2, Fig. 1). In all cases, the reflectors were constructed of CFPR and had six layers made of the flattened plain weave ASPRO fabric and epoxy binder ED-20 with a thickness of 0.4-1.2 mm. The plies were positioned at ± 90° / ± 90° / ± 90°.

Table 1. Variants of ribbed reflectors: five-point star and six-point star arrangement.

Name of variant	Ribs arrangement	Height of ribs, mm	Thickness of ribs, mm	Thickness of shell, mm
Star6_25_12_06	Six pointed star	25	1.2	0.6
Star6_60_06_06	Six pointed star	60	0.6	0.6
Star6_90_06_04	Six pointed star	90	0.6	0.4
Star5_90_06_04	Five pointed star	90	0.6	0.4
Star5_53_06_06	Five pointed star	53	0.6	0.6

Computer modelling demonstrated that the six-pointed star ribbed arrangement possessed superior weight characteristics and high dimension stability (Table 2).

Table 2. Comparison of ribbed design layouts.

Ribs arrangement	Shell and ribs thickness, mm	Mass of the reflector, kg	Maximum displacement of reflecting surface, mm	Linear density, kg/m^2
Six-pointed star	0.6	1.523	0.14	1.354
Five-pointed star	0.6	1.502	0.12	1.328
Izogrid with permanent height of ribs	0.6	1.549	0.185	1.369
Izogrid with variable height of ribs	0.6	1.479	0.18	1.308

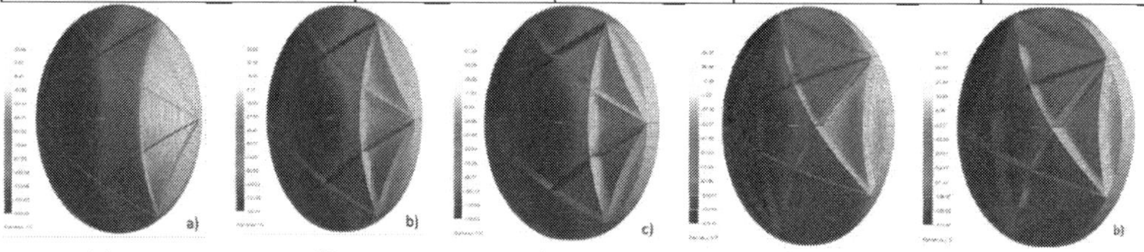

(a) (b) (c) (d) (e)

Fig. 1. Temperature state (degrees Celsius) for: a – "Star6_25_12_06", b – "Star6_60_06_06", c – "Star6_90_06_04", d – "Star5_53_06_06", e – "Star5_90_06_04" with 270° angle between reflector rotation axis and the Sun – Earth line.

2 Reusable space vehicles

2.1 Thermal barrier on the way to reusable spacecraft and rockets

Space vehicles were disposable only for the first 15 years after the First Sputnik had been launched. First-generation launch vehicles were modernized guided ballistic missiles designed to fly in one direction with small coefficients of safety. In 1972, the United States began creating the Space Shuttles, and the USSR in 1976 began working on a similar "Buran" – "Energy" system. Both types of reusable space vehicles (RSV) were designed for 100 flights to the low Earth orbit. The problem of thermal protection (TP) was one of the most difficult because with the RSV surface area of about 1000 m^2, the use of traditional polymer CM was unacceptable for weight reasons. The way out was to utilize combinations of heat-shielding C-C elements, ceramic tiles, flexible coatings from heat-resistant fabrics [5]. After 30 years of the Space Shuttle operation, some of the most significant drawbacks were the extremely high labor costs associated with preflight preparation of the tiled TP, reaching about 40,000 man-hours.

New generation of RSV required elements that could be replaced quickly and had larger dimensions than tens of thousands of 150x150 mm and 200x200 mm tiles in the first shuttles [6, 7]. At the same time, more lightweight TP-materials were investigated consisting of sintered heat-resistant materials or lightweight wool-like (felt) alumina materials inside metal or composite containers or combinations with silica glass fibers with less than 100 kg/m^3 density. These materials comprise randomly oriented thin fibers (about 2 μm in diameter) with a boron oxide binder. Porosity of the material exceeds 94%. The heat transfer process in such materials is combined. The front surface of the TP element has several functions. It prevents erosion of fibers by high-speed air flows and precipitation, has low catalytic activity and high emissivity, which reduces the heat flux to the fibrous insulation. Nevertheless, as a result of aerodynamic heating, this surface temperature reaches more than 1500 K with the excess heat is radiated into the outer space and into the interior of the porous material, where it is absorbed and scattered. Such a complex picture of heat transfer in a highly porous material is crucial for determining its effective coefficient of thermal conductivity, which is used in designing of the TP.

2.2 Heat transfer in high-porosity materials for thermal protective system

The TP of RSV requires technically feasible and inexpensive materials with low density, low heat conductivity, high thermal stability. The create TP with optimum weight-geometric and economic characteristics it is of high significance to determine the main performance characteristics of porous materials under operating conditions. However, these data for the vast majority of materials, especially the coefficient of thermal conductivity, only exist in a limited temperature range or do not exist at all. At the same time, with the current laboratory equipment we can measure only the effective coefficient of thermal conductivity of semi-transparent materials, characterizing the total heat transfer, whose values depend strongly on the measuring conditions.

In terms of thermal design, it is extremely challenging to rationally choose the mathematical model of heat exchange in a layer of semi-transparent scattering material, such as high porosity ceramics. The problem is in the close interconnection of the temperature and radiation fields in these materials which means solving essentially nonlinear problems of radiative-conductive heat transfer, that include the heat conduction and radiation transfer equations. Solving the transfer equation directly, for example using the discrete ordinates method, will lead to excessive computational burden, which unsuitable at the thermal design stage. The previous research [8] showed that the necessary balance of performance and accuracy is achieved when using the diffusion approximation for the radiation transfer equation. In the future, when computer technology is better developed (multi-core and multiprocessor systems, GPGPU), it may be reasonable to use a more accurate method for solving the radiation transfer equation: the half moment method.

The aim of the work is to investigate the combined heat transfer regularities in porous TP materials for RSV at wide temperatures interval (300-2000 K) and at high rates of heating (up to 100 K/s).

2.3 Methodology of theoretical and experimental investigation

The main approaches to this problem are: mathematical simulation and design of thermal test conditions; thermal tests of new porous materials samples and measurements of temperature distribution inside the samples; heating methods – contact, radiative or convective; solution of a 1-d and 2-d inverse problems of effective thermal conductivity (IP ETC) and radiative and conductive heat transfer (IP RCHT).

Modern methods for solving IP RCHT are based on the comparison of calculated and experimentally measured temperature values in a sample of semi-transparent material. Obviously, IP ETCs are a special case of IP RCHT. Probably, the pioneering work in this direction is [8], where the problem of determining the temperature dependence of the thermal conductivity in a semi-transparent scattering material was solved for the first time using experimental data corresponding to nonstationary heat transfer in the Buran TPC. Later this approach developed into more complex multidimensional problems [9, 10] and methods for determining the coefficient of semi-transparent materials thermal conductivity adapted for processing the investigation results using laboratory facilities [11-13].

Our theoretical and experimental investigations have proved that for highly porous semi-transparent materials, the effective thermal conductivity coefficient depends most significantly on the rate and period of heating. It has been established that for the optical layer thickness exceeding 100, the effective thermal conductivity coefficient at 3 K/s surface heating rate (curve 3 in Fig. 2) practically coincides with the value of the effective thermal conductivity coefficient obtained in the Rosseland approximation (curve 2 in Fig. 1), and at the surface heating rate exceeding 300 K/s it is close to the molecular coefficient of thermal conductivity (curves 1 and 5 in Fig. 2). At the heating rate of more than 3 K/ and less than 300 K/s, the effective coefficient of thermal conductivity lies between the values of the effective thermal conductivity coefficient in the Rosseland approximation and the molecular coefficient of thermal conductivity, tending to the latter as the heating rate increases (curves 4 and 5 in Fig. 2) [14].

The study of the temperature dependencies of the effective thermal conductivity coefficient of heat-insulating materials with varying densities [15] has shown that at the temperature of up to 600 K, when the heat transfer is mainly conducted along a solid framework and a gaseous medium, the material with the lowest density has the lowest effective heat conductivity (curve 1 in Fig. 3). However, in the region of high temperatures, when the share of radiative heat transfer to the total heat transfer increases, the effective thermal conductivity is lower for denser materials (curve 3 in Fig. 3). This is explained by the fact that as the density of the material increases, the concentration of fibers per unit volume increases, which causes an increase in the scattering coefficient and, as a consequence, a decrease in the radiative energy transfer. This result opens up possibilities for optimizing the constructions of TP multilayered elements using heat-insulating materials of variable density depending on the specific conditions of thermal loading (Fig. 3).

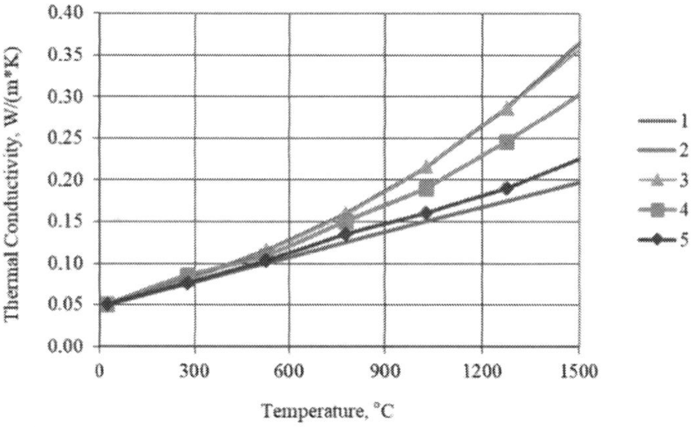

Fig. 2. Temperature dependences of the thermal conductivity of a semi-transparent scattering material: 1 – conductive coefficient of thermal conductivity; 2 – effective coefficient of thermal conductivity (Rosseland); 3 – effective coefficient of thermal conductivity at 3 K/s rate of heating of the frontal surface; 4 – the same for 30 K/s; 5 – the same for 300 K/s.

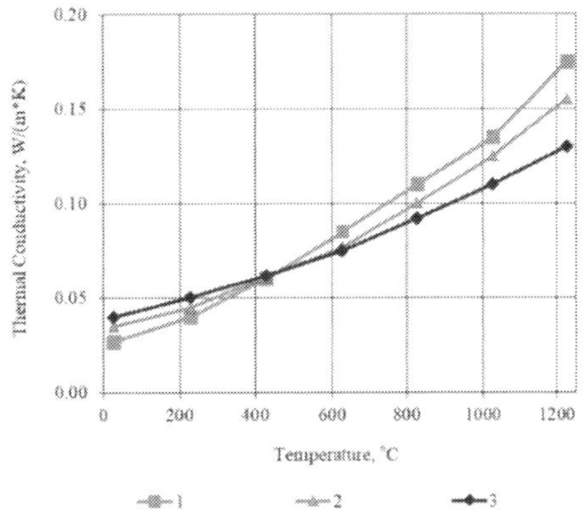

Fig. 3. Temperature dependencies of the ETC coefficient of silica fibrous materials: 1 – with 90 kg/m^3 density; 2 – 105 kg/m^3 density; 3 – 120 kg/m^3 density.

Conclusion

The increasingly complex and ambitious challenges of space exploration stimulate composite structures development. Experience shows that the design layouts of any spacecraft must be selected in view of the expected thermal regimes. Even though there is an ample variety of space composite structures, there exist universal approaches to thermal design based on direct and inverse problems of combined heat transfer.

The author would like to acknowledge A.A. Alekseev, A.V. Azarov, W.P.P. Fisher, M.L. German, P.S. Grinchuk, S.A. Gusev, O.V. Denisov, A.D. Novikov, K.V. Mikhailovsky, A.M. Mikhalev, L.Ya. Paderin, N.V. Pavlyukevich, I.R. Shafikova and M.S. Tret'yak for their scientific cooperation.
Separate parts of this work are fulfilled with financial support on contract from Russian Ministry of Education and Science on project on Agreement 14.574.21.0146 from 26 September 2017 (Unique identifier of the project RFMEFI57417X0146).

References

1. S.V. Reznik, P.V. Prosuntsov, A.V. Azarov, J Eng Phys Thermophys. **88**, 3 (2015)
2. S.V. Reznik, P.V. Prosuntsov, A.V. Azarov, J Eng Phys Thermophys. **88**, 4 (2015)
3. P.V. Prosuntsov, S.V. Reznik, K.V. Mikhailovsky, A.D. Novikov, Zaw Ye Aung, IOP Conf. Series: Materials Science and Engineering **153**, (1), 012012 (2016)
4. S.V. Reznik, A.D. Novikov, MATEC Web Conf. **92**, 01012 (2017)
5. M.Ya. Gofin, *Heat-resistant and thermal protection structures of reusable aerospace vehicles* (ZAO "TF MIR", Moscow, 2003)
6. M.L. Blosser, J Spacecraft Rockets **41**, 2 (2004)
7. J.T. Dorsey, C.C. Poteet, K.E. Wurster, R.R. Chen, J Spacecraft Rockets **41**, 2 (2004)
8. P.V. Prosuntsov, S.V. Reznik, JOEP **49**, 6 (1985)
9. P.V. Prosuntsov, Heat Transfer Research 36, 6 (2005)
10. P.V. Prosuntsov, I.A. Mayorova, A.V. Zuev, Thermal Processes Technology 7 (2014)
11. P.V. Prosuntsov, S.V. Reznik, Siberian Phys.-Tech. J. **1** (1991)
12. P.V. Prosuntsov, S.V. Reznik, V.E. Tyutyunnik, Siberian Phys.-Tech. J. **4** (1992)
13. P.V. Prosuntsov, S.V. Reznik, Siberian Phys.-Tech. J. **4** (1992)
14. S.V. Reznik, P.V. Prosuntsov, A.V. Zuev, *Proc. 2-nd Int. Symp. on Inverse Problems, Design and Optimization* (Miami, FL, USA, 2017)
15. S.V. Reznik, *Abstracts of the Reports and Communications of 5-th Int. Forum on Heat and Mass Transfer* (Minsk, Belarus, 2004)

An experimental study of the effect of water bodies temperature on water heat pump performance

Amer Saloum[1,*], and *Vyacheslav* I. Maksimov[1]

[1]National Research Tomsk Polytechnic University, 634050 Tomsk, Russia

Abstract. The purpose of this research is to study the effect of water temperature change in the reservoir on water heat pump performance. An experimental analyse is done on the heat pump station which depends on water as a heat source. The experimental results confirm that the ice formation has a negative effect on the heat exchange efficiency, and the increase in water source temperature slow down the ice formation rate which causes an appositive effect on the system performance.

1 Introduction

Traditional heating systems are the basic source of pollution, that's why the use of heat pumps grows nowadays. The heat pumps depend on the water-air-and ground as an energy source. the heat pump which depends on water as the energy source (rivers-lakes-sea water) is a promising technology because even when the air temperature drops to -5 °C in many areas in the world, the water surfaces temperatures keep in the range (7-13) °C. There is a lack of information about water source heat pump systems.

The thermodynamic analysis shows that the temperature of the heat source has the highest effect on the heat pump performance, the heating time for water in condenser and the ice formation characteristics. The velocity of ice formation ratio on the evaporator tube has a high effect on the system performance and heat exchange rate between the Freon and water in the evaporator.

2 Experimental setup and study technique

An experimental study is done to analyze the influence of water surface temperature on heat pump performance. The experimental method is explained in [4], But in these experiments 15 thermocouples are fixed in the water reservoir with step 1.53cm by height between every two thermocouples to measure the change in water temperature around the evaporator and condenser, and 15 thermocouples were fixed along the evaporator and condenser tube as shown in Table 1 and Fig. 1.

* Corresponding author: amer-salom@hotmail.com

Table 1. Thermocouples location along the heat exchanger pipes.

*	1	2	3	4	5	6	7	8	9	10	11	12	13	14	15
x, m	0	0.1	-0.075	0.03	0.1	-0.075	0.03	0.11	-0.075	0.055	0.085	-0.075	0.08	0.055	0
y	0	0.095	0.125	0	0.145	0.1	0	0.145	0.1	0	0.145	0.075	0	0.145	0.075
z	0.02	0.045	0.045	0.09	0.09	0.09	0.135	0.135	0.135	0.18	0.18	0.18	0.225	0.225	0.02

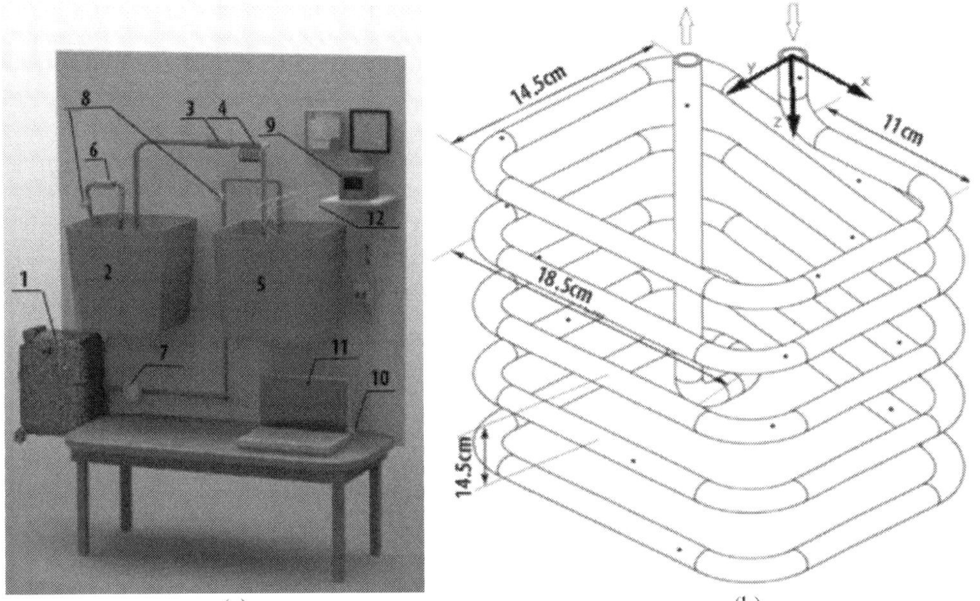

(a) (b)

Fig. 1. (a) Schematic diagram of the experimental setup: (1) compressor, (2) condenser, (3) desiccant, (4) capillary tube, (5) the evaporator, (6) flowmeter, (7) Manometer, (8) valve, (9) Analog-to-digital converter, (10) the network adapter, (11) the computer, (12) thermocouples; (b) Tubes of the heat exchanger(evaporator-condenser). The points represent the thermocouple locations.

3 Experimental results

The initial drop in Freon temperature, is the same for both cases when start water reservoir temperature 7 °C (-15 °C) and the case when it is 13 °C, that means that the drop-in water temperature in the reservoir does not affect the Freon temperature in the tubes. The initial increase in the system efficiency Fig. 2 is explained by the sharp decrease in freon temperature in evaporator because according to thermodynamics the drop in Freon temperature in the evaporator leads to increase in system efficiency.

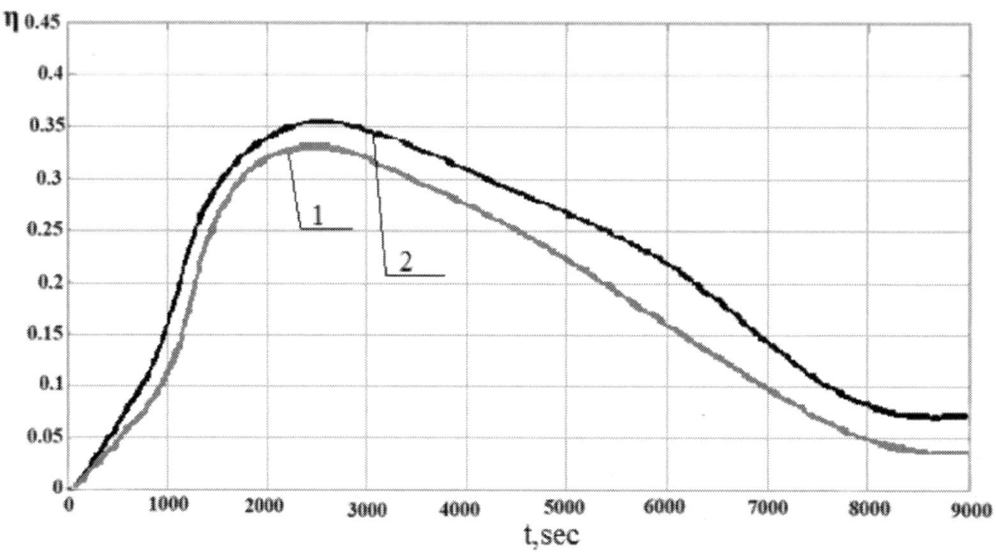

Fig. 2. Changes in the coefficient of cooling with time. 1 – For the case where the initial temperature is 280K. 2 – For the case when the initial temperature is +13 °C.

The water temperature does not fall down the value +4 °C for the studied time, when the initial water temperature is 13 °C, but this temperature drops to the value +2 °C in case, the initial water temperature is 7 °C and the water behavior change like it was explained in [4], but how to answer the question if the drop in Freon temperature is the same why the system efficiency is less , this is caused by ice formation rate, because the ice reduces the efficiency of heat exchange, and the ice layer on the evaporator tube is thicker for the first case as can be seen in Fig. 3 and Fig. 4 (the maximum ice thickness for the first case is more than 10 mm where it is less than 7 mm for the second case.

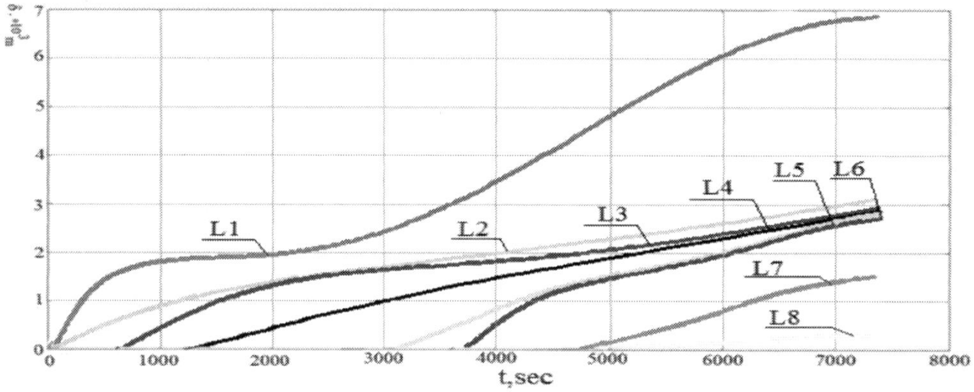

Fig. 3. Ice thickness change with time in different locations along the evaporator tube (in the case where the water initial temperature is +7 °C K in the reservoir around the evaporator). (L1..L8) the distances for the start point of the evaporator which are shown in Table.2.

Table 2. The location of measuring points for ice thickness along the evaporator pipe.

*		L1	L2	L3	L4	L5	L6	L7	L8
m		0	0.115	0.26	0.445	0.59	0.775	0.92	0.1105
Cartesian coordinates	x	0	0.115	0.115	-0.07	-0.07	0.115	0.115	-0.07
	y	0	0	0.145	0.145	0	0	0.145	0.145
	z	0	0.045	0.045	0.045	0.045	0.09	0.09	0.09

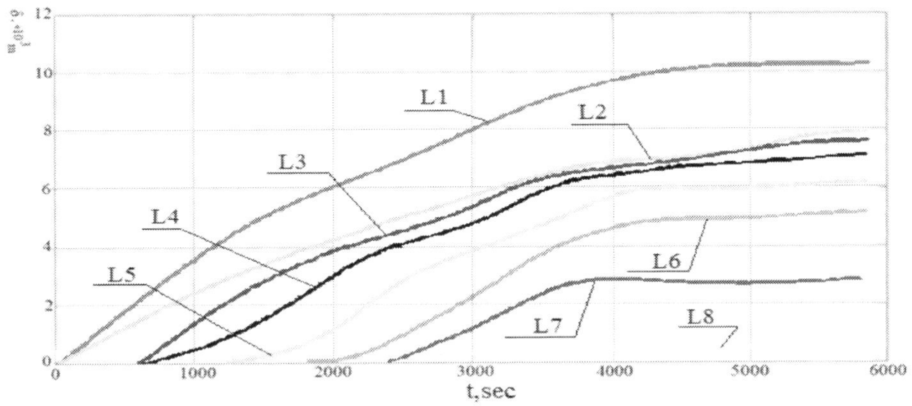

Fig. 4. Ice thickness change with time at different locations along the evaporator tube (in the case where the water initial temperature is +13 °C in the reservoir around the evaporator). (L1...L8) the distances for the start point of the evaporator which are shown in Table 2.

4 Conclusion

The experimental results show that the Freon temperature in the heat pump system, does not change when the temperature of the water in reservoir is less, but the ice growth rate on the evaporator pipe is higher, and thicker layer of ice in the range 5 % for every1 °C, the existence of ice layer on the evaporator tube reduced the heat exchange efficiency and the system efficiency to collect energy from the water is reduced to 1.4 % for every1 °C.

It must be pointed out that the experiment is done with the use of the same capillary tube, which affect the drop in the Freon temperature and pressure, in the next research the effect of the enthalpy on the entry of the capillary tube on the system performance must be studied to choose the best dimensions for this tube which are compatible with the system working conditions.

The results of this research can be used to assess the water heat pump work under the conditions of ice formation, in different conditions of water source temperatures, and how that affects the system efficiency.

References

1. S. Zou, X. Xie, Appl. Therm. Eng. **7**, 200 (2016)
2. A.C. Morton, *Assessing the performance of a reservoir-based water source heat pump* (Strathclyde university, Glasgow, 2013)
3. Y. Wang, *Analysis on the practical use of open-loop surface water source heat pump systems (SWHP)*(Chongqing University, Oklahoma, 2011)
4. V.I. Maksimov, A. Saloum, MATEC Web Conf. **110**, 01074 (2017)
5. L. Schibuola, Massimiliano Scarpa Energy and Buildings, **6**, 182 (2015)
6. N.V. Demyanovich, V. I. Maksimov, T.A. Nagornova, MATEC Web Conf. **92**, 01067 (2017)
7. G.P.Vasilyeva, N.V. Peskova, V.F. Gornova, M.V. Kolesova, Geothermics **10**, 93 (2016)
8. T.A. Kawalecki, A. Cybulski, Polish Academy of Science **10** (1995)
9. V.I. Maksimov, T.A. Nagornova, V.S. Chernyshev, MATEC Web Conf. **23,** 01051 (2015)

Developing English language competence for specialists in rocket and space composite structures

Inna Shafikova[1,*], *Galina* Gurova[1], and *Andrey* Novikov[1]

[1]Bauman Moscow State Technical University, 105005 Moscow, Russia

Abstract. Composite materials is a rapidly developing field of science and technology. Training specialists in composite materials requires special approach with regard to their English language competence. The paper looks at the proposed English language course for Rocket and space structures. The course in question must comply with the requirements of the Bologna process, curriculum and education standards and at the same time meet the learners' needs in terms of their professional competence. A multi-level system is proposed that includes academics apart from undergraduate and postgraduate students.

Composite materials continue to find new applications in rocket and space engineering. Valuable combination of properties make composites a material of choice when it comes to demanding applications in extreme conditions. Innovative, energy-efficient manufacturing technologies make composites an economically feasible solution. Current education trends in Russia and abroad manifest increased interest in training specialists to work in the fields of composite materials design, manufacture and testing.

Existing teaching methods and courses must be reevaluated within regard to the contemporary requirements. The most important factor is that since 2003, after signing the convention on the mutual recognition of diplomas and certificates, Russia is in the process of integration into the global education system, the so-called Bologna process. The Bologna doctrine prescribes system of credit units for the learning outcomes assessment, adequate accreditation procedures, two-tier architecture of academic degrees, transparency of information flows for all education modules, and introduction of the new diploma supplements. All of the above must lead to academic mobility, as well as compatibility and comparability of diplomas and qualifications, and, ultimately, through integration, increase competitiveness of the European higher education system. The aim of this initiative is to unify the European education system as a response to globalization, to ensure its quality and uniformity worldwide.

In spite of the numerous organizational challenges associated with the transitional period, Russia can expect significant benefits for its education system, once the Bologna doctrine is fully implemented. For one, Russian universities will be a part of the international education market. Additionally, recognition of Russian academic degrees globally will stimulate

* Corresponding author: shafikova_inna@mail.ru

academic exchange, thus fostering scientific progress. However, the achievements of the past must be preserved, while the Russian academic standards are brought in line with the European requirements [1].

The English language teaching is one of the spheres most seriously affected by the Bologna process. It is evident that higher proficiency in one or several European languages is a prerequisite to academic mobility. Consequently, the approach to the English language teaching courses needs to be modified. The main trends are:

- courses in English for Specific Purposes (ESP) tailored for the specific professional activity;
- life-long learning, i.e. designing a range of courses to enable continuous improvement of the professional communicative competence;
- diversification of the second language learning system, so that the needs of students with different aims and different abilities are provided for;
- equal access to the foreign language learning for the students from various backgrounds;
- global unification of the second language learning and increased academic mobility;
- information technology skills enabling autonomous learning;
- learner-centered approach; i.e. accounting for the learners' various needs and planning the course accordingly [2-4].

Another important factor is the insufficient English-language competence of the university academics. Some sources suggest that this aspect is hindering the internalization of the Russian higher education into the global education market [5, 6].

New generation of the educational standards and curricula for the Bachelor's and Master's degree programs in rocket complexes and space studies (24.03.01; 24.04.01 majors) contains requirements with regard to the English (or another second) language competence. The graduates are expected to be capable of conducting literature research, preparing reports and presentations for scientific conferences and communicating in the target language in the academic and professional contexts.

All of the above considerations were taken into account when designing an English language curriculum for the students and teachers of Rocket and Space composite structures department at Bauman MSTU (Fig. 1). The choice of the course content is governed by the demands of the industry specified in the professional standards and by the learners' needs, which are diagnosed via needs analysis.

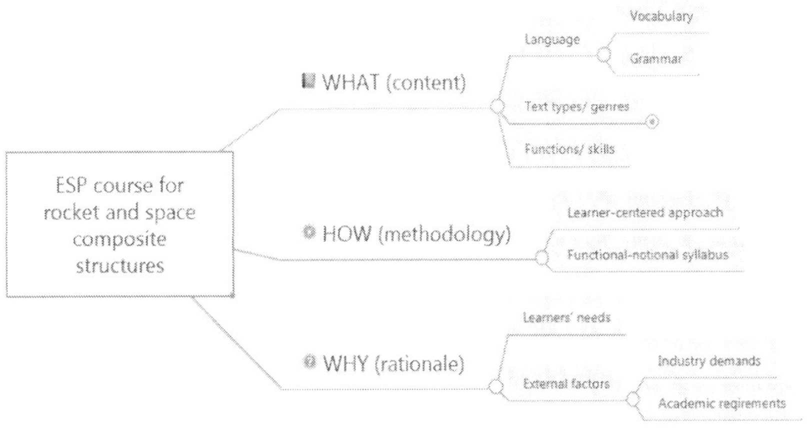

Fig. 1. ESP course for rocket and space composite structures' content, methodology and rationale.

The syllabus of an ESP course must be consistent with the specialized courses taught at the department under consideration. This principle ensures that the students are aware of the importance of the English language for their future profession. It also helps to supplement and support the specialized content. Below (Table 1) is a proposed syllabus of English for rocket and space composite structure course. It takes into account the content of the specialized courses taught to the undergraduate students [7]. The same principle should equally be applied to a Master's degree course, since some Master's degree students hold a Bachelor's degree in a different field of engineering and thus would especially benefit from an ESP course tailored for the department curriculum.

Table 1. Proposed syllabus for English for rocket and space composite structure course.

Course	Function	Vocabulary	Grammar
Technology of processing and modifying new materials	Explaining needs, problems and solutions; stating objectives;	Composites manufacturing processes verbs	Modals, Infinitives of purposes
Fundamentals of physics, chemistry, and fabrication of composites	Describing a process	Physics, chemistry terms	Conditionals
CAD basics	Explaining dimensions; working with drawings	Specifications of dimensions	
History of research in composites	Sequence of events		Past tenses
Composite structures mechanics	Explaining forces	Physical forces	
Principles of scientific research	Describing methods, giving definitions		Relative clauses
Polymer composite structures development	Describing use or function; describing shape and appearance	Composite materials, properties. Shapes of composite parts	Noun attributes
Technology of reusable space vehicles	Describing parallel processes	Stages in a composites fabrication process	Passive voice
Composite media mechanics	Describing motion, forces	Verbs to denote movement	
Optimization of composite structures and procedures	Expressing purpose; expressing strengths/ weaknesses	Processes sequence/ simultaneity; Types of composites fabrication equipment	Purpose clause

A multi-level system was proposed for the Rocket and space composite structures department at Bauman MSTU. The novelty of such system is that apart from undergraduate and postgraduate students it also includes academics. The assumption was that the academics will need to increase their proficiency in English for their own purposes, such as dealing with international information sources, participating in international conferences, conducting business and academic correspondence, and interacting with students from abroad [8,9]. At the same time, academics who are proficient in English will stimulate their students' English language-related activity. Another stimulating factor was the participation of teachers and postgraduate students in scientific conferences in the UK, organized jointly by the Bauman MSTU and the University of Glyndwr in North Wales. As a result, the publication activity increased noticeably. The number of foreign students is constantly growing at the department, with academic communication conducted not only in Russian, but also in English. All this testifies to the effectiveness of the developed methods.

References

1. N.Yu. Pimenova, Universitetskoye upravleniye: praktika i analiz **1**, 15 (2006)
2. T.D. Margaryan, I.H..Margaryan, Eur. Social. Sc. J. **3**, 256 (2017)
3. N.N. Nikolaeva, Humanities Bulletin of BMSTU **7** (2017)
4. T.Y. Polyakova, Vestnik Moscow State Ling. Univ. **12** (618), 9 (2011) [in Russian]
5. E. Frumina, R. West, *Internationalisation of Russian Higher Education: The English Language Dimension* (British Council, Moscow, 2012)
6. K.M. Inozemtseva, Higher ed. Russia, **5**, 145 (2014)
7. I. R. Shafikova Advanced Composite Materials and Technologies for Aerospace Applications, 83 (2013)
8. M.V. Zosimov, V.A. Moiseyev, I.R. Shafikova, S.A. Shorokhov, Vestnik Kostromskogo gos. univ. **23** (3), 127 (2017)
9. I. R. Shafikova, Vestnik Tomsk. gos. ped. univ. **10**, 160 (2017)

Interdroplet distance in a structured monolayer of liquid microdroplets levitating over hot liquid surface

Almira Shatekova[1, 2,*]

[1]Kutateladze Institute Institute of Thermophysics, 630090 Novosibirsk, Russia
[2]Novosibirsk State University, 630090 Novosibirsk, Russia

Abstract. Structured arrays of liquid microdroplets levitating over a heated liquid film have been observed in experimental works, but the nature of this phenomena is not completely understood. The aim of this work is to measure the distance between drops and the size of monolayer depending on the main parameters. We found that the average distance between drops increases with time from 40 to 100 μm. And the size of monolayer increases with time from approximately 1 mm to 5 mm. The distance between drops increases when approaching the boundary of monolayer.

1 Introduction

It is known that an ordered "monolayer" structure can be formed over a heated liquid surface [1, 2]. A monolayer is a highly ordered triangular structure consisting of hundreds and thousands microdroplets (Fig.1).

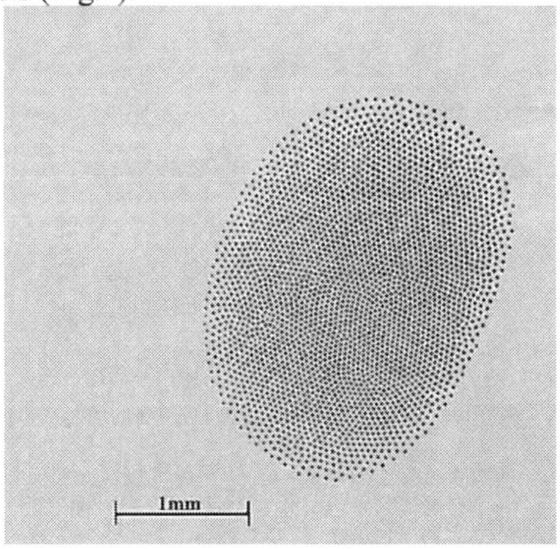

Fig. 1. The monolayer (top view). The thickness of liquid film h= 0.5 mm and temperature of substrate T = 83 °C.

* Corresponding author: shatekova@list.ru

The microdroplets are levitating above the liquid surface at a distance comparable to the droplet diameter. The typical size of the microdroplets in a monolayer varies from 10-100 μm. The study of monolayer can be used for practical applications, for example, spray cooling, when a steam of droplets is directed to the heated surface, it is important to know under what conditions the microdroplets levitate [3]. It is also important to know conditions of levitating for delivering drugs via sprays [4]. Moreover, microdroplets can be used as tracers. On trajectories, it is possible to restore the microscale gas-vapor flows near the liquid-gas contact line [5, 6]. Initially the levitating microdroplets are drops of condensate arising from the evaporation of the heated liquid. After drops gain enough mass and then fall down on the different parts of the liquid surface (supported by the steam originated at the surface of liquid, as discussed in [2]) and collected in an ordered structure. Experimentally, it is known that the structured arrays of microdroplets can be observed over tea, coffee, tap water, distilled water and other kinds of liquid [7, 8]. Coalescence of a monolayer is a chain reaction: one drop touches the water layer, creating a capillary wave on the surface, and that destroys adjacent drops [9]. The process takes a split second and visually looks like an instant disappearance of the monolayer. However, using a high-speed video camera (at least 30000 frames per second), it is possible to visualize the details of this process. In the papers [5, 6, 10], the behavior of levitating liquid microdroplets near the contact line was studied. It was found that when approaching the contact line, the droplet levitation height increases several times, which possibly indicates the existence of intense evaporation near the contact line. In addition, over the dry surface near the contact line exist a "forbidden zone", where drops are not observed. The width of the forbidden zone varies from 25 to 75 μm. Moreover, three scenarios of the migration of liquid microdroplets over the contact line to the dry area were revealed: 1) droplets fly through the contact line with a change in the levitation altitude, 2) the drops travel along a circular path near the contact line with one or more complete rotations and fall onto a dry surface, 3) some drops levitate at a fixed location. Levitation of drops over a heated dry surface was observed in many experimental studies on the Leidenfrost effect [13, 14], but self-organization of microdroplets into structures, as described in [2, 11, 12], was not researched before. Leidenfrost's levitation model to the case of [2, 11, 12], is not applicable because the Leidenfrost temperature (a 200 °C) is higher than substrate temperature in [2, 11, 12] (a 55 °C). In addition, a theoretical model that satisfactorily describes the dependence of the levitation height on the droplet size was offered.

The aim of this work is to determine the size of monolayer and the distance between drops depending on the main parameters.

2 Experimental setup

The test section is a plate of stainless steel base in the centre of which a copper rod with size 10x10 mm^2 is pressed, which serves as a heating element. Thermocouples were installed to measure the temperature of the copper rod surface. The temperature of heater surface is 50-100 °C. The entire experimental setup was placed in a box that minimized the external airflows. The test section is oriented horizontally and open to the atmosphere. The ambient temperature is 20-25 ° C. As a working fluid ultra-pure, nanofiltered water Milli-Q is used. The initial temperature of working liquid is 25±1 °C. The specified volume of the working fluid is applied by means of a syringe to the working surface with the formation of a liquid film. The thickness of the film is calculated by the volume of the liquid to be filled. The film thickness varied from 0.3 to 0.9 mm. The optical system consisting of lenses, semi-transparent mirror and digital camera was used. The optical resolurion – 10 μm/pix. The scheme of the experimental setup is shown in Fig. 2.

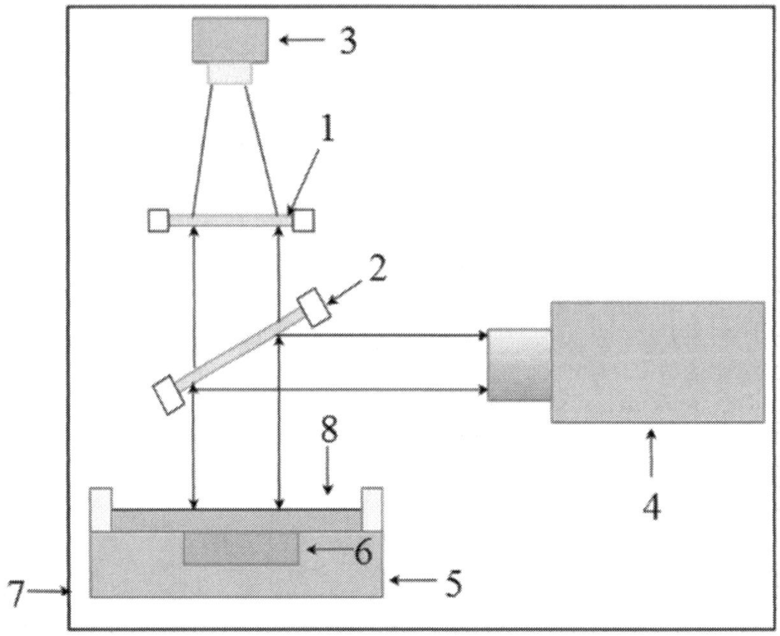

Fig. 2. The scheme of the experimental setup: 1-lens, 2- semi-transparent mirror, 3- light source, 4-camera, 5-stainless steel plate, 6-heater, 7-box.

3 Experimental results

In most cases, the monolayer had an elliptical shape (Fig. 1). Fig. 3. shows the dependence of average diameter on time.

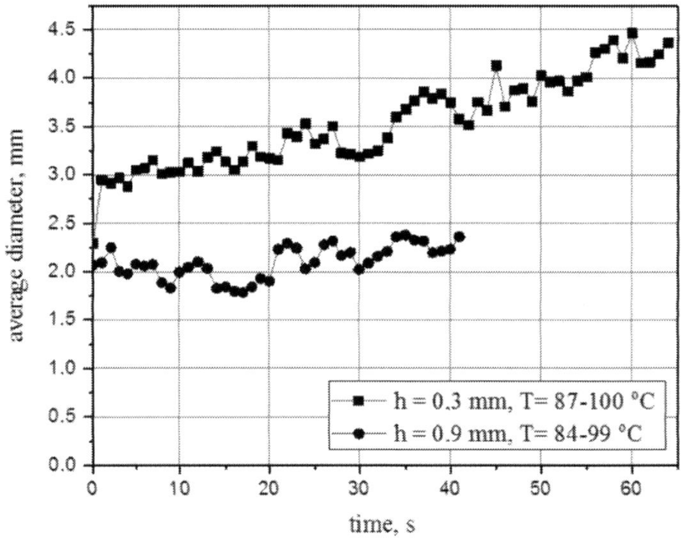

Fig. 3. The average diameter vs. time.

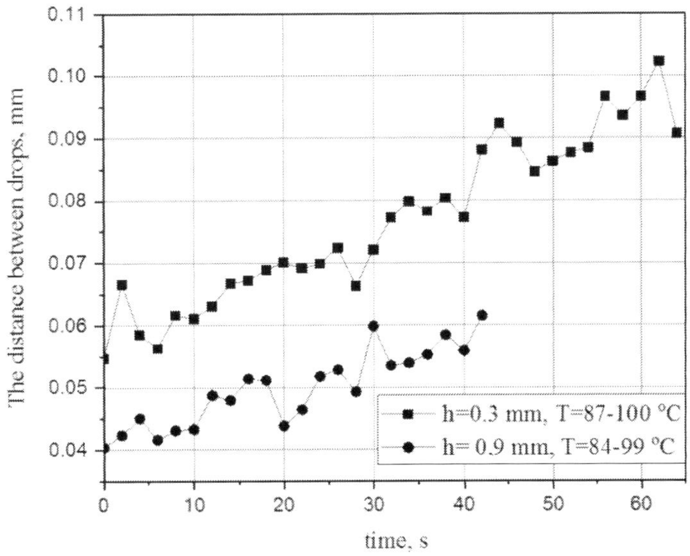

Fig. 4. The distance between drops vs. time.

The average diameter is defined as sum of semi-minor axis and semi-major axis. The dependence was obtained for two different monolayers with varying thickness of liquid layer (h = 0.3 or 0.9 mm) and temperature (T = 84-100 °C). As can be seen from Fig.3 the average diameter increases with decreasing thickness of the liquid layer. The average diameter increases with time from 1 to 5 mm. Fig. 4 shows the dependence of the distance between drops on time. The interdroplet distance increases with decreasing the thickness of the liquid film.

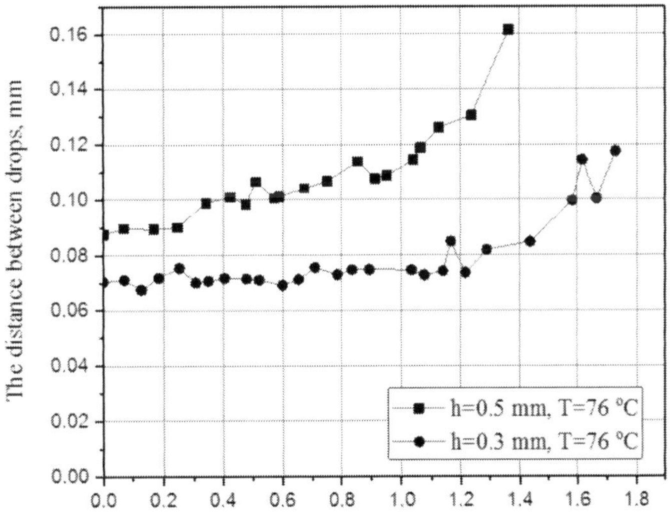

Fig. 5. The distance between drops vs. The distance from the center of monolayer.

The distance between drops varies from 0.04 to 0.1 mm. The lines in Figure 4 correspond to the same two monolayers as in Figure 3. Figure 5 shows a plot of the distance between the drops depending on the distance of the center of monolayer. The dependence is presented for two different monolayers at different thicknesses of the liquid layer (h = 0.3 or h = 0.5 mm) but at the same temperatures (T = 76 °C). The peculiarity of this graph is that the processing was carried out using one image only, i.e. all parameters are the same. As can be seen from the Fig. 5. the distance between the drops increases when approach the boundary of monolayer.

4 Summary

The monolayer parameters such as the size of monolayer, the distance between the drops depends on many parameters such as temperature, the thickness of liquid layer. The average diameter and interdroplet distance increases with decreasing the thickness of the liquid layer. The average diameter of monolayer increase from 1 to 5 mm. Moreover, the distance between drops varies from 40 to 100 μm. The interdroplet distance increase when approaching the periphery of the monolayer.

The research was supported by the Russian Science Foundation, Project No. 1419-01755.

References

1. A. Fedorets JETP Lett. **79**, 372 (2004)
2. D. V. Zaitsev, D. P. Kirichenko, V. S. Ajaev, O. A. Kabov, Phys. Rev. Lett. **19**, 094503 (2017)
3. J. Kim, Int. J. Heat Fluid Flow **28**, 753 (2007)
4. Z. Zhang, C. Kleinstreuer, J. Donohue, C. Kim, Aerosol Sci. **36**, 211 (2005)
5. O. A. Kabov, D. V. Zaitsev, D. P. Kirichenko, V. S. Ajaev, Nanosc. Microsc. Thermophys. Eng. **21**, 60 (2017)
6. O. A. Kabov, D. V. Zaitsev, D. P. Kirichenko, V. S. Ajaev, Interfacial Phenomena and Heat Transfer **4**, 207 (2016)
7. T. Umeki, M. Ohata, H. Nakanishi, M. Ichikawa, Sci. Rep. **5** 8046 (2015)
8. V. J. Schaefer Am. Sci. **59**, 534 (1971)
9. A. Fedorets, I. Marchuk, P. Stryzhak, O. Kabov, Thermophys. Aeromechanics **22**, 535 (2015)
10. D. V. Zaitsev, D. P. Kirichenko, O. A. Kabov, MATEC Web Conf. **92**, 01026 (2017)
11. D. V. Zaitsev, D. P. Kirichenko, O. A. Kabov, MATEC Web Conf. **84**, 00044 (2016)
12. D. P. Kirichenko, D. V. Zaitsev, O. A. Kabov, J. Phys. Conf. Ser. **754**, 042008 (2016)
13. D. Quere, Annu. Rev. Fluid Mech. **45**, 197 (2013)
14. F. Celestini, T. Frisch, Y. Pomeau, Phys. Rev. Lett. **109**, 034501 (2012)

MATEC Web of Conferences **194**, 01053 (2018) https://doi.org/10.1051/matecconf/201819401053
HMTTSC-2018

Analysis of the effect of water steam concentration on water drop cooling process efficiency in conditions of water cooling towers of thermal power plants

Sergei Shevelev[1,*]

[1]National Research Tomsk Polytechnic University, 634050 Tomsk, Russia

Abstract. Numerical investigation of transient distribution of temperature in water drops cooled in cool blast in conditions typical for high capacity evaporative water cooling towers has been done. Water drops cooling through thermal conductivity and water evaporation has been taken into consideration. It has been proved that the analysis of temperature distribution along the radius of water drop increases reliability of evaluation results of average temperature of water drops when they are cooled. Effect of water steam concentration in steam-gas mixture layer surrounding a water drop on water cooling towers efficiency has been evaluated.

1 Introduction

Water cooling towers are an important part of water treatment system at thermal power plants (TPP) [1, 2]. However, process cycle characteristics of recycled water evaporative cooling are selected experimentally and mainly with the help of integral process characteristics (temperature of large amounts of water after passing through water cooling towers) until recently. It is not possible to experimentally determine the temperature of water drops moving in comparatively cold (in comparison with recycled water) and dry air flow. At the same time the results of theoretical [3, 4] and experimental [5, 6] researches of water drops movement processes through gas flow have been known. The authors [3-6] have proved that temperature field of water drops when heated is not homogeneous (under certain conditions water drop temperature gradient is 20-30 K). The analysis of water drops cooling processes during their movement through the cold air of water cooling towers is of interest. Moreover [4], water evaporation intensity is known to be dependent on its steam concentration near evaporation boundary (air humidity). Water cooling towers of TPP operate in conditions of atmospheric air intake for recycled water cooling without dehydration of the so called refrigerant. Therefore, analysis of air humidity effect on water drops cooling efficiency in operating paths of water cooling towers has acquired practical importance.

[*] Corresponding author: wevelev@tpu.ru

The purpose of the paper is mathematical simulation of temperature fields of single water drops under water steam concentration variation in cold air.

2 Research objective

A spherical water drop is slow-speed under the gravity of approach cold air flow which temperature is significantly less than the initial temperature of water. Water drop cooling is due to convective heat transfer and evaporation heat absorption on the surface of the water drop. Water steam formed as a result of water evaporation is blown into near-wall region and mixed with air, and increases its humidity. The formed steam-gas mixture surrounds a drop. The dimensions of the drop decrease over time due to the removal of the part of the mixture with water steam.

It was assumed that the diameter of water drops is significantly smaller than the distance between them. Therefore, it is supposed that heat removal from the water drop has no significant effect on the temperature of the surrounding steam-gas mixture. Heat transfer coefficient was determined for conditions of flow around the sphere by a flow with low velocities due to low velocity of water drops.

Thermophysical characteristics of components of 'drop-air-water steam' system are taken as depending on temperature and pressure (for water they were determined according to the data from [7], for air according to figures provided by [8]).

The scheme of problem solution region is presented in Fig. 1.

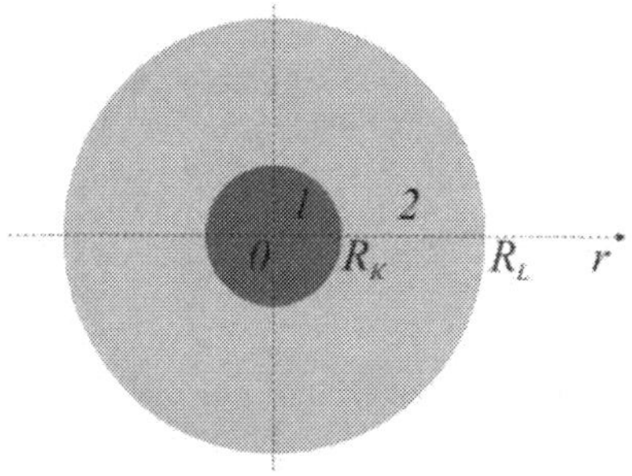

Fig. 1. Scheme of problem solution region. 1 – water drop, 2 – humid air.

The problem of heat and mass transfer set above is described with the help of the system of differential equations in partial differential coefficients written in spherical system of coordinates.

$$\rho c \frac{\partial T}{\partial t} = \lambda \left(\frac{\partial^2 T}{\partial r^2} + \frac{2}{r} \frac{\partial T}{\partial r} \right), 0 < r < R_K \tag{1}$$

Here: T, K – temperature; ρ, kg / m^3 – density, c, J / (kg · K) – heating capacity; λ, W / (m · K) – thermal conductivity, C_{ws} – water steam concentration, D, m^2 / s – water steam diffusion coefficient.

Boundary data:

$$r = 0 : \frac{\partial T}{\partial r} = 0 \tag{5}$$

$$r = R_K : -\lambda \frac{\partial T}{\partial r} = \alpha \left(T_\Gamma - T_B \right) + QW \tag{6}$$

$$r = R_K : -D\frac{\partial C_{ws.}}{\partial r} = \frac{W}{\rho_2} \tag{7}$$

$$r = R_L : \frac{\partial C_{ws}}{\partial r} = 0 \tag{8}$$

Here, Q, J / kg – heat of phase change; W, $kg / (m^2 \cdot s)$ – mass rate of vaporization, defined with mathematical expression of Hertz-Knudsen law, in accordance with [8].

The system of equations (1) – (2) presented above, with the initial and boundary data (3) – (8) was solved with the finite difference method using a tacit finite-difference scheme. The method of catching the front into a special grid node was used to take into account the movable boundary of the phase transfer. Non-linier problem of heat and mass transfer (1) – (8) was solved with the iteration method similar to [9]. The solution of differential rates of differential equations was carried out with the sweep method.

3 Results and discussion

Numerical modelling was carried out at rather typical initial data values: T_0=323 K, T_B=288 K, D=2.16·10^{-5} m^2/s.

Fig. 2 shows temperature distribution along the radius of the water drop over time. The radius of the water drop at the initial time is 5 mm.

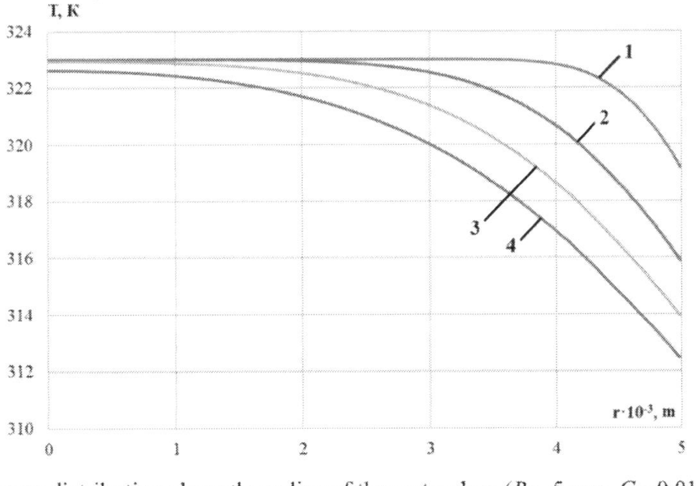

Fig. 2. Temperature distribution along the radius of the water drop (R_K=5 mm, C_0=0.01) over time: 1 – t=1 s; 2 – t=4 s; 3 – t=7 s; 4 – t=10 s.

Based on the dependency analysis presented in Fig. 2, we can con-clude about the significant temperature difference along the radius of the water drop: in one second the temperature of the surface of the water drop is less than the temperature of its center by 3.8 K, and in ten seconds by 10.2 K. Herewith, the average temperature of the water drop is 321.6 and 316.4 K respectively. Thus, recycled water temperature evaluation at the outlet of the water cooling tower by water drop surface temperature may cause incorrect results. Therefore, in mathematical simulation of water drops cooling in cold air flow under conditions typical for modern water cooling towers, temperature changes along the radius of the water drop must be taken into account.

Fig. 3 shows temperature distribution along the radius of the water drop over time for different water steam concentration values.

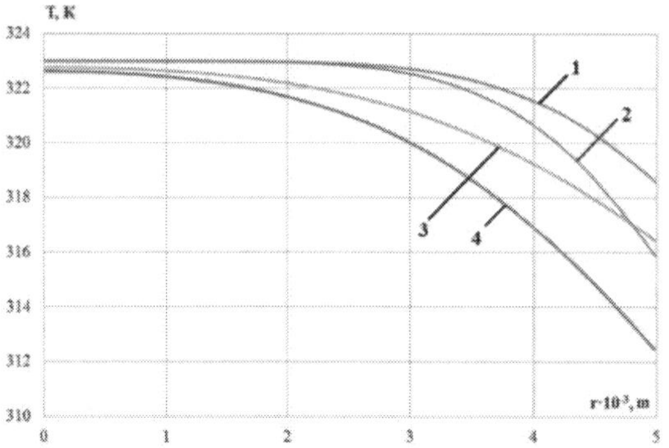

Fig. 3. Temperature distribution along the radius of the water drop over time for two water steam concentration values (R_K=5 mm): C_0=0.05 (1 – t=1 s; 3 – t=10 s); C_0=0.01 (2 – t=4 s; 4 – t=10 s).

According to the result of the analysis of dependencies presented in Fig. 3 we can conclude about the significant effect of initial water steam concentration on the temperature distribution along the radius of the water drop. So increase of C0 from 0.01 to 0.05 after four seconds of flight causes average integrated temperature increase of water drops by 1.2 K and by 2.4 K after ten seconds accordingly. Consequently, when creating a mathematical model designed for recycled water cooling efficiency evaluation in a water cooling tower both the value of water steam concentration and the temperature and air velocity must be considered.

4 Conclusion

As a result of the theoretical research the necessity and reasonability of the account of the effect of water steam concentration on temperature of drops at calculation of water cooling in technical water supply systems have been shown. The established results and common factors can be used for development of mathematical models of heat and mass transfer processes in the movement of water drops heated to high temperatures when they are cooled in water cooling towers.

References

1. S. R. Alavi, M. Rahmati, Energy Convers. Manage **122**, 504 (2016)
2. L. D. Berman, *Evaporative cooling of circulating water* (Pergamon, London, 1961)
3. G. V. Kuznetsov, P. A. Stizhak, J. Eng. Phys. Thermophys. **87**, 1 (2014)
4. G. V. Kuznetsov, P. A. Strizhak, High Temp. **52**, 4 (2014)
5. R. S. Volkov, G. V. Kuznetsov, P. A. Stizhak, Tech. Phys. **59**, 7 (2014)
6. G. V. Kuznetsov, P. A. Stizhak, Tech. Phys. Lett. **40**, 6 (2014).
7. Y. Sun, Z. Guan, K. Hooman, Renew. Sust. Energ. Rev. **79**, 618 (2017)
8. Revised Release on the IAPWS Industrial Formulation 1997 for the Thermodynamic Properties of Water and Steam. The International Association for the Properties of Water and Steam www.iapws.org (Lucerne, Switzerland, 1997)
9. A. A. Zukauskas, *High-performance single-phase heat exchangers* (Hemisphere Publishing Corporation, London, 1989)
10. V. L. Strakhov, A. N. Garashchenko, G. V. Kuznetsov, V. P. Rudzinskii, Combust. Explos. **37**, 2 (2001)

MATEC Web of Conferences **194**, 01054 (2018) https://doi.org/10.1051/matecconf/201819401054
HMTTSC-2018

Dynamics of the vapor cavity in the vertical tube of a small diameter after boiling-up in aqueous lithium bromide solutions

Ekaterina Slesareva[1,2,*], *Ruslan* Dekhtyar[1,2], and *Valeriy* Ovchinnikov[1]

[1]Kutateladze Institute of Thermophysics, 630090 Novosibirsk, Russia
[2]Novosibirsk State University, 630090 Novosibirsk, Russia

Abstract. Dynamics of a single vapor cavity, moving under gravity in a vertical tube of a small diameter, has been studied. The experiment was carried out in water and aqueous LiBr solution in a tube with the length of 400 mm and diameter of 16 mm. Experiments have shown that dynamics of the rising vapor cavity after-the-fact boiling-up of aqueous LiBr solution in a vertical tube of a small diameter for some initial conditions differs qualitatively from dynamics of a gas bubble rising in a pool. Dynamics of the vapor cavity in aqueous LiBr solution and in water demonstrates quantitative but not qualitative differences. During bubble rising in a small diameter tube to time dependence for volume and mass of the vapor bubble, the regimes with pulsations were detected.

1 Introduction

Vapor bubbles are the extremely labile objects. They quickly respond to even slight changes in the ambient pressure or temperature of liquid with which they come into contact. [1-3]. Thus, in the aqueous solution of lithium bromide (LiBr), the velocity of interphase boundary depends on distribution of absorbent concentration. Because of this, experiments with the vapor bubbles rising in the lithium bromide solution are not less significant than experiments with gas bubbles in ordinary liquids. In particular, the importance of influence of various factors on dynamics of the rising vapor bubbles in absorbent solution should be estimated.

When rising, the vapor bubble deforms strongly and moves along an intricate trajectory. For example, Figure 1 shows the video frames of vapor bubble rising in aqueous LiBr solution of 56 wt% at temperature $T_0 = 297$ K. Vapor at temperature $T_{V0} = 332$ K was fed through the capillary tube in the volume of liquid. Data on volume *V 1* of the bubble and liquid pressure *P 2* near the bubble, obtained by processing bubble position and shape in the video frames, are shown in a graph of Figure 1 as a function of time. Knowing the dependence of vapor density on pressure and temperature in the bubble, it is possible to obtain the calculated values of the bubble volume at constant *3* and variable, due to absorption *4*, vapor mass *m*. In the graph, we can see that there is significant discrepancy

* Corresponding author: styuardessa@yandex.ru

between experimental data and output computation by simplified model, when the bubble approaches the liquid surface in the tube ($t > 500$ ms). To understand the mechanisms affecting the dynamics of a lifting vapor bubble, a series of experiments in a small-diameter vertical tube with the vapor bubbles produced by boiling-up in water and aqueous LiBr solution was carried out in this work.

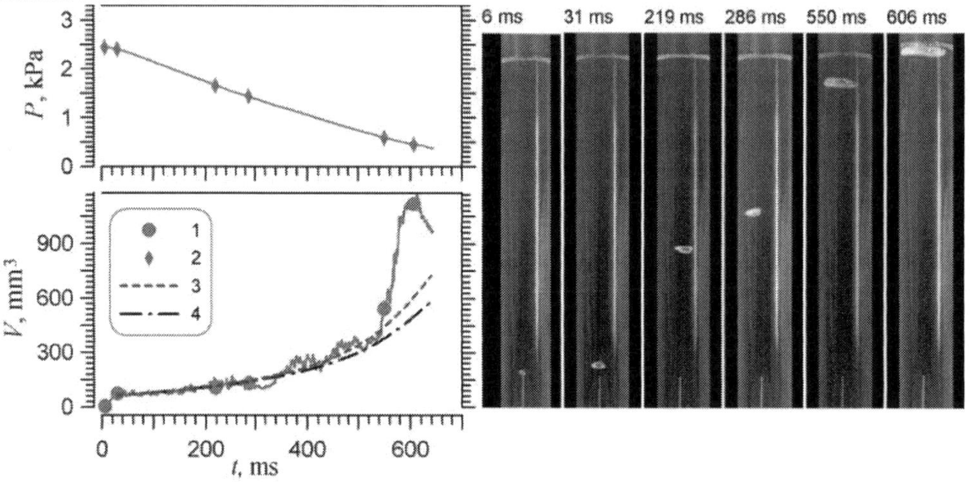

Fig. 1. The video frames of a rising vapor bubble in aqueous LiBr solution of 56 wt% at temperature $T_0 = 297$ K, vapor temperature $T_{V0} = 332$ K. The volume of bubble and liquid pressure near the bubble vs. time: 1 – volume V, 2 – pressure P, 3, 4 – results of calculations of bubble volume (3 – with a constant mass of vapor, 4 – with variable mass of vapor reduced because of absorption by 20%).

2 Experimental part

The schematic diagram of experimental setup is shown in Figure 2. Calibrated glass tube *1* of length $L = 400$ mm, inner diameter $D = 16$ mm was filled with test liquid. Height H of the tube filled with liquid varied from 100 to 200 mm. The setup was vacuumed by vacuum pump *3* through a vacuum trap. The pressure above liquid level was measured by vacuum manometer *2* (precision class 0.25). At the height of 10 mm from the bottom of the tube, electric heater *8* was installed with galvanic isolation. The heater is made of a nichrome tube with the diameter of 6 mm and wall thickness of 0.5 mm. Tubular heater *9* of 4-mm diameter and length of 16 mm with copper leads *10* of 3-mm diameter was placed in the nichrome tube. In the central part of the nichrome tube, a chromel-alumel thermocouple was placed. The space between the tube and the heater was filled with finely dispersed electrically insulating paste with a high coefficient of thermal conductivity. The level of heat generation at the heater was set by source of constant current *6*. The voltage drop on the heater varied from 0 to 30 V, electric current varied from 0 to 100 A. To create a hermetic seal between the heater and the working volume, vacuum rubber bushing *7* was used. The temperature of liquid before boiling-up was controlled by infrared imager 5. For determine the volume of V_B, surface area S_B, and position of the vapor bubble by digital video camera *4* at the rate of up to 1000 fps video recording of the vapor bubble and of the liquid level in the tube was performed. The volume of the vapor cavity for different times t was determined from the change in the liquid level height in tube h_L and estimated by the formulas taking into account the shape and characteristic dimensions of the vapor bubble.

$$V_B(t) = \pi \cdot \frac{D^2}{4} \cdot \left[h_L(t) - h_L(0) \right] \tag{1}$$

The pressure in the vapor cavity P was calculated taking into account the height of the liquid column above the bubble and gas pressure in the tube P_0 in the form:

$$P(t) = P_0 + \rho_L(T_L) \cdot g \cdot [h_L(t) - h_B(t)] \quad (2)$$

where h_B is the height of the bubble top, ρ_L is the density of liquid at temperature T_L, g is acceleration of gravity. The vapor mass in bubble m was calculated using reference data for vapor density ρ_V as a function of pressure P and temperature T [4]:

$$m(t) = V_B \cdot \rho_v(T, P) \quad (3)$$

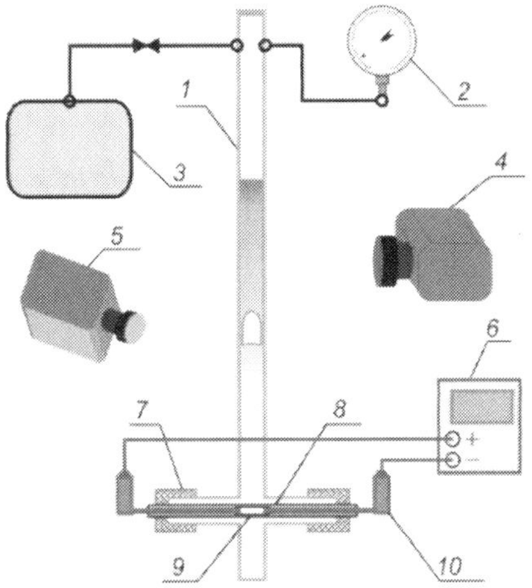

Fig. 2. The scheme of setup: 1 – glass calibrated tube (internal diameter D = 16 mm, length L = 400 mm); 2 – a vacuum manometer; 3 – membrane vacuum pump; 4 – digital video camera; 5 – infrared imager; 6 – DC power supply (voltage 0 ÷ 30 V, current strength 0 ÷ 100 A); 7 – vacuum rubber bushing; 8, 9, 10 – electric heater with galvanic isolation (8 – nichrome tube with heat conducting paste and thermocouple temperature sensor, 9 – tubular heater, 10 – copper leads).

3 Results and discussion

It was shown that dynamics of the rising vapor cavity after-the-fact boiling-up of aqueous LiBr solution in a vertical tube of a small diameter for some initial conditions has a qualitative distinction from dynamics of a gas bubble rising in a pool. It should be noted that dynamics of the vapor cavity in aqueous LiBr solution and water demonstrates quantitative but not qualitative differences. During bubble rising in a small-diameter tube to time dependence of the cavity volume and of the vapor mass, the regimes with pulsation were detected (Figures 3, 4). The video frames of the rising vapor bubble in a LiBr solution of 58 wt% at liquid level above the heater H_0 = 128 mm, pressure in the volume P_0 = 518 Pa, solution temperature T_0 = 297.4 K, and heater temperature before boiling-up T_h = 413.4 K are presented in Figure 3. In these frames, one can see that when floating up, the height of the vapor cavity has nonmonotonic type of relation in time. A graph of varying vapor mass of the vapor cavity in time is shown in Figure 4 for the rising bubble in water and in LiBr solution. Differences in dynamics of the vapor cavity for water and LiBr solution can be correlated with parameter P/P_0 (the ratio of pressure at the heater level to pressure above the liquid surface); this parameter is about 1.5 for water and 4.5 for LiBr solution (because of a higher density of liquid phase and of a lower density of gas phase). Therefore, differences in duration and amplitude of pulsations in experiments with the rising vapor cavity in tube with water and LiBr solution are noticeable. For a narrow tube, in contrast to pulsations of the vapor bubble in a pool [5], no regimes with shock waves were observed. This is probably due to the fact that the mass of the bubble rising in the tube differs in order of magnitude from the vapor mass in the bubble, which is formed by the action of laser radiation, when shock waves are observed.

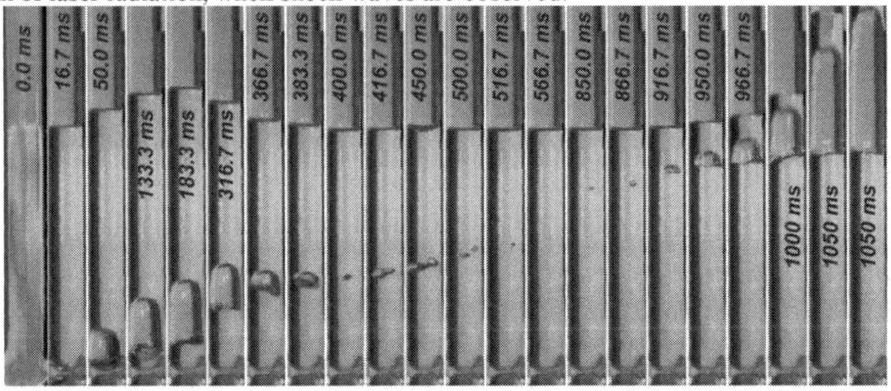

Fig. 3. Video frames after boiling-up of 58 % of aqueous LiBr solution in tube $D = 16$ mm, $H_0 = 128$ mm, $P_0 = 518$ Pa, $T_0 = 297.4$ K, $T_h = 413.4$ K.

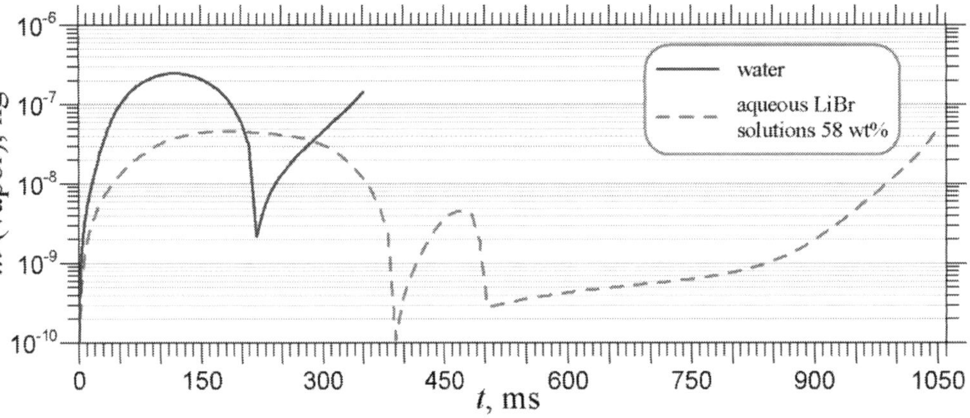

Fig. 4. Variation vapor mass in the vapor cavern in time for the rising bubble in water ($H_0 = 98$ mm, $P_0 = 2730$ Pa, $T_0 = 295.6$ K, $T_h = 361.1$ K) and in aqueous LiBr solution 58 wt% ($H_0 = 128$ mm, $P_0 = 518$ Pa, $T_0 = 297.4$ K, $T_h = 413.4$ K).

The acknowledgements: the study was funded by FASO of Russia.

References

1. A. Prosperetti, Ann. Rev. Fluid Mech. **49**, 221 (2017)
2. V .E. Nakoryakov, S. Y. Misyura, J. Engng. Thermophys. **25**, 24 (2016)
3. St. Kiirner, L. Friedel, J. Loss, Prev. Process Ind. **10**, 221 (1997)
4. H. Lover, *Thermodynamischen und physikalische Eigenschaften der wässrigen Lithiumbromid Lösung* (Dissertation, Karlsruhe, 1960)
5. I. Akhatov, O. Lindau, A. Topolnikov, R. Mettin, N. Vakhitova, W. Lauterborn, Phys. Fluids. **13**, 2805 (2001)

Ignition and combustion of high-energy materials containing aluminum, boron and aluminum diboride

Alexander Korotkikh[1,2,*], *Ivan* Sorokin[1], and *Ekaterina* Selikhova[1]

[1]National Research Tomsk Polytechnic University, 634050 Tomsk, Russia
[2]Tomsk State University, 634050 Tomsk, Russia

Abstract. Boron and its compounds are among the most promising metal fuel components to be used in solid propellants for solid fuel rocket engine and ramjet engine. Papers studying boron oxidation mostly focus on two areas: oxidation of single particles and powders of boron, as well as boron-containing composite solid propellants. This paper presents the results of an experimental study of the ignition and combustion of the high-energy material samples based on ammonium perchlorate, ammonium nitrate, and an energetic combustible binder. Powders of aluminum, amorphous boron and aluminum diboride, obtained by the SHS method, were used as the metallic fuels. It was found that the use of aluminum diboride in the solid propellant composition makes it possible to reduce the ignition delay time by 1.7–2.2 times and significantly increase the burning rate of the sample (by 4.8 times) as compared to the solid propellant containing aluminum powder. The use of amorphous boron in the solid propellant composition leads to a decrease in the ignition delay time of the sample by a factor of 2.2–2.8 due to high chemical activity and a difference in the oxidation mechanism of boron particles. The burning rate of this sample does not increase significantly.

1 Introduction

Boron is one of the most promising metals for rocket fuels due to high specific combustion heat value (among the highest per unit of mass and the maximum possible per unit of volume [1]). However, its application is significantly complicated by the fact that inert layer of boron is formed on the surface of boron particles during storage and combustion. It prevents the access of oxidizer [2] to pure boron, significantly reduces the ignition time and combustion of particles [3].

Devoted to the investigation of boron oxidation processes articles are mainly represented in literature by three types of works: the study on the oxidation of single particles [4, 5] and boron powders [6–9], as well as boron-containing composite fuels [9, 10].

* Corresponding author: korotkikh@tpu.ru

Numerous studies presented in literature on the characteristics of boron propellants ignition and combustion with different particle size [11], binder (most commonly used NH_4ClO_4, KNO_3) and the ratio of fuel to oxidizer [12]. In studies aimed at studying ignition and combustion at high pressures [13] or under conditions close to real ones [1, 13] the conductive initiation (for example, heating with nichrome filament or hot gases in TGA chamber) was used. It was established that with an increase in the pressure and content of boron particles to 20 mass. % in the fuel the linear burning rate and the surface temperature of the sample increase with the pressure dependence form close to linear [9].

Various techniques have been used to improve the characteristics of composite boron-based fuels. The one popular solution is using of coatings to prevent the formation of an oxide film, another is using bimetallic powders [10] with an ignition temperature of second metal lower than the evaporation temperature of the boron oxide film.

In this paper, we present the results of the experimental study of the ignition, combustion and corresponding surface temperature of the high-energy material samples based on ammonium perchlorate, ammonium nitrate and an energetic combustion binder with various types of boride powders. The study was carried out using a laser of continuous action.

2 Experimental sections

2.1 Sample characterization

Three HEM samples were used for ignition delay time and burning rate definition experiments. The first is the basic composition of HEM, containing 15 wt.% ammonium perchlorate (PCA) (fraction 160–315 μm), 35 wt.% ammonium nitrate (NA) (fraction less than 100 μm), 20 wt.% active fuel-binder MPVT-ASP and 30 wt.% of metal fuel – microsized aluminum powder ASD-4 with an average particle diameter d_{43} = 10.8 μm. Other two formulations of HEM aluminum powder is completely substituted by boron powders B (d_{43} = 2.0 μm) and aluminum diboride AlB_2 (d_{43} = 6.2 μm) obtained by the SHS-method (self-propagating high-temperature synthesis).

2.2 Experimental setup

The HEM ignition process was carried out using setup based on the CO_2 laser of continuous action with a wavelength of 10.6 microns and a maximum power of 200 W (Fig. 1). The diameter of the laser beam at the exit from the translucent mirror of the CO_2 laser was approximately equal to the diameter of the sample. HEM samples were cut into 5 mm high tablets before the experiment. The flat end surface of the samples was visually monitored for the absence of pores, cavities and cracks.

The test sample (5) of the HEM was attached to the holder (6). When the shutter (3) was opened, exposure time ~5 ms, the heat flux was directed to the investigated HEM sample (5). The signals from the photodiodes (7) were transmitted through the L-card E-14-440 ADC (10) and recorded using personal computer (11), and then processed using the LGraph2 software. The ignition delay time t_{ign} of HEM was defined as a difference between the times of signal from photodiodes (7), one of which registered the occurrence of flame on the flat end surface of the sample. Relative error for delay time values t_{ign} was 5–12% with confidence level 0.9.

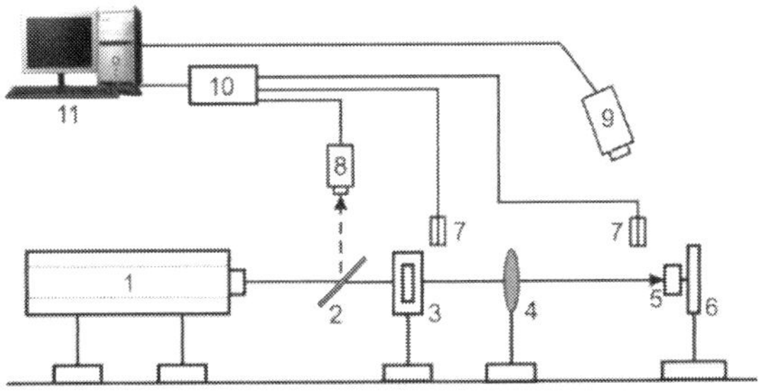

Fig. 1. The scheme of experimental setup based on CO_2 laser: 1 – CO_2 laser; 2 – beam-splitting mirror; 3 – shutter; 4 – lens; 5 – HEM sample; 6 – holder; 7 – photodiodes; 8 – thermoelectric sensor of radiation power; 9 – video camera; 10 – ADCs; 11 – PC.

The surface temperature of the fuel sample during the heating was recorded using a Jade J530 SB thermal imaging camera at a frequency of 50 Hz in the temperature range 370–1800 °C. The thermal imaging data and the photodiode parameters were synchronized by a characteristic click produced by shutter opening. It corresponds to the beginning of the HEM sample heating.

The power of the laser radiation heat flux on the HEM sample was measured with a thermoelectric sensor Ophir FL400A (8). A maximum value of the radiation power at the center of the laser beam was determined through 2 mm diameter diaphragm. The ratio of the maximum value to the mean value over the area of the beam diameter of the radiation power was ~ 1.7.

2.3 The burning rate

The burning rate of the HEM was measured under atmospheric pressure. We used HEM with a diameter of 10 mm and a height of 25–30 mm. The flat end surface of the samples was visually monitored for the absence of pores, cavities and cracks. The sample was placed on the holder and ignited with a nichrome spiral. The burning rate of the HEM was calculated from the known length of the sample and the burning time determined by the stopwatch.

2.4 Activation energy calculation

The activation energy value was determined by solving the inverse heat conduction problem of the heated by radiant flux sample according to the given in [14] method. It is based on the following equation:

$$\lg\left(\frac{t_{ign}}{\left(1-T_0/T_{ign}\right)^2}\right)=\lg\left(\frac{0.359\cdot Ec}{(1-0.8\cdot\beta)R\cdot Qz}\right)+\frac{E}{RT_{ign}}\lg e, \quad (1)$$

where t_{ign} – ignition delay time, s; T_0 – initial temperature of solid fuel sample, K; T_{ign} – quasistationary ignition temperature of the HEM sample surface, K; E – effective activation energy of the reaction, J/mole; c – specific heat of the solid fuel sample, J/(kg K); $\beta=\frac{RT_{ign}}{E}$ – unitless parameter which expresses the degree of dependence of the reaction rate on temperature; R – universal gas constant, J/(mole K); Q – the thermal effect of the reaction, J/kg; z – reaction pre-exponent, 1/s.

The value of the characteristic temperature can be determined by the following formula:

$$T_{ign} = T_0 + \sqrt{\frac{t_{ign} \cdot (1 - 0.8 \cdot \beta)}{\lambda c \rho}} \cdot q \ , \quad (2)$$

where q – heat flux density, W/cm^2, λ, ρ – thermal conductivity and density of the sample, W/(m·K) и kg/m^3.

3 Results and discussion

3.1 Ignition delay time

The dependence of the ignition delay times of solid fuel samples on the radiation flux density were obtained using setup for radiant heating. These dependences are shown in Fig. 2.

The graph in Fig. 2 shows that a complete replacement of the ASD-4 aluminum powder with boron powder results in a decrease in the ignition delay time of solid fuel samples by 2.2–2.8 times. The use of aluminum borides makes it possible to reduce the ignition delay time of fuels in comparison with the base composition by 1.7–2.2 times. An analysis of these results shows that the difference in the ignition delay time of the HEM during radiant heating is associated with oxidation reaction rate and chemical reaction heat during heating and ignition of the used metal additives.

The large values of t_{ign} for the basic composition containing aluminum ASD-4 are explained by the higher average diameter and the presence of a churlish oxide layer on the particles surface.

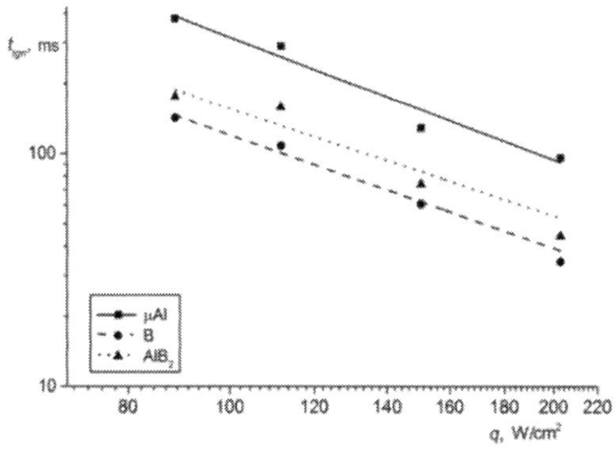

Fig. 2. HEM ignition delay time vs laser heat flux density: $t_{ign} = 9.90 \cdot 10^5 q^{-1.75}$ – sample with Al; $t_{ign} = 2.02 \cdot 10^5 q^{-1.61}$ – sample with B; $t_{ign} = 1.78 \cdot 10^5 q^{-1.53}$ – sample with AlB_2.

3.2 Ignition temperature

We obtained the synchronized video frames and thermal imaging corresponding to the main stages of ignition for the investigated samples. When processing the measured data, the temperatures on the reaction layer surface of HEM samples were obtained at a constant emissivity factor of 0.9 at the time of steady ignition (Table 1).

Table 1. Sample surface characteristic temperatures at the moment of ignition.

Sample	Al	B	AlB_2
Surface maximal temperature, °C	1150	1260	1240
Surface average temperature, °C	690	560	570

It is worth mentioning, that the maximum surface temperatures of sample ignition with aluminum boride and amorphous boron powders are fairly close. This may indicate the leading role of particles B and AlB_2 oxidation reaction during heating and ignition of HEM.

Fig. 3 shows the experimental maximum temperature values of the HEM surface as a function of the heating time. The heating and ignition process of the HEM can be divided into two stages differing in the rates of temperature increasing on the surface of the samples reaction layer. This can be caused by an increased rate of exothermic chemical reactions at higher decomposition temperatures of the oxidant, active fuel-binder and oxidation of metal particles.

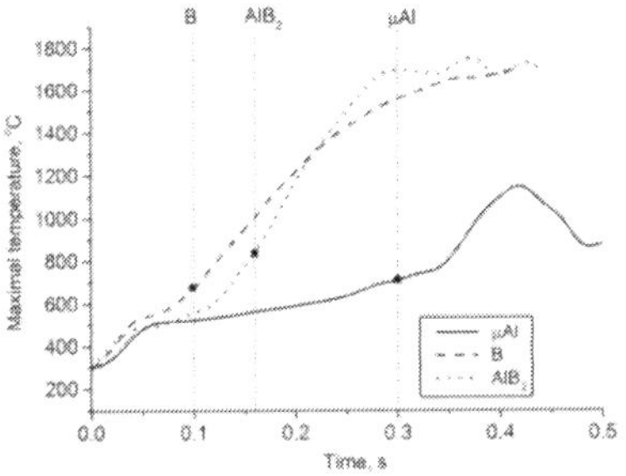

Fig. 3. Maximum surface temperature of the HEM samples as a function of the heating time.

3.3 The burning rate

We performed three parallel experiments for each of HEM samples and determined average values of the burning rate that is shown in Fig. 4.

Complete replacement of microsized aluminum powder ASD-4 with amorphous boron powder allows increasing the burning rate by 1.32 times. The burning rate of the HEM composition with aluminum diboride is 4.83 times higher than that for the base composition with ASD-4.

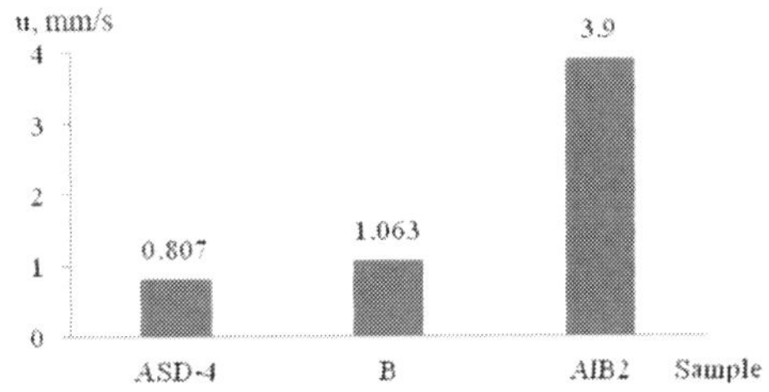

Fig. 4. Burning rate of HEM containing metal powder.

3.4 Effective activation energy

We determined the values of activation energy, reaction thermal effect, pre-exponent and characteristic ignition temperature of the HEM samples (Table 2). According to Table 2, samples with Al additive have the highest activation energy, while for samples with AlB_2 it is the lowest. Based on these data it can be concluded that the addition of AlB_2 into composition of HEM is preferred.

Table 2. Calculation results of kinetic parameters and ignition temperature of samples.

Sample	Al	B	AlB_2
E, kJ/mole	159	90	74
Qz, W/g	$9.92 \cdot 10^{13}$	$3.31 \cdot 10^{11}$	$4.24 \cdot 10^{9}$
T_{ign}, °C at q = 115 W/cm²	846	641	687

4 Conclusions

The paper presents an experimental study of ignition and combustion of the three solid fuel compositions based on perchlorate and ammonium nitrate, active binder and additives of metal powders ASD-4, B and AlB_2. Ignition delay times, reaction layer surface temperature, ignition temperature, and linear burning rate under atmospheric pressure were determined. It is established that ignition delay times of HEM containing boron powder are 2.2 to 2.8 times smaller than the sample with aluminum ASD-4 in the heat flux density range of 90–200 W/cm^2. At the same time, the sample burning rate increased insignificantly (by 1.3 times) due to incomplete burning of boron and the heat released.

The use of aluminum diboride powder in the HEM composition allows to reduce the ignition delay time by 1.7–2.2 times in the indicated range q and significant increase the burning rate (by 4.8 times) in comparison with HEM containing aluminum powder ASD-4 by increasing the efficiency and heat combustion of boron in the chemical reactions zone.

The reported study was supported by RFBR according to the research project No. 16-03-00630 a and by The Tomsk State University competitiveness improvement programme.

References

1. D. S. Sundaram, V. Yang, V. E. Zarko, CESW **51**, 2 (2015)
2. W. Ao, Tz.-Kh. Zhou, Tz.-C. Liu, W.-Cz. Jan, Yu Wang, H.-R. Li, Phys. Combust. Explos. **50**, 3 (2014)
3. K.-L. Chintersingh, M. Schoenitz, E. L. Dreizin, Combust. Flame **173**, 288 (2016)
4. R. O. Foelsche, R. l. Burton, H. Krier, Combust. Flame **117**, 32 (1999)
5. J. Xi, J. Liu, Y. Wang, Y. Hu, Y. Zhang, J. Zhou, J. Propul. Power **30**, 1 (2014)
6. J-Z. Liu, J-F. Xi, W-J. Yang, Y-R. Hu, Y-W. Zhang, Y. Wang, J-H. Zhou, Acta Astranaut. **96**, 89 (2014)
7. L-J. Liu, G-Q. He, Y-H. Wang, J. Therm. Anal. Calorim. **114**, 3 (2012)
8. V. Ao, T. Kh. Zhou, V. C. Yan, C. C. Liu, Yu. Wang, Phys. Combust. Explos. **50**, 6 (2014)
9. T. Kuwahara, N. Kubota, Propellants Explos. Pyrotech. **14**, 43 (1989)
10. A. G. Korotkikh, V. A. Arkhipov, O. G. Glotov, V. E. Zarko, A. B. Kiskin, Combust. Flame **178**, 195(2017)
11. F. Maggi, A. Bandera, L. Galfetti, L. T. De Luca, T. L. Jackson, Acta Astronaut. **66**, 1563 (2010)
12. V. A. Arkhipov, A. B. Kiskin, V. E. Zarko, A. G. Korotkikh, Combust. Explos. Shock Waves **50**, 622 (2014)
13. T. L. Connell Jr., G. A. Risha, R. A. Yetter, C. W. Roberts, G. Young, J. Propul. Power **31**, 1 (2014)
14. V. N. Vilyunov, V. E. Zarko, *Ignition theory of condensed matter* (EDP Science, Novosibirsk, 1984)

Research of efficiency of application of external physical fields for sedimentation of liquid-droplets aerosols of different concentration

Maria Yu. Stepkina[1,*], *Olga* B.Kudryashova[1], and *Natalia* V. Korovina[1]

[1]Institute for Problems of Chemical and Energetic Technologies of the SB RAS, 659322 Biisk, Russia

Abstract. A theoretical estimation of extent of influence of physical and chemical properties and concentration of the weighed particles on evolution of parameters of a liquid-droplets aerosol is carried out in the work. A series of experiments on a research of parameters of aerosol media with a various structure and concentration characteristics of a dispersive phase under external (ultrasonic and electrostatic) fields were made.

1 Introduction

In the industry, the modern equipment, technology and everyday life it is necessary to consider the processes and the phenomena proceeding with the participation of aerosol systems and powdery materials. The research of aerosols and aero-disperse systems is extremely important for the organization of environmental protection and health safety. Formation of an aerosol happens with mechanical fragmentation and dispersion of solids or liquids, crushing, friction, explosions, burning, dispersion by sprayers.

Now the problem of aerosols creation for needs of fire extinguishing, disinfection, deactivation, the creation of barriers on the way of distribution of toxic gases, non-lethal weapon, inhalation in medicine, etc. is relevant.

No matter how the liquid-droplets aerosol has been received, the cloud evolves further: liquid droplets evaporate or increase in the course of condensation of vapors, particles stick together or split up, fall down. Acceleration of process of sedimentation of the considered aerosol media is increased under external physical fields [1,2].

2 Theoretical part

An evaporation of the condensed phase of a liquid-droplets aerosol makes the important contribution to the process of the transformation of the size range of aerosol-cloud particles. Most strongly evaporation processes influence on particles less than 10 microns in size, therefore, these processes need to be taking into account by consideration of ensemble of such particles. Such influence is caused by the active heat exchange with the environment and energy ablation from the large surface of droplets.

* Corresponding author: bvborisov@tpu.ru

For calculation of time of evaporation of t on the basis of Maxwell's equation the following formula is used [3]:

$$t = \frac{r^2}{\dfrac{6D_f M}{\rho_\kappa R T_\infty} e^{-\frac{MH}{R}\left(\frac{1}{T_K}-\frac{1}{293}\right)} 2338.8 \left(e^{\frac{2\sigma M}{\rho_\kappa R T_K \cdot r}} - v \right)}, \tag{1}$$

where ρ_κ is the droplets density; D_f is diffusion coefficient; M is molecular weight of liquid; r is the particle radius; R is universal gas constant, σ is coefficient of a surface tension of substance of a droplet; T_κ is the droplet temperature; v is humidity; T_∞ is the temperature of the air.

An example of the calculation is presented for the dependence of evaporation time of aerosol particles with a diameter 1-15 micrometers for substances – water, acetone, ethyl alcohol (Fig.1), hydrochloric acid and glycerin (Fig.2) at an ambient temperature of 20 o C and humidity of 70 %.

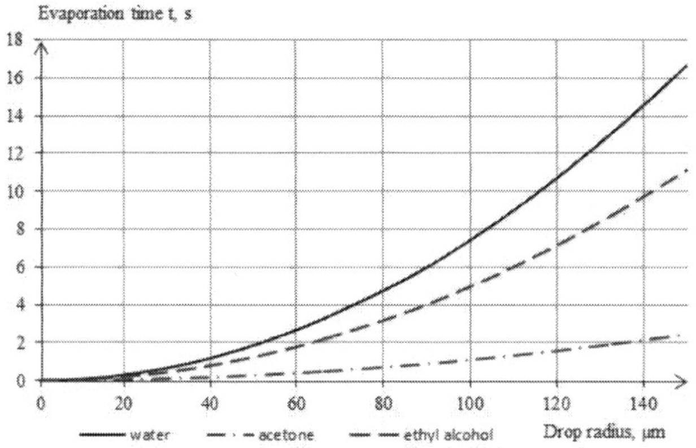

Fig. 1. Time of evaporation of aerosol particles of water, acetone, ethyl alcohol at ambient temperature 20°C.

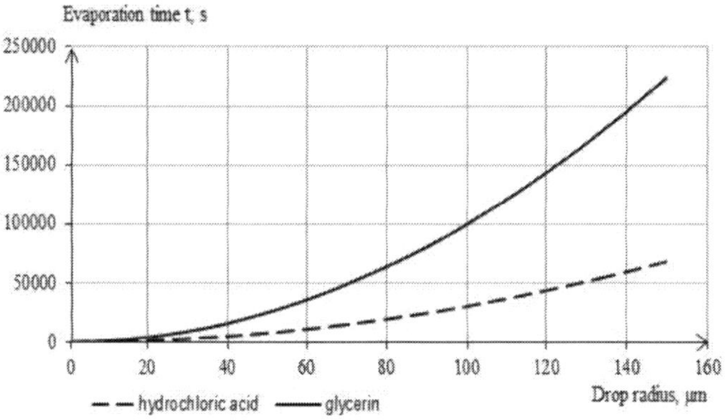

Fig. 2. Time of evaporation of aerosol particles of hydrochloric acid and glycerin at ambient temperature 20°C.

Transformation of distribution of particles by the sizes is based on the balance equation (an integrated variant of the equation of Smolukhovsky), as following [6]:

$$\frac{\partial f(D,t)}{\partial t} = I_1 + I_2 + I_3, \tag{2}$$

where I_1 is a decrease in mass of droplets with a diameter of D in a unit of time, in unit of volume due to collision of a droplet of diameter D with any droplet of diameter D'; I_2 describes formation of particles of diameter of D due to collision of droplets with diameters D' and $D–D'$; I_3 is the reduction of mass of droplets due to their evaporation.

The term of I_1 is calculated by a formula:

$$I_1 = -f(D,t)\int_0^{\infty} K(D,D')f(D',t)dD', \tag{3}$$

where $K(D,D') = b_k(D^2 + D'^2)$ is the probability of collision of drops with diameters of D and D' in the unit of time.

For the calculation of I_2 a formula is used:

$$I_2 = \frac{1}{2}\int_0^{D} K(D-D',D')f(D',t)f(D-D',t)dD'. \tag{4}$$

The term is defined by Maxwell (1) equation with Thomson formula differentiated on the mass of particles:

$$I_3 = \frac{\partial}{\partial D}\left[4\pi D_f M p_{pl}\left(\exp\left(\frac{4\sigma M}{\rho_l RTD}\right)-1\right)\frac{f(D)}{RTD\rho_l}\right]. \tag{5}$$

It is necessary to consider initial conditions for the equation (1) at $t=t_0$ $f(D,t_0)=f_0(D)$ – initial distribution particles by the sizes.

In the course of sedimentation of an aerosol the particle, which weight exceeds critical value instantly drop out in a deposit and don't take part in coagulation. However, to track dynamics of the process of sedimentation it is necessary to consider the dependence of the critical size on time, for this purpose it is necessary to write down the equation for sedimentation speed as particle derivative of height above the ground of h on time:

$$V_o = \frac{dh}{dt} = \frac{2}{9}\frac{D^2 g(\rho_k - \rho_0)}{\mu_g} \tag{6}$$

Then in the time of t all the droplets with the diameter more than D_{cr} will drop out in a deposit. The value of D_{cr} can be expressed from the equation (6) as:

$$D_{cr} = \sqrt{\frac{9\mu_g h}{2g(\rho_K - \rho_0)t}}, \tag{7}$$

Then the equation (3) taking into account of D_{cr} will take a form:

$$I_1 = -f(D,t)\int_0^{D_{cr}(t)} K(D,D')f(D',t)dD' \tag{8}$$

Thus, considering sedimentation of large particles, the range of the particle sizes on each time point will be cut off on the right and gradually this border is displaced towards smaller particles [4-5].

3 Experimental part

Experimental study of the influence of external fields on the evolution of parameters of a liquid-droplets aerosol was carried out in the measuring volume of 1 m3. For the creation of a liquid-droplets aerosol, the airbrush "CRATON R 200 LVLP-02S" was used. The research of evolution of parameters of a liquid-droplets aerosol in the experimental camera was measured by "LID-2M" system because this system is intended for measurement of dispersion and concentration of aerosol particles from the moment of aerosol generation till the total disappearance of it. The principle of action of the measurement system is based on the modified method of low-angle scattering representing a contactless method of determination of the dispersing characteristics of the aerosol based on a classical method of low-angle scattering [7].

For the creation of external fields in experimental researches, the electro precipitator and the ultrasonic radiator are used. Technical characteristics of the ultrasonic device are following: diameter of the radiator – 320 mm, the level of sound pressure – not less than 140 dB, the frequency of fluctuations – 28 kHz. Design data of the electric precipitator used in an experiment are the following: the area of sedimentation is 1000 mm2; length of a discharge element, length of an active zone – 0.085 m; distance between discharge electrodes – 0.01 m; distance between the planes of discharge and precipitation electrodes – 0.005 m; radius of a discharge electrode – 0.004 m; average tension – 8000 V [8].

The mass of the sprayed water solution of 20% of glycerin in experiments was 9.5 g; the time of the spraying is 35 seconds; initial humidity is 40% and the temperature is 20 o C. The time of registration of process is more than one hour. After the spraying we start the registration of the process of sedimentation within 30 minutes, and then we turn on the pre-cipitator or the ultrasonic radiator for 1 hour in experiments with external fields.

The values of a decrease of relative concentration of Cm for glycerin solution are received as a result of an experiment using the software of the "LID-2M" laser system. The first series of experiments were carried out without any influences under normal conditions (Fig.3). Also calculations of time of sedimentation taking into account evaporation, coagulation and sedimentations have been carried out (the equation (2) taking into account or without the member of I3). The following series of experiments were carried out in the presence of acoustic and electrostatic field (Fig.4) in identical conditions of an experiment.

The calculated curve taking into account the only sedimentation of particles predicts sedimentation of an aerosol already to 1500 second. Evaporation of droplets reduces their size and they remain in the air longer, without sedimentation. Both calculation and an experiment show this fact (Fig.3).

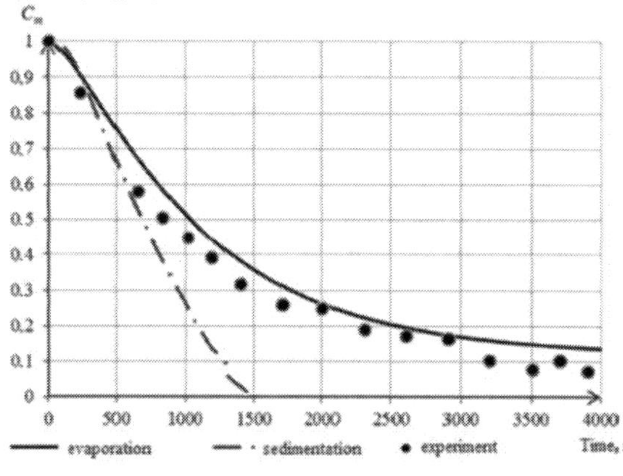

Fig. 3. Dependence of relative concentration of aerosol particles of 20% solution of glycerin on time at a humidity of 40 % and temperature of the disperse media of 20°C.

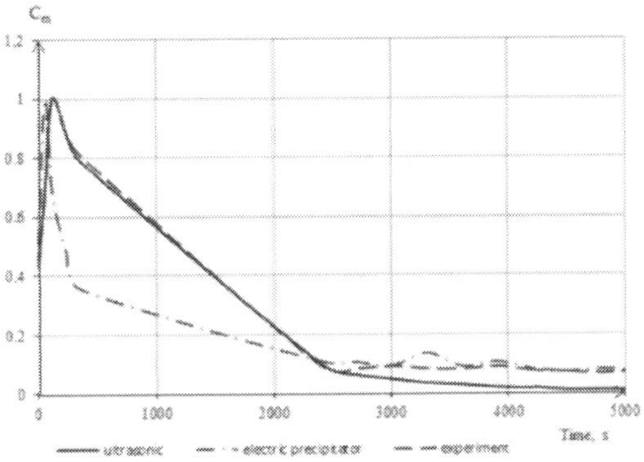

Fig. 4. Experimental dependences of the relative concentration of aerosol particles of 20% water solution of glycerin on time without influences and at under external fields with humidity of 40% and temperature of the disperse media of 20°C.

Under the external field of ultrasound (US) the experimental curve of concentration tends to zero after 4000 sec., that allows confirming a good result of sedimentation of an aerosol of the small sizes of particles under the field. Influence of electrostatic field allows to remove quickly from the air the large particles, but smaller ones continue to remain in the air for a long time.

4 Conclusion

The theoretical estimation of the extent of influence of physical and chemical properties of a dispersed phase on the evolution of parameters of liquid-droplets aerosols allowing to predict the evaporation time of an aerosol of different substances is carried out. It has been as a result established (Fig.1-2) that time of evaporation of large particles (140 micrometers) of glycerin will be more than 55 hours. Among a number of model liquid-droplets particles, the acetone is the most easily evaporated substance. The experimental trends of a decrease of relative concentration received by means of laser measuring system have allowed revealing the most effective mechanism of sedimentation on the example of an aerosol of 20% of a water solution of glycerin: it turned out the ultrasonic sedimentation.

This study was supported by the Russian Science Foundation (grant No. 17-79-10209).
The work was conducted on equipment of the Biysk Regional Center of Collective Use of the SB RAS (IPCET SB RAS, Biysk).

References

1. V.I. Roldugin, *Fizikokhimiya poverkhnosti* (Intellekt, Moscow, 2008) [In Russian]
2. V.A. Arkhipov, A.S. Usanina, *The movement of aerosol particles in a stream* (Publishing house of the Tomsk state university, Tomsk, 2013)
3. M.Y. Stepkina, O.B. Kudryashova, B.V. Borisov, MATEC Web Conf. 23 **(2015)**
4. A.A. Antonnikova, N.V. Korovina, O.B. Kudryashova, I.M. Vasenin, Polzunovsky vestnik **1** (2013) [In Russian]
5. O.B. Kudryashova, A.A. Antonnikova, Tomsk Polytechnic University News **322** (2013)
6. A.A. Antonnikova, N.V. Korovina, O. B. Kudryashova, OJA E **3**, 3 (2013)
7. O. Kudryashova, *Photodetectors* (InTech, Croatia, 2012)
8. M.Y. Stepkina, O.B. Kudryashova, A.A. Antonnikova, Arch Acoust. **43 (**2018)

Analysis of the vapor-oxygen oxidizer in the synthesis gas production from solid fuel

Elena S. Popova[1], and *Alexander* N. Subbotin[1,*]

[1]National Research Tomsk Polytechnic University, 634050 Tomsk, Russia

Abstract. A universal method of processing low-grade coals, as well as energy waste, is their gasification. A mathematical model of the conversion of coke, placed in a cylindrical capsule, through which a steam-oxygen blow is carried out, is considered. The determining parameter is the proportion of oxygen in the oxidizer. The values of the oxygen concentration in a steam-oxygen oxidizer are found to produce a synthesis gas with a predetermined percentage of combustible components.

1 Introduction

A huge amount of waste from coal mining and coking (slimes and stubble) have no application. The combustion of coal adversely affects the environment. In this connection, interest in alternative methods of processing is growing. Special attention is now being paid to the gasification of solid fuels. There is an opportunity to process low-grade coal, from power inputs with a significant reduction in emissions to the atmosphere [1]. The resulting synthesis gas is applicable both in power engineering and in the chemical industry. In particular, a sufficiently large amount of hydrogen is required in the production of ammonia and methanol [2]. Gasification is possible with different modes of solid fuel conversion. As reacting agents, steam or oxygen (or a mixture thereof), air may act. Involving oxygen in the process of gasification makes it possible to accelerate and increase the stability of reactions occurring during the conversion [1]. Currently, mathematical modeling of various physical processes is widely used [3]-[6]. Consider the mathematical model of the solid fuel conversion process.

2 Physical and mathematical model

The investigated object is a cylindrical capsule filled with coke. Fuel is ignited due to the steam-oxygen mixture heated to a certain temperature. In the modeling of fuel conversion, the following chemical reactions are considered:

$$C + H_2O = CO + H_2 - 130{,}5 \text{ kJ/mol},$$
$$C + CO_2 = 2CO - 175{,}5 \text{ kJ/mol},$$
$$C + O_2 = CO_2 + 395 \text{ kJ/mol},$$
$$2C + O_2 = 2CO + 219 \text{ kJ/mol},$$

* Corresponding author: subbot@tpu.ru

$$2H_2 + O_2 = 2\,H_2O + 219\ \text{kJ/mol},$$
$$2CO + O_2 = 2CO_2 + 571\ \text{kJ/mol},$$
$$CO + H_2O = CO_2 + H_2 + 40{,}4\ \text{kJ/mol}.$$

The task is to determine the composition of the synthesis gas during the conversion of coke, the analysis of its composition, depending on the quantitative oxygen content in the steam-oxygen oxidizer. The research was carried out using a mathematical model. According to [3], a system of equations is developed:

- conservation of coke mass

$$\rho_3 \cdot \frac{\partial \phi_3}{\partial \tau} = -R_C\,, \quad R_C = R_{C1} + R_{C2} + R_{C3} + R_{C4}\,; \tag{1}$$

- continuity of the gas phase

$$\frac{\partial\, \rho_5 \phi_5}{\partial \tau} + \frac{\partial\, \rho_5 \phi_5 v}{\partial z} = R_C\,; \tag{2}$$

- continuity α – of the gas phase component

$$\frac{\partial(\rho_5\,\varphi_5\,c_\alpha)}{\partial \tau} + v\,\frac{\partial(\rho_5\,\varphi_5\,c_\alpha)}{\partial z} = \frac{\partial}{\partial z}(\rho_5\,\varphi_5\,D_\alpha\,\frac{\partial c_\alpha}{\partial z}) + R_{5\alpha} + R_{S\alpha},\ \ \alpha = \overline{1,4}\ , \tag{3}$$

here parameter with an index $\alpha = 1$ corresponds to the oxygen (O_2), $\alpha = 2$ – CO, $\alpha = 3$ –CO_2, $\alpha = 4$ – H_2, $\alpha = 5$ – H_2O;

- conservation of energy of the gas phase

$$\begin{aligned}\rho_5\,\varphi_5\,c_{P5}\left(\frac{\partial T}{\partial \tau} + v \cdot \frac{\partial T}{\partial z}\right) &= \frac{\partial}{\partial z}\left((\lambda_5 + \lambda_R)\,\varphi_5\,\frac{\partial T}{\partial z}\right) - A_s(T - T_S)\\ &+ q^1_{CO}\,R^1_{CO} + q^2_{CO}\,R^2_{CO} + q_{H_2} R_{H_2} - \frac{2}{r_o}\alpha_1\,\varphi_5\,(T - T_S)\,;\end{aligned} \tag{4}$$

- conservation of energy of the condensed phase

$$\begin{aligned}\sum_{i=3}^{4} \rho_i\,\varphi_i\,c_{p_i}\,\frac{\partial T_S}{\partial \tau} &= \frac{\partial}{\partial z}(\lambda_S\,\frac{\partial T_S}{\partial z}) + q_{C1}\,R_{C1} + q_{C2}\,R_{C2} - q_{C3}\,R_{C3} - q_{C4}\,R_{C4} +\\ A_S\,(T - T_S) &+ \frac{2 \cdot r_1}{r_0^2}\,\alpha_2\,(T_S - T_E) + q_R\,;\end{aligned} \tag{5}$$

- gas motion in the pores (in the form of a quadratic filtration law) and the equation of state

$$v = \frac{k}{\mu + k \cdot \beta \cdot \rho_5\,|V|} \cdot \frac{\partial p}{\partial z},\quad p = \frac{\rho_5 \cdot R \cdot T}{M_5}\,. \tag{6}$$

The system of equations (1) - (6) was solved under the boundary conditions:

$$\tau = 0:\quad \varphi_i = \varphi_{iH},\ i = \overline{3,4};\ \ c_\alpha = c_{\alpha H},\ \alpha = \overline{1,4},\ \ T = T_S = T_H,\ \ p = p_H\,; \tag{7}$$

$$z=0:\quad T=T_G,\quad (pv)_w=const,\quad \lambda_S\cdot\frac{\partial T_S}{\partial z}=\alpha_1\cdot(T_S-T_G),\quad c_\alpha=c_{\alpha G},$$
$$z=h:\quad \frac{\partial c_\alpha}{\partial z}=\frac{\partial T}{\partial z}=0,\quad p=p_H,\quad \lambda_S\frac{\partial T_S}{\partial z}=0. \qquad (8)$$

In the equations (2) - (7) the following notation is introduced: $c_{p5}=\sum_{\alpha=1}^{5}c_{p\alpha}c_\alpha$,

$\lambda_S=\sum_{i=3}^{4}\lambda_i\varphi_i$, $\lambda_R=\frac{16\sigma T^3}{s}$, $M_5=1/\sum_{\alpha=1}^{7}\frac{c_\alpha}{M_{4\alpha}}$, $R_{C1}=s\,\rho_5\varphi_5c_7k_{S1}\varphi_3\exp\left(\frac{-E_{C1}}{R\cdot T_S}\right)$,

$R_{CO}^2=k_{CO}^2(c_1M/M_{51})^{0.25}(c_2M/M_{52})\cdot T^{-2.25}\exp\left(\frac{-E_{CO}^2}{R\cdot T}\right)$, $R_{C2}=s\,\rho_5\varphi_5c_3k_{S2}\varphi_3\exp\left(\frac{-E_{C2}}{R\cdot T_S}\right)$,

$R_{C3}=s\,\rho_5\varphi_5c_1k_{S3}\varphi_3\exp\left(\frac{-E_{C3}}{R\cdot T_S}\right)$, $R_{CO}^1=k_{CO}^1c_2c_7\exp\left(\frac{-E_{CO}^1}{R\cdot T}\right)$, $R_{H_2}=k_{H_2}c_1c_4\cdot\exp\left(\frac{-E_{H_2}}{R\cdot T}\right)$,

$R_{52}=\gamma_5R_{C1}+\gamma_6R_{C2}+\gamma_7R_{C4}$, $R_{54}=\gamma_{13}R_{C1}+\gamma_{14}R_{CO}^1-\gamma_{15}R_{H_2}$, $R_{57}=\gamma_{16}R_{H2}-\gamma_{17}R_{C1}-\gamma_{18}R_{CO}^1$,
$R_{51}=-(\gamma_1R_{C2}+\gamma_2R_{C3}+\gamma_3R_{H_2}+\gamma_4R_{CO}^2)$, $R_{53}=\gamma_8R_{C3}-\gamma_{10}R_{C2}+\gamma_{11}R_{CO}^1+\gamma_{12}R_{CO}^2$,

$R_{C4}=s\,\rho_5\varphi_5c_1k_{S4}\varphi_3\exp\left(\frac{-E_{C4}}{R\cdot T_S}\right)$. All other notations coincide with the notation introduced in [5].

3 Analysis of numerical results

According to [6], during steam gasification of solid fuel, the composition of the synthesis gas is affected by the temperature of the oxidizer, the specific surface area of the pores and the porosity. In [7], the influence of the oxygen concentration in the vapor-oxygen oxidant on the percentage composition of the synthesized gas formed is considered. As shown by numerical calculations, an important parameter in this case is the fraction of oxygen in the oxidant. Table 1 shows the studies on the composition of the resulting synthesis gas during the conversion of coke in a heated vapor-oxygen oxidizer with various mass concentrations of oxygen in it.

The synthesis gas compositions shown in Table 1 were obtained at: T_e = 873 K, T_e is the temperature of the oxidant supplied to the capsule; $\varphi_{3H}=0.8$, φ_{3H} - initial volume fraction of coke placed in the capsule $S_*=5\cdot10^4$, 1/m, S_* is the initial specific surface area of the pores.

Table 1. Composition of synthesis gas and temperature of gas and condensed phases during steam-gasification of coke.

Concentration O_2		0.04	0.06	0.08	0.1	0.15	0.2	0.25	0.3	0.35
Volumetric composition synthesis gas , %	CO	8.0	14.3	22.2	33.0	55.4	60.6	64.1	67.0	70.2
	H_2	13.6	23.6	33.7	42.9	44.0	39.4	35.9	33.0	29.8
	O_2	–	–	–	–	–	–	–	–	–
	CO_2	2.9	4.4	5.5	5.2	0.4	–	–	–	–
	H_2O	75.5	57.7	38.6	18.9	0.2	–	–	–	–
Maximum temperature, K	gas	910	950	996	1040	1175	1327	1405	1478	1510
	coke	918	964	1018	1067	1248	1415	1510	1621	1703

From the analysis of the data presented in Table 1, it follows that at a mass concentration of oxygen up to 0.1, the volume fraction of the combustible gas is about 22 to

75%, there is a lot of water vapor, the volume fraction of which is approximately 75-19%.The products of gasification contain a small amount of carbon dioxide. Under a mass concentration of oxygen in a steam oxygen oxidizer of more than 0.15, a synthesis gas consisting of hydrogen and carbon monoxide is produced.
The lower lines of the table show the maximum values of the temperature of synthesis gas and coke in the combustion zone. Wesee that when the oxygen concentration in the oxidant is less than 0.15, the temperature in the conversion region is low, does not exceed 1100 K. Therefore, the process proceeds slowly and the gasification products contain a lot of water vapor. As follows from the table, with an increase in the oxygen concentration, the temperature of both the gas and condensed phases increases.

With the increase in the oxygen concentration to 0.15, the volume content of carbon dioxide and hydrogen in the synthesis gas increases. With a mass concentration of oxygen in the oxidizer of more than 0.15 in the synthesis gas, the volume fraction of hydrogen begins to decrease, and carbon dioxide continues to grow.

Table 2. Characteristics of synthesis gas for steam-oxygen gasification of coke ($\varphi_{зн}$ = 0.5, S_* = $5\cdot10^3$, 1/m).

Concentration O_2		0.1	0.15	0.2	0.25	0.3	0.35	0.4	0.45
Volumetric composition synthesis gas, %	CO	2.7	7.0	17.1	33.7	46.9	55.7	61.0	64.6
	H_2	19.0	36.4	49.0	49.0	44.0	39.9	36.8	34.3
	O_2	0.2	0.1	0.1	–	–	–	–	–
	CO_2	12.0	20.3	22.8	16.1	9.1	4.4	2.2	1.1
	H_2O	66.1	36.2	11.0	1.2	–	–	–	–
Maximum temperature, K	gas	934	979	1015	1051	1085	1122	1153	1186
	coke	984	1065	1149	1245	1346	1482	1604	1722

Table 2 shows the characteristics of the synthesis gas for various concentrations of oxygen in the oxidant during the conversion of coke to a larger fraction. Analyzing the characteristics of the synthesis gas obtained in the two tables, we conclude that when the coke is converted into a larger fraction, more oxygen must be added to the water vapor. In particular, in order to obtain a synthesis gas with substantially the same characteristics, in the first case the oxygen concentration in the oxidant should be 0.15 and 0.25, and for the larger fraction coke, respectively, 0.35 and 0.45.

Table 3. Characteristics of synthesis gas with decreasing oxidizer temperature (S_* = 5 · 103.1 / m).

Coke volume fraction		$\varphi_{зн}$ = 0,2		$\varphi_{зн}$ = 0,5		$\varphi_{зн}$ = 0,8	
Concentration of O_2 (C1)		0.40		0.35		0.25	
Oxidizing agent temperature		873 K	373 K	873 K	373 K	873 K	373 K
Volume synthesis of gas,%	CO	59.3	56.1	55.7	53.6	58.8	58.2
	H_2	36.6	37.3	39.9	40.3	39.9	40.2
	O_2	–	–	–	–	–	
	CO_2	4.1	6.6	4.4	6.1	1.3	1.6
	H_2O	–	–	–	–	–	–
Maximum temperature, K	gas	1235	1207	1122	1073	1066	1022
	coke	1806	1759	1482	1461	1394	1363

Table 3 presents numerical calculations for different porosities and initial concentrations of oxygen in a steam-oxygen oxidizer. The gas composition and maximum temperatures in the combustion zone, given in the first columns of the table for different values of $\varphi_{зн}$(volume fraction of coke), were obtained at the temperature of a steam oxygen oxidizer of 873 K. The second columns contain the same values of the synthesis gas parameters when the oxidant temperature decreases after ignition up to 373 K.

Analyzing the data given in Table 3, we come to the conclusion that after the ignition of the fuel, the conversion process can be carried out at a lower oxidant temperature. The composition of the resulting combustible gas varies slightly. The change in the composition of the resulting synthesis gas at different grinds (large, medium fine) is shown in Table 4.

Table 4. Dependence of characteristics of synthesis gas on the specific surface of the pores (S_*) at two initial porosity and oxygen concentrations of $C_1 = 0.35$.

Initial porosity , φ_{5H}		0,8			0,5		
S_*, 1/m		$5 \cdot 10^2$	$5 \cdot 10^3$	$5 \cdot 10^4$	$5 \cdot 10^2$	$5 \cdot 10^3$	$5 \cdot 10^4$
Volumetric synthesis of gas,%	CO	30.3	52.6	61.8	34.6	55.7	61.1
	H_2	43.4	39.9	38.0	43.1	39.9	38.9
	O_2	3.4	–	–	1.4	–	–
	CO_2	21.2	7.5	0.2	19.2	4.4	–
	H_2O	1.7	–	–	1.7	–	–
Maximum temperature, K	gas	887	1199	1425	886	1122	1342
	coke	1648	1592	1551	1517	1482	1457

Analyzing the data of Table 4, it can be argued that when the fraction of converted fuel decreases, the composition of the gas and the temperature of the process change. The temperature difference between the gas and solid phases decreases with decreasing fuel fraction.

4 Conclusions

The analysis of the synthesized gas compositions obtained allows us to state that:

a) the oxygen concentration in the steam-oxygen oxidant significantly changes the composition of the gasification products, therefore, it is always possible to select the oxidant necessary to produce a combustible gas of a certain composition;

b) the smaller the fraction of convertible solid fuel, the less oxygen must be added to the oxidizer during conversion;

c) it is always possible to select the composition of a steam oxygen oxidizer, in which a synthesis gas consisting of carbon monoxide and hydrogen is obtained.

References

1. B. I. Bondarenko, E. A. Khadzhinov, Dnepropetrovsk: NPVK "Triacon" **3** (2011)
2. B. K. Nefedov, Catalysis in the chemical and petrochemical industry **4** (2008)
3. P. S. Gergelizhiu, S. A. Khaustov, R. B. Tabakaev, P. Y. Novoseltsev, A. V. Kazakov, A.S. Zavorin, *Proc. of 2014 Int. Conf. on MEACS* (2014)
4. S. A. Khaustov, Y. A. Belousova, K. V. Buvakov, A. Y. Dolgih, R. N. Kulesh, 11th International Forum on Strategic Technology (IFOST - 2016) Proceedings **2** (2016)
5. A. N. Subbotin, Izvestiya Tomsk Polytechnic University **4** (2008)
6. R. N. Kulesh, A. S. Mazanik, A. N. Subbotin, Izvestiya Tomsk Polytechnic University. Technics and technologies in power engineering **4** (2014)
7. R. N. Aryukov, A. N. Subbotin, *Collection of materials II All-Russian youth scientific-practical conference "Introduction to energy"* (2016) [in Russian]

Kinetics of biomass low-temperature pyrolysis by coats–redfern method

Dariga Altynbaeva[1,*], *Alexander* Astafev[1], and *Roman* Tabakaev[1]

[1]National Research Tomsk Polytechnic University, 634050 Tomsk, Russia

Abstract. Energy is one of the main sources of environmental pollution. In this regard, the world scientific community seeks to reduce its negative impact on human health and nature. To do this, some of the organic raw materials, which are fuel for power plants, are gradually replaced by renewable resources, in particular - biomass. The combustion of biomass by traditional methods is accompanied by high operating costs, which leads to the conclusion that it is necessary to process it into energy-valuable products. The development of one of the most promising such directions - thermal processing - requires studying the thermophysical properties of biomass and studying its kinetics of decomposition. The aim of this work is to estimate the activation energy of thermal decomposition of biomass by the method of Coats-Redfern.

1 Introduction

As is known [1], the power engineering is one of the main sources of environmental pollution. In which connection, the power stations, used coal and brown coals, anthracites as fuel, are cause damage to the environment. According to the World Health Organization [1, 2], air pollution leads to premature deaths of several million people per year, what is comparable to the size of a large metropolis.

In this regard, developed and developing countries (for example, the USA, Finland, France, Canada, etc. [3]) announced a policy by gradually fossil organic fuels use reducing for energy production. Replacement of organic fuel should be carried out by increasing the share of renewable energy sources (RES), such as sun, wind, biomass, etc. The use of RES will increase the ecological purity of energy production; will reduce the external energy dependence of countries, which not have own fuel resources; will stimulates the development of local and regional production; will contribute to the development of the agro-industrial complex and creation of new jobs [4]. In 2015, 7.1% of electricity was generated on the basis of renewable energy (excluding hydropower) [5]. At the same time, the share of such output will increase to 20% by 2020 at the plan of the European Union [6].

Biomass is considered as one of the most important and promising RES, what is due to its ubiquity and large volumes of formation. It is recognized that biomass is neutral in the carbon dioxide circulation: when burning biomass, the same amount of carbon dioxide is

* Corresponding author: dba10@tpu.ru

formed, which was absorbed during her life as a result of photosynthesis. The sulfur is practically absent in biomass content, what ensures the low SO_x emissions at the time of its burning [7]. At present time, the share of biomass in the world fuel and energy balance is about 9.7% [5].

However, the burning of biomass by traditional methods is accompanied by a number of problems, which prevent its wide using in the energy sector. For example, the biomass in a natural state has a high humidity, which reaches to 30.4-79.7% [8], what leads to a low calorific value in terms of the operating state. In addition, such a high humidity of biomass leads to its caking, sticking during a transportation, and in winter – to freezing, what additionally complicates the process of transportation and unloading of raw materials [9]. Some types of biomass (for example, bran) are prone to slagging of heating surfaces during combustion [10], as a result of which the service life of the fuel-burning equipment is reduced and the cost of repair works is increased. The heterogeneous fractional composition of the biomass leads to a high value of the failure through the grate at layer combustion, increasing the losses with mechanical underburning (q_4) and thereby reducing the overall boiler efficiency. Thus, the energy use of biomass by traditional methods of combustion is accompanied by high operating costs, which prevents its widespread use as fuel.

The previous studies [11] are shown the prospects for thermal processing of biomass into energy-valuable products, such as high-calorific solid fuel, liquid biofuels or flammable gas. The combustion of these products, even by traditional methods, is highly efficient and requires significantly less cost than using of raw materials in their natural state.

The various studies of biomass thermal processing by physical experiment is a labour-intensive process required a large time and material resources. The construction of a mathematical model and the conduction on its basis of numerical studies is contributed to increasing of productivity and study economy. The development of model for thermal processing is required of study a large number of physical and chemical quantities, which necessary to obtain of reliable and adequate results, in particular the activation energy.

In this regard, the purpose of this paper is evaluation of biomass thermal decomposition activation energy by Coats-Redfern method.

2 Material and methods

Several samples of biomass were considered in this work: straw, chips from various types of wood and low-lying peat (Sukhovskoe deposit, Tomsk region). The samples were previously dried to air-dry state (Table 1).

Table 1. The humidity of researched biomass samples

Biomass	Straw	Chips	Peat
Humidity W^a, %	8.5	8.5	11.5

The differential thermal (DTA) and thermogravimetric (TGA) analyzes with thermal analyzer STA 449C Jupiter (Netzsch, Germany) were performed to evaluation of the activation energy. The samples were heated at a rate of 10°C per minute (averaged heating rate according to the generally accepted standard of fuels semi-coking of ISO 647-74), the temperature range of the experiment was from 20°C to 600°C. The weight of samples was 50 mg. Analyzes were performed in an inert medium (helium) in order to exclude the influence of oxidative reactions on the results of the study.

The activation energy was calculated by the method of A. Coats and J. Redfern [12]. The method is based on the hypothesis about that the single-stage first-order reactions, submitted to the Arrhenius laws occur, are procced at the time of solid fuels thermal decomposition. The activation energy equation according to Coats-Redfern has the following form:

$$\ln\left[\frac{-\ln(1-\alpha)}{T^2}\right] = ln\left[\frac{AR}{\beta E_a}\left(1-\frac{2RT}{E_a}\right)\right] - \frac{E_a}{RT} \quad (1)$$

where $\alpha = \frac{m_0 - m_i}{m_0 - m_K}$

m_0 – mass of reaction start, kg;
m_K – mass of reaction finish, kg;
m_i – mass in the period of i, kg;
α – conversion level;
T – temperature of heating, K;
A – pre-exponential coefficient;
R – gas constant, J/(mole·K);
E_a – activation energy, kJ/mole;
β – reaction rate, K/min.

Further, the dependency diagram $ln(-ln(1-\alpha)/T^2)$ from $1/T$ is constructed in accordance with the formula for the first-order reaction, its result is a straight line, the slope of this line is E_a/R, and the intersection point of this line with the axis $ln(-ln(1-\alpha)/T^2)$ is equal to $ln(AR/\beta E_a)$.

3 Results and discussion

The results of DTA and TGA for researched samples are shown in Fig. 1. It can be seen that during heating from 20 to 140-150°C there occurs moisture evaporation from the samples, which was accompanied by an endothermic peak. The thermal decomposition of biomass starts at a temperature of 160-180°C and accompanied by its mass decrease (Fig. 1, TG curve). The temperature 240°C was the beginning for main exothermic reactions, terminating at 580°C, for straw and chips. With the thermal decomposition of peat samples, two exothermic peaks were observed: first – 200-330°C, second – 380-560°C.

The graphical construction of the Coats-Redfern equation is shown in Fig. 2. It can be seen that the biomass samples decomposition proceeds in two stages. The calculation of the activation energy for each stage on dependence of biomass type is given in Table 2.

Table 2. Activation energy of thermal decomposition

Biomass	Activation energy E_a, kJ/mole		Start of decomposition t_0, °C	Transition temperature t_i, °C	Finish of decomposition t_K, °C	Pre-exponential coefficient, c^{-1}	
	I	II				I	II
straw	74.2	29.1	162	350	595	5919	8213333
chips	85.0	11.7	180	366	575	61773	5480814
peat	68.9	18.9	180	340	565	2537	3629208

The analysis of obtained data (Table 2) is shown that the activation energy for the first stage have a greatest value for the sample of chips, a smallest – for sample of peat. At the second stage, the straw has a greatest value of the activation energy, and the chips – a smallest. Considering that the activation energy is the energy, which necessary for breaking bonds between molecules, the most energy-consuming raw material for processing (including both stages) is straw, and less energy-consuming – peat.

The obtained data are accorded with the results of previous studies by other authors [13].

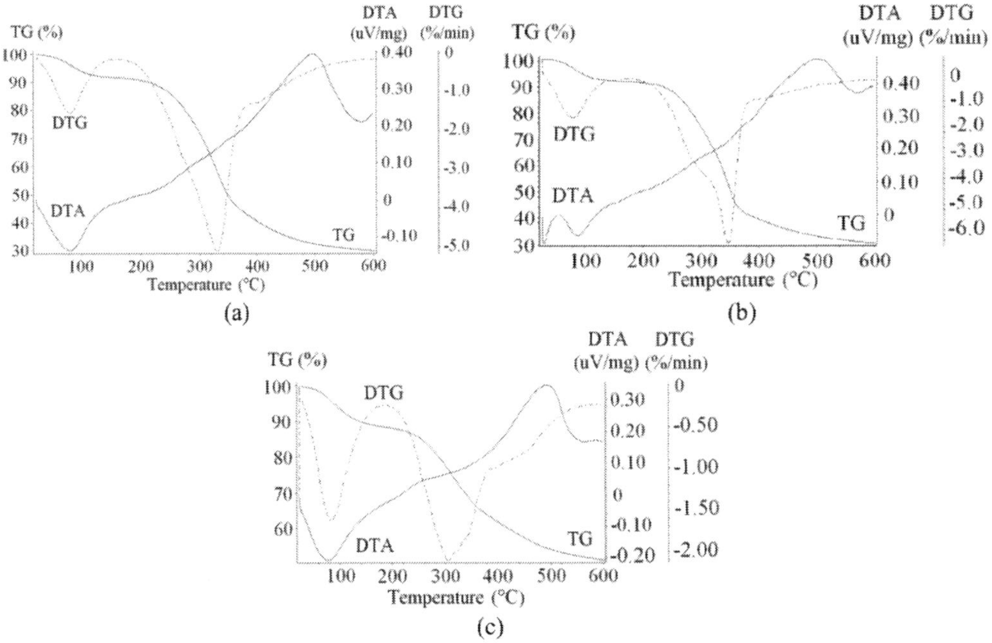

Fig. 1. Curves of DTA analysis: a – straw, b – chips, c – peat.

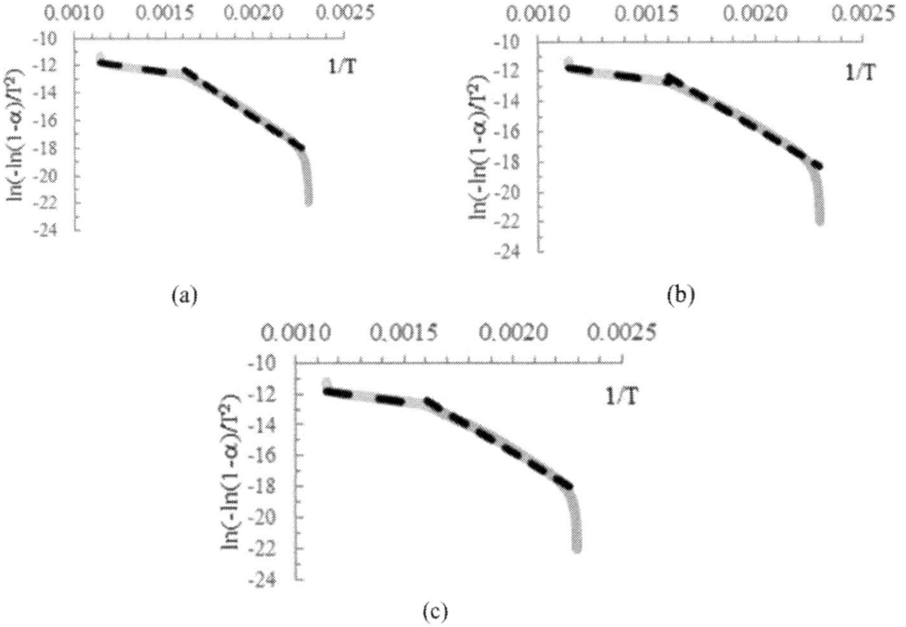

Fig. 2. The dependency ln(-ln(1-α)/T^2) from 1/T: a – straw, b – chips, c – peat.

4 Conclusions

The results of DTA and TGA for biomass samples of several types (straw, chips and peat) are given in this paper. The activation energy of the biomass thermal decomposition is calculated with using of the Coats-Redfern method, developed for evaluation of the thermal analysis results. It has been established that the thermal decomposition of biomass proceeds in two stages: the first stage start at temperature 162-180°C and finish at temperature 340-

366°C; the second stage – from 340-366°C to 565-595°C. Wherein, the most energy-consuming raw material for thermal processing is straw and less energy-consuming is peat.

The reported study was funded by RFBR according to the research project No. 18-38-00648.

References

1. *World Energy Outlook Special Report. Energy and Air Pollution* (International Energy Agency, Paris, 2016)
2. J. Lelieveld, J.S. Evans, M. Fnais, D. Giannadaki, A. Pozzer, Nature **525**, 367 (2015)
3. O.S. Popel, B.F. Reutov, A.P. Antropov, Therm. Eng. **57**, 909 (2010)
4. J.L. Miguez, L.M. Lopez-Gonzalez, J.M. Sala, J. Porteiro, E. Granada, J. C. Moran, M.C. Juarez, Renew. Sust. Energ. Rev. **10**, 225 (2006)
5. International Energy Agency (IEA): Key world energy statistics (Chirat, France, 2017)
6. EU Road Map: http://europa.eu/legislation_summaries/energy/renewable_energy/l27065_en.htm 2007 (Accessed 01.03.2018)
7. S. Sofer, O. Zaborski, *Biomass conversion processes for energy and fuels* (Plenum Press, New York, 1981)
8. S. V. Vassilev, C. G. Vassileva, V. S. Vassilev, Fuel **158**, 330 (2015)
9. A. D. Uchitel, M. V. Kormer, V. P. Lyalyuk, I. A. Lyakhova, E. O. Shmel'tser, Y.I. Vititnev, Coke Chem. **56**, 167 (2013)
10. M. Yu. Chernetskiy, A. A. Dekterev, A. Burdukov, Journal of Siberian Federal University. Engineering & Technologies **6**, 625 (2013)
11. R. Tabakaev, I. Shanenkov, A. Kazakov, A. Zavorin, J. Anal. Appl. Pyrolys. **124**, 94 (2017)
12. A. W. Coats, J. P. Redfern, Nature **201,** 68 (1964)
13. C. Lu, W. Song, W. Lin, Biotechnol Adv. **27**, 583 (2009)

Dynamics of dry spots in the liquid film moved by the gas flow in the mini-channel under intensive local heating

Egor Tkachenko[1,2,*]

[1]Kutateladze Institute Institute of Thermophysics, 630090 Novosibirsk, Russia
[2]Novosibirsk State University, 630090 Novosibirsk, Russia

Abstract. Experimental studies of hydrodynamics and the heat transfer crisis were carried out for a two-phase stratified flow in a mini-channel with intensive heating from a heat source of 1x1 cm^2. It has been established that as the heat flow increases, the total area of dry spots on the heater increases, but when a certain temperature of the heater surface reaches ≈100 °C, the area of dry spots begins to decrease. With the help of high-speed visualization (shooting speed 100000 frames per second), several stages of formation of a dry spot (a typical size of the order of 100 microns) were isolated. It was found that at a heat flux of 450 W/cm^2 about 1 million dry spots per 1 second are formed and washed on the surface of the heater (1 cm^2). The speed of the contact line when dry spot is forming reaches 10 m/s.

1 Introduction

One of the most important problems of thermal physics today is the cooling of microelectronic equipment. At present, the production of processors varies from 14 nm technology to 7 nm technology. New IBM processors use 7-nm transistors. This was made possible by using a silicon-germanium alloy in the construction of a transistor instead of pure silicon, a traditional material for the manufacture of microchips. A normal-sized processor can accommodate more than 20 billion transistors having a size of 7 nm. In addition to reducing the size of the microprocessor unit, the use of a silicon-germanium alloy has improved the switching speed of transistors and reduced energy consumption. The average density of heat flow on the chips of commercially available computers and other electronic devices is up to 200-300 W/cm^2. In local areas from 100 μm to several square millimeters, the heat flux density reaches values of 1 kW/cm^2 or even higher (IBM). Next-generation devices can have even higher heat fluxes and impulse loads [1], which requires the output of thermal regulation to a new level.

There are three "classical" methods of heat removal from localized heat sources of high intensity: 1) boiling in microchannels [2], 2) spray cooling [3], and 3) micro-jet cooling. In [4,5], the fourth effective cooling method was proposed, in which the heat removal is

* Corresponding author: egor_tkachenko@mail.ru

caused by the intensive evaporation of a thin liquid film moving in a flat micro- and mini-channel under the action of a gas flow. The method turned out to be an order of magnitude more efficient, in comparison with the falling liquid film [6-8]. In [9-11], studies were carried out that showed the possibility of removing heat fluxes of density up to 1200 W/cm^2 from the heating area 1x1 cm^2 by means of a thin film moving under the action of the gas flow in the channel. In [12, 13] it was shown that such a system can work steadily in a wide range of channel height changes (0.17-2.00 mm) and channel inclination angles to the horizon (0-360°). In [14], heat transfer was studied in a liquid film moving in a channel with an extended heater (at relatively low heat fluxes). In the present work, systematic experimental studies of the flow and rupture of a water film moving in a channel are considered when heated from a local heat source 1x1 cm^2. The dynamics of dry spots is investigated using a high-speed camera with the frequency of up to 100 000 frames per second.

2 Experimental stand

Fig. 1 shows the scheme of the experimental stand. The stand includes a test section, measuring systems, an open gas line and a closed fluid circuit, as well as a system for supply and monitoring of working phases. The stand also includes an automated National Instruments ADC (not shown in Fig. 1).

Working gas – atmospheric air – with a relative humidity of 20-40% and an initial temperature of 25 °C is used, which is supplied to the working area by means of an air compressor. The gas flow rate is regulated by the flowmeter and the Bronkhorst flow controller and ranges from 0 to 200 liters/minute.

As working fluid in the experiment, Ultra-pure distilled Milli-Q water is used at an initial temperature of 25 ° C. The liquid is supplied to the working area by means of a gear pump Reglo-Z 183. The flow rate of the liquid is regulated by means of a pump in the range from 4.2 to 420 ml/minute. The design of the test section allows to create flows in the range $Re_l = 0 - 210$.

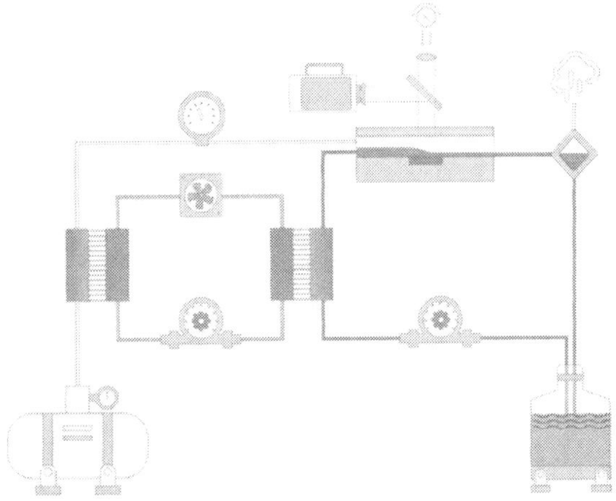

Fig. 1. The scheme of the experimental stand.

Fig. 2 shows the design of the test section – sealed channel with openings for the entry and exit of liquid and gas. The lower part of the channel is a plate made of stainless steel, located on a substrate made of textolite. In the steel plate, a copper cylinder is inserted, acting as a heater. The surface of the copper heater is a square area of 10x10 mm^2. The upper part of the channel is an optical glass, installed to observe the processes in the channel.

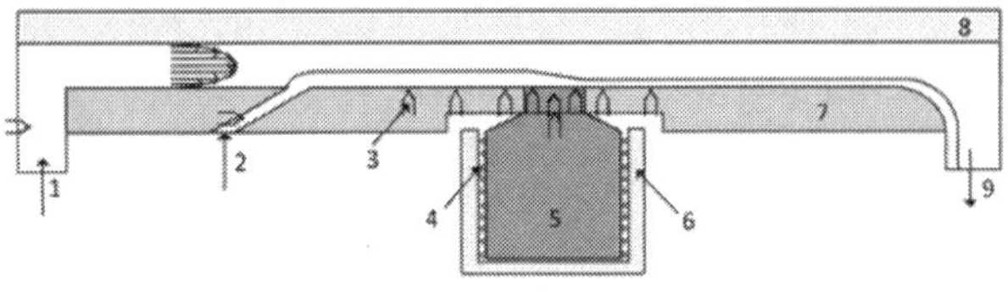

Fig. 2. Design of the test section. 1 – gas nozzle; 2 – liquid nozzle; 3 – thermocouples; 4 – nichrome tape; 5 – copper heater; 6 – thermal insulation; 7 – steel plate; 8 – optical glass; 9 – outlet of liquid and gas.

Thermocouples built into the stainless steel plate and into the copper rod allow to determine the temperature of the working surface. The heat flux is determined by the electrical power released on the heating coil. The thermal conductivity of copper is 400 W/mK, which is more than an order of magnitude higher than the thermal conductivity of stainless steel (15 W/mK). This provides moderate heat flow from the heater to the steel plate. According to estimates, using the measurements of thermocouples embedded in a steel plate, the heat spreading is about 15% at $q>200$ W/cm^2. To reduce heat losses to the atmosphere, the heater was wrapped with a layer of heat-insulating material. According to estimates using thermocouple measurements embedded in the heater, the thermal losses to the atmosphere do not exceed 10% at $q>400$ W/cm^2. Consequently, thermal congestions and losses do not total more than 25% at $q>400$ W/cm^2.

3 Optical visualization

The optical visualization of the experiment was carried out by two methods:

1) Visualization using a Nikon D200 digital camera with a resolution of 3872x2592 pixels (10 Megapixel), with a Micro-Nikkor 105 mm F2.8G VR micro-lens, which allows shooting with a spatial resolution of up to 6 μm / pixel. The camera is installed above the surface of the working section perpendicular to the plane of the flow. The survey was conducted at different frequencies, but not less than 3 frames for each stationary mode. Photographs obtained by shooting are one of the main subjects of analysis and comparison.

2) Visualization using high-speed camera FASTCAM SA1.1 CCD. The speed of shooting this camera can vary from 5400 frames per second at a resolution of 1024x1024 pixels and up to 675000 frames per second at resolutions of less than 1024x1024 pixels. The camera is equipped with an optical system of high spatial resolution (up to 500 nm per 1 pixel of the camera sensor).

4 Results

In the visual analysis the following stages of the liquid film rupture were identified with an increase in the heat flux. 1 – formation of a dry spot below the heater. 2 – increase in the dimensions of the first dry spot and the achievement of the surface of the heater. 3 – increase of dry spots on the surface of the heater. 4 – decrease in the area of dry spots on the surface of the heater; 5 – in the pre-crisis moment, the liquid film again covers practically the entire surface of the heater with rapidly appearing and vanishing dry spots on it. And 6 – the moment of crisis, in which the heater is completely dry outed.

An analysis of the experimental photographs was carried out using the ImageJ application. During processing, the places of the heater on which the dry spots formed were assigned a white color and their area in pixels was further considered, knowing the characteristic scale (the length of the heater was 10 mm), the area of the dry spots was translated into mm^2, then the total area of dry spots, S_{spot}, on the surface of heater was plotted in relation to the surface area of the heater, S_{heater}, from the heat flux density q. Fig. 3 is a plot of the area of dry spots from the heat flux at various flow rates of liquid and gas.

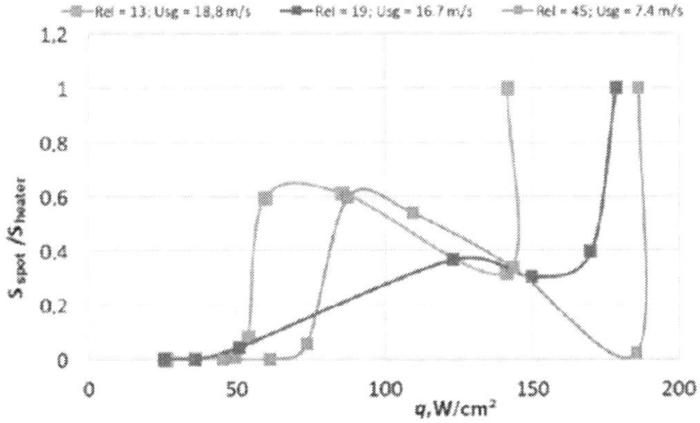

Fig. 3. Dependence of the area of dry spots on the heater surface from the heat flux at various flow rates of liquid and gas.

It can be concluded that the behavior of the liquid film does not qualitatively differ with different heat fluxes. Also there is a qualitative correspondence of the data obtained as a result of processing photos with the help of the ImageJ program and visual observations of the behavior of the liquid film with increasing heat flux density.

With the help of high-speed visualization, the dynamics of formation and washing of a dry spot was obtained. At the initial moment, we can observe a continuous film of liquid, then the film is nonuniform, then it ruptures very quickly, which leads to the formation of a dry spot, further development of this spot, and then after a short period of time, this spot blurs, so the lifetime of one spot is of the order of 1 ms. The size of such spots is about 100 microns. It was also found that with a heat flux of 450 W/cm^2, about 1 million dry spots per 1 second are formed and washed on the surface of the heater (1 cm^2).

For a dry spot the speed of the contact line was calculated (Fig. 4, 5), the maximum speed of the contact line is observed at the time of formation of the dry spot and reaches 10 m/s (because of such high speeds, the frequency of shooting in our experiment does not allow us to understand in detail the mechanism of formation of a dry spot). Further, with an increase in the dimensions of the dry spot, the velocity of the contact line on average takes a value of about 10 cm/s, the speed of the contact line when the dry spot is washed is an order of magnitude greater than increasing, and is of the order of 100 cm/s.

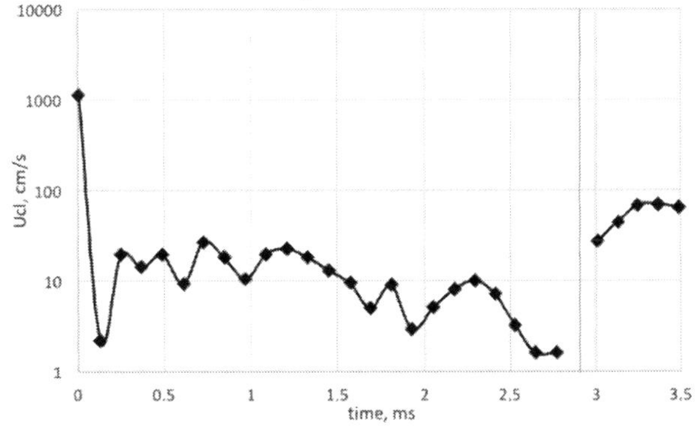

Fig. 4. Speed of the contact line during the formation, increase and washing of the dry spot No 1 (the red line separates the process of increase from the washing of the dry spot).

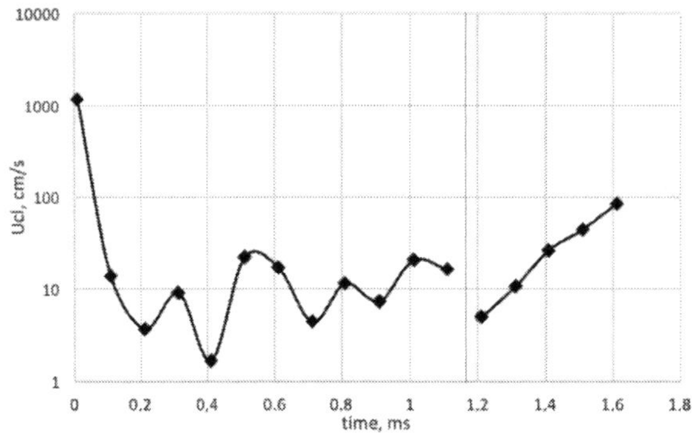

Fig. 5. Speed of the contact line during the formation, increase and washing of the dry spot No 2 (the red line separates the process of increase from the washing of the dry spot).

5 Conclusions

During the experiment, the total area of dry spots increases with increasing heat flow and heater temperature, but when the heater reaches a certain temperature (≈100 °C), the total area begins to decrease. The speed of the contact line when forming a dry spot reaches 10 m/s. With an increase in the size of the dry spot, the velocity of the contact line on average takes on the order of 10 cm/s, and when the dry spot is washed, the speed of the contact line is an order of magnitude larger than when increasing, and is of the order of 100 cm/s. It was also found that with a heat flux of 450 W/cm^2, about 1 million dry spots per 1 second are formed and washed on the surface of the heater (1 cm^2).

The work was supported by the Russian Science Foundation, Project No. 14-19-01755.

References

1. A. Bar-Cohen, C. Holloway, J. Phys. Conf. Ser. **745**, 022002 (2016)
2. I. Mudawar, W. Qu, Int. J. Heat Mass Transf. **47**, 2045 (2004)
3. J. Kim, Int. J. Heat Fluid Flow **28**, 753 (2006)
4. O. A. Kabov, Yu. V. Lyulin, I. V. Marchuk, D. V. Zaitsev, I Int. J. Heat Fluid Flow **28**, 103 (2007)
5. D. V. Zaitsev, D. A. Rodionov, O. A. Kabov, Tech. Phys. Lett. **35**, 680 (2009)
6. O. A. Kabov, D. V. Zaitsev, Multiphase Science and Technology **21**, 249 (2009)
7. O. A. Kabov, D. V. Zaitsev, V. V. Cheverda, A. Bar-Cohen, Exp. Therm. Fluid Sci. **35**, 825 (2011)
8. E. A. Chinnov, O. A. Kabov, A. V. Muzykantov, D. V. Zaitsev, Int. J. Heat Technol. **19**, 31 (2001)
9. E. M. Tkachenko, D. V. Zaitsev, E. V. Orlik, J. Phys. Conf. Ser. **754**, 032019 (2016)
10. D. Zaitsev, E Tkachenko, E. Orlik, O Kabov, MATEC Web Conf. **92**, 01037 (2016)
11. D. Zaitsev, E. Tkachenko, O. Kabov, EPJ Web Conf. **159**, 0054 (2017)
12. D. Zaitsev, O Kabov, MATEC Web Conf. **84**, 00043 (2016)
13. E. M. Tkachenko, D.V. Zaitsev, MATEC Web Conf. **72**, 01114 (2016)
14. T. Hirokawa, H. Ohta, O. Kabov, Interfacial Phenomena and Heat Transfer **3**, 303 (2015)

Algorithm for calculation of a CCGT of a trinary type with an air condenser

Svyatoslav Tsibulskiy[1,*], *Nikolay* Galashov[1], *Denis* Mel'nikov[1], *Alexandr* Kiselev[1], and *Al'bina* Bannova[1]

[1]National Research Tomsk Polytechnic University, 634050 Tomsk, Russia

Abstract. In this paper, an algorithm and a program have been developed that make it possible in a complex to calculate the combined operation of a combined-cycle gas turbine (CCGT) and air condenser (AC), as a result of which it is possible to investigate their operation and performance, depending on a large number of operating parameters.

1 Introduction

The binary type CCGTs currently have the highest efficiency of generating electricity. The most powerful and perfect CCGT is based of GTU 9NA.02 of GE Power, the efficiency of which reached 63.7%. The working body of the steam turbine unit (STU) of this CCGT is water. The ways of further increasing the efficiency of the combined-cycle plant are given in the paper [1]. This is an increase in the initial temperature of a gas turbine, a decrease in heat losses in a steam turbine cycle and a heat recovery steam generator (HRSG), lower temperature heat rejection from the cycle.

The decrease in the heat removal temperature affects the efficiency of the cycle more than the increase in the temperature of the heat supply, but in the gas turbine it is limited by the process in the gas turbine, and in the STU – by the temperature of the cooling medium and the process in the condenser. When the STU is operating on water steam, and cooling the steam in the condenser with water, the heat removal temperature in the cycle can not be lower than 15 °C. In winter, the heat removal temperature can be lowered below 15 °C if the water steam in the condenser is cooled by air, but at the same time the reliability of the installation is reduced due to the possibility of freezing of water in the condenser tubes. Reduce the temperature of heat removal in the cycle below 0 °C and reliably work when using AC is possible, if low-boiling substances (LBS) used as a working fluid in the medium cycle of STU [2].

At a given steam temperature before the turbine, the increase in the efficiency of the STU is possible only due to the decrease in the heat removal temperature in the cycle, therefore it is proposed in [3] to use a low-boiling substance as the working fluid of the cycle of the STU, and to remove heat from it in an air condenser. This in winter will allow to reliably remove heat in the cycle at negative temperatures and increase the efficiency of the STU and CCGT.

* Corresponding author: stzibulsky@tpu.ru

2 Algorithm for calculation of a CCGT of a trinary type with an air condenser

The algorithm is based on the calculation of the cycle diagram of CCGT of a trinary type with AC. In the CCGT, the first cycle is the Brighton cycle for the gas turbine unit (GTU), the second and third are the Rankine cycles operating on water steam and low-boiling substances (LBS). The cycle diagram of the CCGT is shown in Fig. 1.

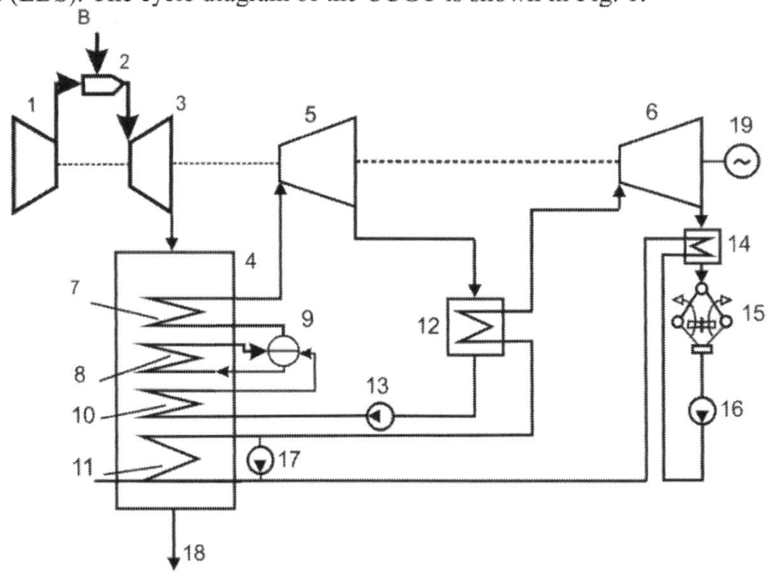

Fig. 1. The cycle diagram of the CCGT trinary type with ondensation low boiling substances in air condenser: 1–compressor; 2–the combustion chamber; 3–gas turbine; 4–HRSG; 5–high pressure cylinder (HPC); 6–low-pressure cylinder (LPC); 7–superheater; 8–evaporator; 9–boiler drum, 10–high-pressure economizer, 11–low-pressure economizer; 12–condenser-evaporator; 13–condensate pump of the second cycle (CP_2); 14–steam cooler; 15–AC; 16–condensate pump of the third cycle (CP_3); 17–recirculation pump, 18–exhaust gases, 19–electric generator.

The initial data:

- type of GTU;
- electric power and termal efficiency GTU;
- combustible fuel composition;
- flow rate and temperature of the combustion products at the inlet to HRSG;
- pressure and temperature of water steam at the entrance to the steam turbine;
- pressure of water steam at the outlet of the steam turbine;
- type of LBS;
- cooling air temperature in the air condenser;
- the geometric characteristics of the air condenser.

For GTU, the heat input to the combustion chamber, the composition and parameters of the combustion products is determined from its initial data. The composition of the combustion products is determined by the following function:

$$c_{C.P.}=f\{N_2;\ Ar;\ O_2;\ H_2O;\ CO_2\},\qquad (1)$$

which depends on the nitrogen content of N_2, argon Ar, oxygen O_2, H_2O steam, CO_2 carbon dioxide CO_2.

The process of calculating the steam-turbine part has an iterative form [5]. For the cycle on water steam are determined the initial parameters at the outlet of the HRSG, is constructed the process of expansion in the high-pressure cylinder and is determined its power:

$$N_E^{HPC}=D_0^{HPC}\cdot H_i^{HPC}\cdot \eta_{EM}\cdot 10^{-3},\ MW, \quad (2)$$

where D_0^{HPC} – water steam flow rate in HPC, kg/s – for the first iteration is given an arbitrary value subsequently clarified; H_i^{HPC} – effective heat drop in the high pressure cylinder, kJ/kg; η_{EM} – electromechanical efficiency of steam-turbine plant.

The properties of the condensation of water steam and the conditions for the transfer of heat to the low-boiling matter in the condenser-evaporator are determined from the pressure at the outlet of the high-pressure cylinder. Calculate the water parameters after the condensate pump – at the inlet to the HRSG:

$$h_{IN}^{EC1} = (h'_W)^{CE}+\Delta h_{CP2},\ kJ/kg, \quad (3)$$

where $(h'_W)^{CE}$ – enthalpy of water condensate in the CE, kJ/kg; Δh_{CP2} – increase in enthalpy in a high-pressure condensate pump, kJ/kg.

The properties of water, steam and combustion products in the waste heat boiler are calculated. Temperature and enthalpy of combustion products at the cold end of the evaporator:

$$\theta_E = t_S^E+\delta t_E,\ ^\circ C; \quad (4)$$

$$I_E = f(c_{C.P.};\ \theta_E;\ p_{C.P.}),\ kJ/kg, \quad (5)$$

where t_S^E – saturation temperature in the evaporator, °C; δt_E – temperature difference at a pinch-point of evaporator, °C; $p_{C.P.}$ – combustion pressure in the path of the HRSG.

The flow rate of water vapor is determined [4]:

$$D_0^{HPC}(h_{0W.S.} - (h'_W)^E) = G_G\cdot(I_{IN}^{HRSG} - I_E), \quad (6)$$

where $h_{0W.S.}$ – enthalpy of water steam at the outlet of the high-pressure superheater, kJ/kg; $(h'_W)^E$ – enthalpy of water at the inlet to the evaporator, kJ/kg; G_G – consumption of combustion products, kg/s; I_{IN}^{HRSG} – enthalpy of combustion products at the outlet from the GTU – inlet to the HRSG, kJ/kg; I_E – enthalpy of combustion products at the evaporator outlet, kJ/kg.

If obtained in (1.6) flow rate of water steam has a predetermined divergence in (1.2) is more than 0.5 %, then the calculation is repeated by substituting found in (1.2) flow rate.

For given type of LBS and water steam property are determined properties of LBS and its flow in CE:

$$(t_S^{LBS})^{CE} = t_S^{CE} - \delta t_{CE},\ ^\circ C; \quad (7)$$

$$p_S^{CE} = f(LBS;\ (t_S^{LBS})^{CE}),\ MPa; \quad (8)$$

$$(h''_{LBS})^{CE} = f(LBS;\ p_S^{CE}),\ kJ/kg; \quad (9)$$

$$D_0^{LPC} = Q_{CE}/((h''_{LBS})^{CE} - h_{OUT}^{EC2}),\ kg/s, \quad (10)$$

where t_S^{CE} – saturation temperature of water vapor in condenser-evaporator, °C; δt_{CE} – temperature head in condenser-evaporator, °C, h_{OUT}^{EC2} – enthalpy at the outlet of the low pressure economizer, in the first approximation is given an arbitrary value subsequently clarified, kJ/kg.

The verification calculation of the air condenser is necessary for determining the condensation temperature t_C^{LBS} and electric energy consumption for the drive of the blowing fans with the cooling air parameters [6].

The condensation temperature t_C^{LBS}, which is determined from the calculation of the air condenser, calculates the pressure at the end of the expansion process into the low-pressure cylinder:

$$p_k^{LBS} = f(LBS;\, t_C^{LBS}),\ \text{MPa}. \tag{11}$$

Construct process LBS steam expansion in LPC:

$$H_0^{LPC} = h_{0LBS} - (h_{kt}^{LPC})^{LBS},\ \text{kJ/kg}; \tag{12}$$

$$H_i^{LPC} = H_0^{LPC} \cdot \eta_{oi}^{LPC},\ \text{kJ/kg}, \tag{13}$$

where $h_{0LBS} = (h''_{LBS})^{CE}$ – the enthalpy of a LBS steam at the inlet to the LPC, kJ/kg; $(h_{kt}^{LPC})^{LBS}$ – the enthalpy of the low-boiling substance at the end of the theoretical expansion process in the low-pressure cylinder, kJ/kg; η_{oi}^{LPC} – internal relative efficiency of the LPC.

The properties of the LBS at the outlet of the LPC are determined and a conclusion is made about the need to install a steam cooler (SC). Thermal calculation is carried out with the purpose of finding parameters of steam and condensate at the output of SC:

$$D_0^{LPC} \cdot ((h_k^{LPC})^{LBS} - h_{SC}^{LBS}) \cdot \eta_{CP} = D_0^{LPC} \cdot (h_{OUT}^{SC})^{LBS} - h_{CP2}), \tag{1.13}$$

where D_0^{LPC} – LBS steam consumption in LPC, kg/s; $(h_k^{LPC})^{LBS}$ – the enthalpy of the LBS at the end of the expansion process in LPC, kJ/kg; h_{SC}^{LBS} – enthalpy of the cooling vapor of the LBS from the SC, kJ/kg; $(h_{OUT}^{SC})^{LBS}$ – enthalpy of liquid LBS from the SC, kJ/kg; h_{CP2} – enthalpy of liquid LBS from the condensate pump of low pressure, kJ/kg.

The heat balance equation is compiled for the recycling system [7]. The heat balance determines the consumption of recycling:

$$D_R = D_0^{LPC} \cdot (h_{IN}^{EC2} - (h_{OUT}^{SC})^{LBS}) / (h_{OUT}^{EC2} - h_{IN}^{EC2}),\ \text{kg/s}, \tag{14}$$

where h_{IN}^{EC2} – the enthalpy of the input to the low-pressure economizer, kJ/kg; h_{OUT}^{EC2} – the enthalpy of output to low-pressure economizer, kJ/kg.

Determined electric power of LPC:

$$N_E^{LPC} = D_0^{LPC} \cdot (h_{0LBS} - (h_k^{LPC})^{LBS}) \cdot \eta_{EM} \cdot 10^{-3},\ \text{MW}. \tag{15}$$

The total power gross of CCGT:

$$N_E^{CCGT}{}_{GR} = N_E^{GTU} + N_E^{HPC} + N_E^{LPC}. \tag{16}$$

Net efficiency of CCGT trinary type with AC:

$$\eta^{CCGT}{}_{NET} = (N_E^{CCGT}{}_{GR} - N_{ON}) \cdot 100/Q_{CC},\ \%, \tag{17}$$

where N_{ON} – electric power own needs, MW.

For to calculate the parameters of the burned fuel, combustion products, water and steam, and low-boiling substances are used the functions of the library REFPROP [8].

3 Conclusion

The proposed algorithm makes it possible to comprehensively calculate the combined operation of a CCGT of a trinary type and an air condenser.

The use of a LBS in the lower Rankine cycle using air condenser allows wintertime reliably remove heat at low outdoor temperatures and low boiling substances and significantly raise the termal efficiency of the CCGT.

On the basis of the algorithm, a program was developed for the calculation of a CCGT with an air condenser in an Excel package, which makes it possible to carry out complex studies of the CCGT operation with changes in the outside air temperature.

References

1. S. Tsibulskiy, N. Galashov, A. Bannova, D. Melnikov, A. Kiselev, MATEC Web Conf. **141**, 01025 (2017)
2. N. N. Galashov, S. A. Tsibulskii, Power Technology and Engineering **48**, 459 (2015)
3. S. Tsibulskiy, N. Galashov, A. Bannova, D. Melnikov, A. Kiselev, MATEC Web Conf. **110**, 01027 (2017)
4. A. G. Kostyuk, V. V. Frolov, A. E. Bulkin, A. D. Trukhny, *Turbines of thermal and nuclear power plants: a textbook for high schools* (Publishing house MEI, Moscow, 2001)
5. A. D. Trukhni, *Steam and gas installations of power plants* (Publishing house MEI, Moscow, 2013)
6. O. O. Milman, V. A. Fedorov, *Air-condensing units* (Publishing house MEI, Moscow, 2002)
7. K. Larionov, D. Gvozdjakov, A. Zenkov, V. Zaytsev, EPJ Web Conf. **110**, 01034 (2016)
8. E. Lemmon, M. Huber, M. McLinden, REFPROP, standard reference database 23, version 8.0, National Institute of Standard and Technology (2007)

MATEC Web of Conferences **194**, 01061 (2018) https://doi.org/10.1051/matecconf/201819401061
HMTTSC-2018

Research of temperature distribution in rooms with radiant heating systems

Ivan Voloshko[1,*], and *Vasiliy* Ushakov[1]

[1]National Research Tomsk Polytechnic University, 634050 Tomsk, Russia

Abstract. The results of temperature measurements in model rooms of a closed type with a system of radiant heating are presented. The temperatures of different designs' surfaces in large-sized rooms, heated by a gas infrared emitter of light type, are established. Conclusions are drawn about the mechanism of the medium's heating, depending on the conditions of heat exchange, and also on providing worker's comfortable working conditions, when the medium is heated by a gas infrared emitter.

1 Introduction

The actual task of the present time is the effective using fuel and energy resources, which is confirmed by the adoption of appropriate regulatory and legal acts at the regional and federal levels.

Modern architectural and constructional solutions of industrial buildings' elements made of lightweight structures require the using of new energy-efficient and energy-saving heating systems. If in the development of project documentation for the construction of new production rooms, energy-efficient heating and ventilation systems are performed with using of modern building structures, then there are many problems with the compensation of heat losses by external walls when using traditional heating systems in large-sized production buildings built many years ago. They are characterized, as a rule, by huge, often empty production areas with thermo-technical properties that do not meet modern requirements. Heat supply of such facilities with the use of traditional heating appliances is inexpedient and impossible since this will lead to irrational energy costs and the inability to ensure a proper microclimate of workers.

Modern practice of heat supply shows that traditional convective heating systems can not effectively heat premises for periodic and short-term use; premises with a partially used area; significant height; distant from heat networks, as well as open and semi-open areas due to their inherent number of fundamental shortcomings.

The solution of tasks to ensure the thermal regime of industrial facilities and create an optimal microclimate in the work area can be the use of autonomous radiant heating systems, for example, gas infrared emitters (GIE) [1-3]. The application of GIE is inextricably linked with the need to study the thermal regimes of working zones in areas with multi-layered enclosing structures, roof and overall equipment in conditions of both open and closed rooms

* Corresponding author: iv_voloshko@mail.ru

[1-6]. The purpose of present research is to register the temperature distributions of the surfaces of the rooms' enclosing structures, to subsequent develop the mathematical model, and, as a consequence, to predict the optimal conditions for the spatial location and angles slope of the GIE.

2 Methods and research's results

To achieve the delivered purpose, experimental plant (Fig. 1) was used, consisting of the main constructive elements: 1) gas infrared emitter – GIE-5 of light type manufactured by "Schwank" s company with a rated heat power of 5 kW; 2) source of gas; 3) model of the heat supply's object; 4) thermocouples chromel-alumel type (junction's thickness is 0.08 mm and measurement's absolute error is 5%, total relative systematic error is 4.6%); 5) ADC - ethernet chassis cDAQ 9181 manufactured by "National Instruments" s company to control the timing, synchronization and transfer of data from a 16-channel, 32-bit isothermal temperature measurement module NI 9214 manufactured by "National Instruments" company; 6) PC.

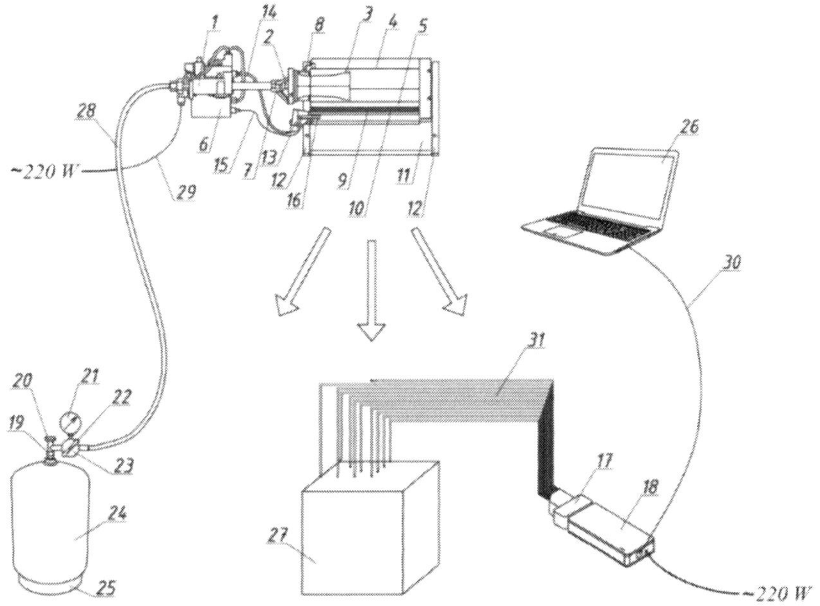

Fig. 1. The scheme of the experimental plant: 1 – valves block, 2 – nozzle, 3 – mixing tube, 4 – gas burner body, 5 – spreaders, 6 – infrared control, 7 – coupling, 8 – connecting tube, 9 – grids, 10 – ceramic tiles, 11 – reflectors, 12 – face fetlock, 13 – protective casing, 14 – high-voltage wire, 15 – flame sensor wire, 16 – spark plug with flame sensor, 17 – temperature measurement module NI-9214, 18 – ethernet chassis CompactDAQ Ethernet cDAQ-9181, 19 – valve, 20 – flywheel, 21 – output pressure manometer, 22 – adjusting screw, 23 – universal regulable reduction gear, 24 – propane gas cylinder's body, 25 – support, 26 – PC, 27 – model of the heat supply's object, 28 – connecting hose, 29 – electrical cable, 30 – connecting cable (twisted pair), 31 – thermocouples.

The experimental plant was located in two enclosed premises with different floor surfaces (white ceramic tiles and black concrete), with overall dimensions 10.2 × 4.9 × 4.4 and 8.7 × 6.0 × 3.4 m, height from floor to the lower edges of the reflector GIE – 2.975 and 2.6 m respectively. The walls were manufactured from brick with a thickness of 70 cm with plastic window openings. The initial air temperature in the room ranged from 7 to 18 °C, while the ambient temperature varied from -15 to -35 °C. During the experiments the GIE was in a stationary position at controlled distances from the walls and in height. The thermocouples were placed on the floor surface and oriented along two coordinate directions (Fig. 2). The beginning of coordinate axes was centre of GIE. The distances (R) between two adjacent thermocouples of the same direction were 0.2 m.

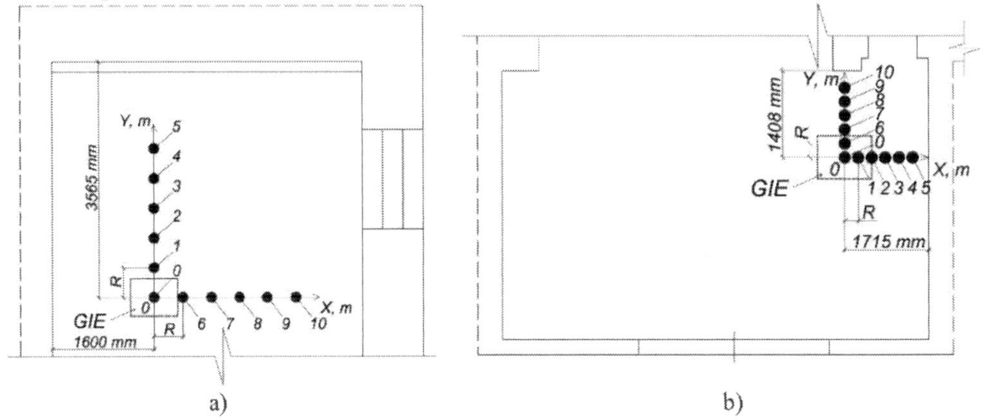

Fig. 2. The schemes of placing GIE and thermocouples in experimental premises with different floor surfaces and geometric dimensions (the numerals correspond to the numbers of the thermocouples): a) - with ceramic tiles and dimensions 10.2 × 4.9 × 4.4 m; b) - with a concrete floor and dimensions 8.7 × 6.0 × 3.4 m.

The typical temperature distributions on different surfaces of the floor in time are shown in Fig. 3, 4.

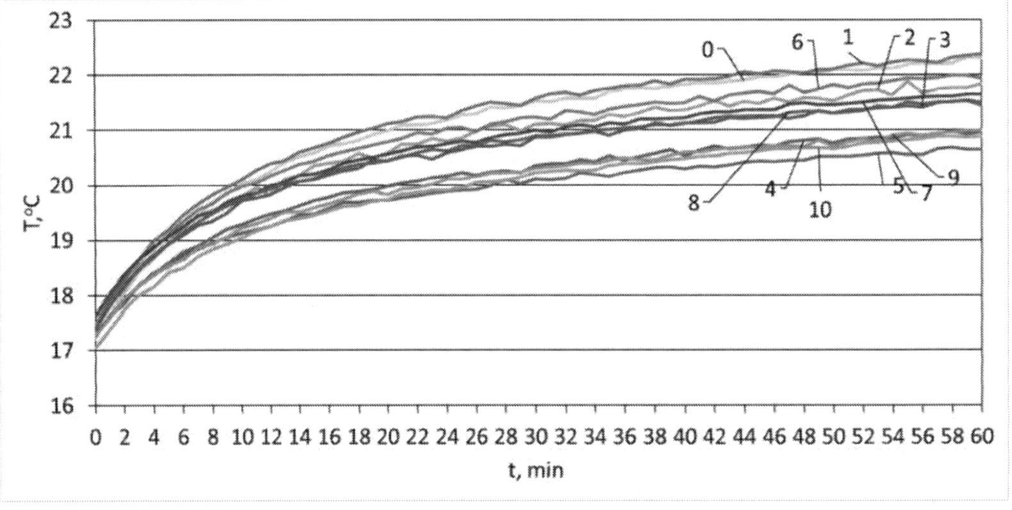

Fig. 3. Change of temperature in time at different points of the floor surface with ceramic tiles (the numerals correspond to the numbers of the thermocouples).

As can be seen from Fig. 3-4, the dynamics of changing the fields' temperatures of floor in time for different premises with different floor surface is the same. The maximum temperature of the floor surface is reached under the centre of the GIE. Then, as the thermocouples are moved away from the centre, the temperature values decrease unevenly. This unevenness is due to the peculiarity of the ceramic matrix's radiation used in the GIE. 50 minutes after the GIE switch-on, the temperature in the central region of the floor in both cases becomes 22 °C.

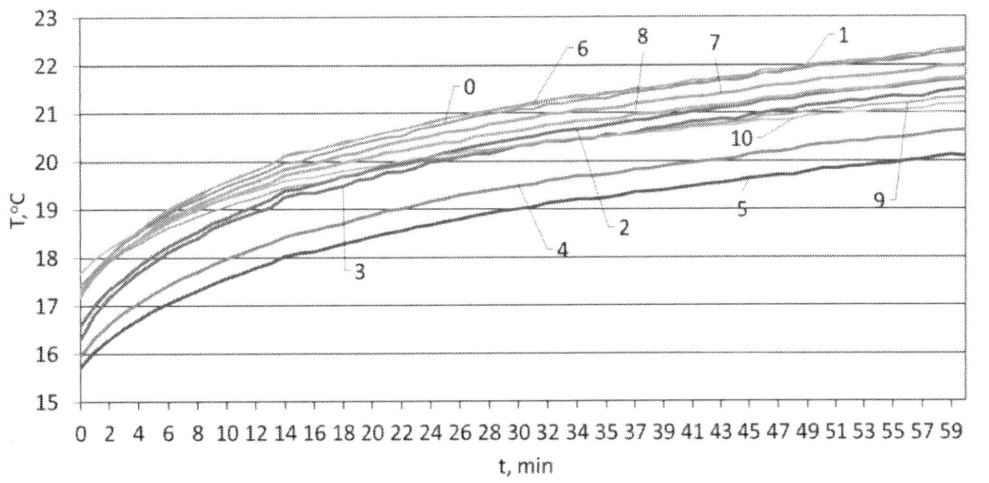

Fig. 4. Change of temperature in time at different points of the concrete floor surface (the numerals correspond to the numbers of the thermocouples).

This allows us to conclude that the temperature distribution on the floor surface of a room heated by a gas infrared emitter does not depend on the characteristic of the radiation of the floor's surface in the optical range. In the infrared range, the emissivity's coefficient of ceramic tiles and concrete is the same and equal to 0.95. The regularities established in the experiments make it possible to formulate the hypothesis that the energy transfer in the "emitter-air-floor surface" system is most likely due to all three mechanisms of heat transfer (radiation, conduction and convection). It can also be concluded that the change in the geometric characteristics of large-sized rooms (volume) by 20% practically does not affect the temperatures' distribution on the floor surface heated by GIE.

The obtained experimental data can be used to improve the methods for calculating thermal regimes of used and designed large-sized premises heated by gas infrared emitters, as well as for developing and verification mathematical models of calculating the thermal regimes of large-sized premises heated by infrared emitters.

References

1. G. V. Kuznetsov, N. I. Kurilenko, V. I. Maksimov, G. Ya. Mamontov, T. A. Nagornova, J. Eng. Phys. Thermophys. **1** (2013)
2. G. V. Kuznetsov, N. I. Kurilenko, G. Ya. Mamontov, L. Yu. Mikhailova, EPJ Web. Conf. **82**, (2015)
3. V. I. Maksimov, T. A. Nagornova, EPJ Web Conf. **76**, (2014)
4. A. N. Ermolaev, S. A. Khaustov, MATEC Web Conf. **110**, (2017)
5. M. Montiel-Gonzales, Appl. Therm. Eng. **75**, 1176 (2015)
6. Q.-q. Li, C. Chen, Y. Zhang, J. Lin, H.-s. Ling, Y. Ma, Build Simul. **7**, 207 (2014)

Ignition of liquid droplets fuels under conditions of radiation-conductive heating in air

Arkadiy V. Zakharevich[1,*], *Mikhail* S. Zygin[1], and *Dmitriy* N. Tsymbalov[1]

[1]National Research Tomsk Polytechnic University, 634050 Tomsk, Russia

Abstract. The results of an experimental investigation of ignition liquid fuel (kerosene, diesel fuel) by the single drops under conditions of radiation-conducting heating in air are presented. The dependences of the ignition delay time of the typical fuels droplets on the initial temperature of the heated oxidant in a limited space are established. The parameters of stable ignition of liquid fuel droplets are set.

1 Introduction

Combustion of sprayed liquid fuels occurs in many energy systems [1,2], installations [3,4] and apparatus [5,6]. The regularities of the droplet combustion of such fuels have been studied for a long time [7,8] and intensively [9]. However, interest in the combustion of liquid droplets is not reduced to the present time [10,11] due to the complex mechanisms of processes of heat and mass transfer and physicochemical transformations occurring in the drop, on its surface and in the near-surface (or adjacent) layer of the vapor-gas mixture. In particular, up to the present time, the temperature field of droplets is often assumed to be uniform when analyzing the conditions and characteristics of the combustion processes of condensed substances [12]. The rate of vaporization of a flammable liquid is generally assumed to depend only on the intensity of the vapor mass transfer process in a thin layer of the external environment adjacent to the surface of the droplet [11]. At the same time the experimental data on the conditions and characteristics of the processes of ignition of fuel droplets and flammable liquids in the medium of high-temperature gases published relatively little. From the review [11] it can be concluded that there are no mathematical models describing the processes of ignition of condensed substance droplets in the oxidizer medium, as well as in the real possible ranges of changes in the main significant factors (ambient temperature, heat exchange intensity, droplet configuration and conditions of its movement, thermal properties of the flammable liquid, etc.). The base in the description of the combustion of drops of liquid is the model of Warsaw formulated many years ago. But the approach [13] does not provide the possibility to analyze the characteristics of the fuel droplet ignition process (primarily the ignition delay time) depending on the conditions of the thermal effect of the environment.

* Corresponding author: bet@tpu.ru

The aim of the work is to establish the conditions and characteristics of typical liquid fuel droplets ignition under conditions of conductive-radiation heating as a result of experimental studies.

2 Experimental method

A specialized experimental stand has been developed to study the integral characteristics of the liquid fuel droplet ignition process (Fig.1), which includes a combustion chamber (1), a high - speed video camera (2) with the "Tema Automotive" software, a motorized coordinate device (3) for moving a drop of fuel at a speed of 0.5 m/s, an electronic dispenser (4) "Finn pipette" (minimum and maximum volumes to be taken is 1 and 10 μl with a variation step of 0.1 μl) for generating drops in the range of 3-8 μl and PC (5).

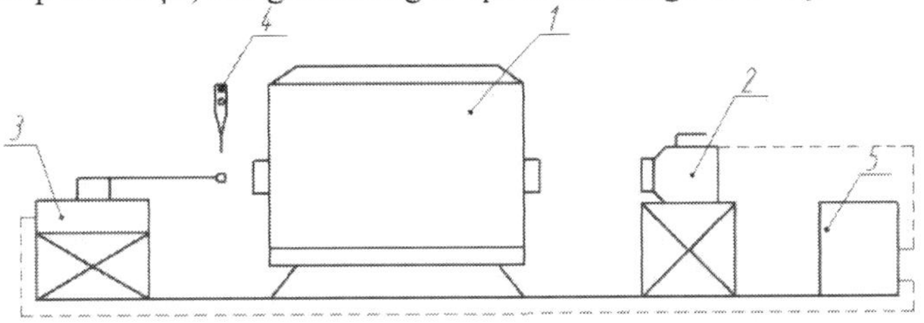

Fig. 1. Schematic diagram of the experimental setup: 1 – tubular oven "Nabertherm R 50/250/13" with integrated temperature controller, 2 – high-speed video camera "Phantom V411", 3 – motorized coordinate device, 4 – electronic dispenser "Finn pipette", 5 – PC.

The objects of the study were widely used liquid fuels – kerosene of the TS-1 brand (GOST 10227-86 with changes 1-3) and winter diesel fuel DT-Z-K3-38 (GOST R 55475-2013). The experiments were carried out under the same and well reproducible conditions (temperature 290-293 K, relative humidity 50-70%, atmospheric pressure 730-760 mmHg.V.). A drop of liquid fuel of a certain volume was dropped by the tip of the Finn pipette dispenser onto the thermocouple junction with a diameter of 0.5 mm. At a fixed speed of 0.5 m/s a drop of liquid fuel was moved by a coordinate device to the center of the combustion chamber. In order to preserve the integrity of the droplet and minimize the time of its movement in the chamber, the optimal speed of movement of the coordinate device was established. The values of the oxidizer temperature in the combustion chamber varied in the range of 663-873 K with a systematic error of ±2 K. The temperature range is due to the need to establish the minimum values at which the stable ignition of a drop of liquid fuel was provided. The processes taking place in the vicinity of the distillate fuel droplet were recorded by a color high-speed camera. After processing the video footage of the experiments, the ignition delay times were determined (τ_{ign}). For each type and volume of the droplet at a constant initial temperature (T_p) in the combustion chamber, a series of (6 – 8) experiments were carried out. The τ_{ign} parameter characterized the time from the moment of introduction of a drop of distillate fuel into the tube space of the combustion chamber until the flame appears. After this was determined the average for each constant value of T_p the values of τ_{ign} and the standard deviation. As a result of the analysis of ignition delay times for the performed experiments, it was found that the coefficients of variation varied from 0.15 sec (at high T_p) to 1.1 sec (at relatively low T_p).

3 Results and discussion

Fig. 2 shows the experimental dependences of the ignition delay time of distillate fuel droplets on the initial temperature in the combustion chamber. The parameters of stable ignition of liquid fuel droplets are set. Each fuel is characterized by a minimum value of the ignition process and stable combustion (T_p). According to the results of the experiments, it can be concluded that the droplet size has little effect on the ignition delay time of distillate fuels.

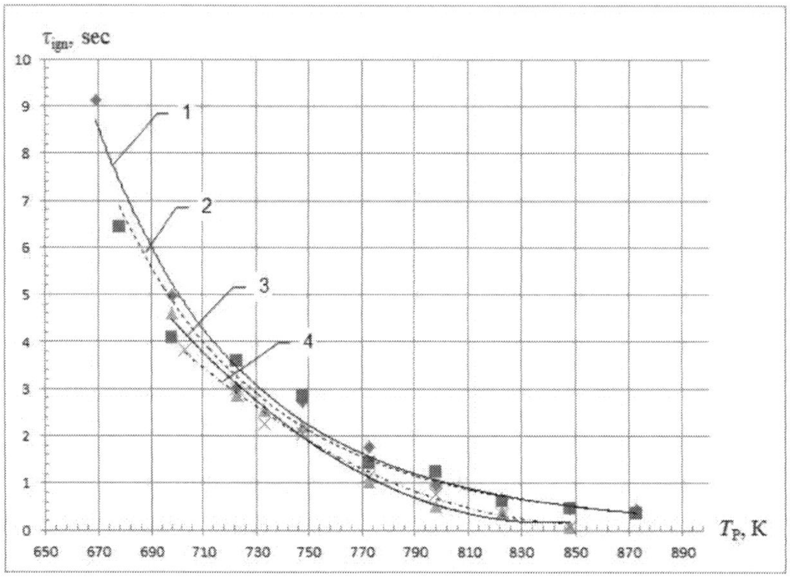

Fig. 2. Experimental dependences of delay time of ignition the drops of distillate fuels from the initial temperature in the combustion chamber: 1,2 – diesel; 3,4 –kerosene; 1,3 – drop volume of 7 μl; 2,4 – drop volume of 4 μl.

Deviations for drops by τ_{ign} of 7 ml and 4 μl do not exceed random measurement errors of the main characteristics of the process. The mechanism of the established regularities should be investigated in the future using the means of registration of a higher level of detail.

4 Conclusions

On the basis of the results of experimental studies, the limiting modes of droplet ignition of fairly typical liquid fuels under conditions of radiation-conductive heating in the air are established. The parameters of stable ignition of liquid fuel droplets are set. The obtained experimental data allow us to conclude that the droplet sizes have little effect on the ignition delay time of distillate fuels.

This work was supported by the Russian Foundation for Basic Research (project No. 17-29-05093).

References

1. S. Y. Misyura, Nature: Scientific Reports **7**, 14759 (2017)
2. S. Y. Misyura, Int. J. Heat Mass Transfer **166**, 667 (2018)
3. O. P. Korobeinichev, A. G. Shmakov, V. M. Shvartsberg, A. A. Chernov, S. A. Yakimov, K. P. Koutsenogii, V. I. Makarov, Fire Safety J. **51**, 102 (2012)
4. A. Yu. Varaksin, High Temp. **51**, 377 (2013)
5. D. Tarlet, C. Allouis, J. Bellettre, Int. J. Therm. Sci. **107**, 179 (2017)
6. B. Kichatov, A. Korshunov, A. Kiverin, E. Son, Fuel Proces. Technol. **166**, 77 (2017)

7. E. A. Salgansky, A. Y. Zaichenko, D. N. Podlesniy, M. V. Salganskaya, M. Toledo, Int. J. Hydrogen Energ. **42**, 11017 (2017)
8. V. E. Zhukov, A. N. Pavlenko, M. I. Moiseev, D. V. Kuznetsov, High Temp. **55**, 79 (2017)
9. S. S. Sazhin, E. A. Shchepakina, V. A. Sobolev, Combust. Flame **187**, 122 (2018)
10. K. Warncke, S. Gepperth, B. Sauer, A. Sadiki, J. Janicka, R. Koch, H.-J. Bauer, Int. J. Multiph. Flow **91**, 208 (2017)
11. S. S. Sazhin, Fuel **196**, 69 (2017)
12. S. S. Sazhin, Prog. Energy Combust. Sci. **2**, 162 (2006)
13. G. A. Varshavsky, Proceedings of NII-1 **6** (1945)

Mathematical modeling of the melting of ice under conditions of motion of a heat drill

Ekaterina Zakharova[1,*], and *Mikhal* Purin[1]

[1]National Research Tomsk Polytechnic University, 634050 Tomsk, Russia

Abstract. The mathematical modeling results of the simultaneous processes of heat and mass transfer under the conditions of intense phase changes (melting of ice) during the movement of cryobot have been given. The spatial unevenness of the melting rate of ice has been taken into account. It has been established that the rate of passage of the cryobot depends essentially on its temperature. According to the results of the numerical simulation, considerable cooling of the cryobot sheath has been established. The latter is due to the high endothermic effect of melting ice.

1 Introduction

One of the more effective ways of studying the structure, the material composition and the dynamics of ice deposits is drilling wells. The main aim of deep drilling was initially to obtain continuous ice-core samples. Its study allows getting the most complete information about the gas composition, the degree of its dustiness and the temperature fluctuations of our planet for the last 900 thousand years [1].

When drilling ice, like any breed, there are problems. As is known, under the influence of gravity the glaciers are in a stressed state. The walls of the well become unstable due to deformation of the ice.[2] There is a narrowing of the well, which in the future can lead to accidents. To prevent this situation, the well is filled with a special fluid that compensates for the pressure of the ice. Recently, during drilling, a mixture of a hydrocarbon liquid is used that is similar to kerosene with difluoroethane. The mixture at low temperatures (up to -60 °C) maintains fluidity, and its density is close to the density of ice. The main disadvantage is the toxicity for living microorganisms. As a consequence, the samples taken can be contaminated [3].

This process can be carried out in three ways: the first variant involves the thermal destruction of ice structures [4]; the second variant involves the use of mechanical action on the layers of ice ("classical" drilling) [5]; the third variant is based on the combined chemical and thermal effects [6].

The third way is to use a melting probe. Its movement is carried out by heating the ice to the formation of melt water. Water flows around the cryobat and freezes behind it, while the cryobot itself continues to move under the influence of gravity. The use of cryobot avoids the previously described problems, since a mine is not created for its operation, and the impact on ice occurs only at the expense of high temperature [7]. Energy costs for its

* Corresponding author: eaz15@tpu.ru

use are relatively small. However, like any method, there are drawbacks: low penetration rate and the need for a liquid phase [5]. The representative of this method is the cryobot specialized probe-robot of deep-glacial penetration – cryobot.

In the summer of 2001, the cryobot was tested on the Longyear Glacier, where a well of 23 meters deep was drilled without complications [8]. At the current stage of the development of science and technology, this depth is maximal. Most likely, this is due to the poor knowledge of the processes of melting and crystallization of massive ice structures by a local source of heating. This is due to the fact that a phase transition front is formed in the melting process, which is characterized by the presence of a powerful and singularly concentrated heat sink, as well as by a jumplike change in the thermophysical characteristics of ice upon passage through it. In addition, it should be noted that the shape of this phase transition is not substantially linear and depends on the divergent distribution of heat fluxes in the ice, respectively, we can say that the processes of phase transformations during the melting of ice are actual unresolved problems of modern physics.

The aim of this work is the mathematical modeling of heat transfer processes that occur together under ice melting conditions during the movement of the cryobot.

2 Physical statement

At a time t = 0, the heat flux is acts by a limited region of ice. Ice heats up and melts. A heterogeneous structure is formed, consisting of a solid and a liquid phase of water [9,10]. In solving the problem, ice has been assumed to be homogeneous. At the initial time, the temperature of the entire volume of ice has been assumed to be the same.

3 Mathematical statement

The equation for "the ice-water" system has the following form [7, 11, 12]:

$$\frac{C\cdot P}{Fo}\cdot\frac{\partial\theta}{\partial\tau}+U\cdot Pe\frac{\partial\theta}{\partial X}+V\cdot Pe\frac{\partial\theta}{\partial Y}=\mathrm{div}(\Lambda\cdot\mathrm{grad}\theta)\mp Pom_{melt}. \quad (1)$$

where Fo – the Fourier's criterion; Pe – the Peclet's criterion; θ – dimensionless temperature; Λ – dimensionless thermal conductivity; Pom_{melt} – Pomerantsev's criterion; U, V – dimensionless water velocities; $C=c/c_0$, $P=\rho/\rho_0$, $\Lambda=\lambda/\lambda_0$, $\theta=T/T_0$; $X=x/l, Y=y/l$ – dimensionless heat capacity, density, coefficient of thermal conductivity, temperature and dimensionless steps in area; c_0, ρ_0, λ_0, T_0, l – parameters of dimensionization.

The Pomerantsev's criterion:

$$Pom_{melt}=\frac{Q\cdot W\cdot l_0^2}{h\cdot\lambda_0\cdot T_0},$$

where Q – melting heat of ice, J/kg; W – melting rate, m/s; h – step in area, m; T_0 – temperature, K.

To solve this equation, the following assumptions are made:

1. The water that forms during melting has a very thin layer that "sticks" to the cryobot and ice. Accordingly, the speed of water movement is not taken into account;

2. The temperature field of the ice at the initial instant of time is homogeneous;

3. Heat transfer from the Z coordinate is not taken into account.

For the convenience of analyzing the information obtained, the calculations were carried out in dimensionless quantities. The basic equation "the ice-water" system is the energy equation [12]:

$$\frac{C \cdot P}{Fo} \cdot \frac{\partial\theta}{\partial\tau} = \frac{\partial}{\partial X}\left(\Lambda \frac{\partial\theta}{\partial X}\right) + \frac{\partial}{\partial Y}\left(\Lambda \frac{\partial\theta}{\partial Y}\right) \mp Pom_{melt}. \quad (2)$$

The equation energy for cryobot:

$$\frac{C_c \cdot P_c}{Fo} \frac{\partial\theta}{\partial\tau} = \Lambda_c \left(\frac{\partial^2\theta}{\partial X^2} + \frac{\partial^2\theta}{\partial Y^2}\right) + Pom_{tv}, \quad (3)$$

where C_c, P_c, Λ_c – dimensionless heat capacity, density and coefficient of thermal conductivity of cryobot; Pom_{tv} – Pomerantsev's criterion of thermal emission. [12] The mathematical statement also includes the corresponding boundary conditions:

$$x = 0 \;\; 0 < y < H \;\; \frac{\partial\theta}{\partial X} = 0, t > 0; \; y = 1 \;\; 0 < y < H \;\; \frac{\partial\theta}{\partial X} = 0, t > 0;$$

$$y = 0 \;\; 0 < x < l \;\; \frac{\partial\theta}{\partial Y} = 0, t > 0; \; y = H \;\; 0 < x < x_1 \;\; \Lambda_{ice}\frac{\partial\theta}{\partial X} = \Lambda_{air}\frac{\partial\theta}{\partial X};$$

$$y = H \;\; x_2 < x < l \;\; \Lambda_{ice}\frac{\partial\theta}{\partial X} = \Lambda_{air}\frac{\partial\theta}{\partial X} \;\; H < y < H_1 \;\; x = x_2 \;\; \Lambda_c\frac{\partial\theta}{\partial X} = \Lambda_{air}\frac{\partial\theta}{\partial X} + Ki;$$

$$H < y < H_1 \;\; x = x_1 \;\; \Lambda_c\frac{\partial\theta}{\partial X} = \Lambda_{air}\frac{\partial\theta}{\partial X} + Ki; \; H < y < H_1 \;\; x = x_2 \;\; \Lambda_c\frac{\partial\theta}{\partial X} = \Lambda_{air}\frac{\partial\theta}{\partial X} + Ki,$$

where Ki – criterion of Kirpichev [13].

4 The results

According to the results of the mathematical modeling of cryobot movement, the formation of a vertical shaft with even edges has been established.

Fig. 1 "a" and 1"b" show the temperature fields in the "cryobot-ice-air" system the moment of immersion of the cryobot at depth 1m. It can be noted that ice heats up slightly and temperature gradients are present only in the region near the device. A nonlinear fusion front is formed, the shape of which is equidistant to the shape of the cryobot. Analysis of Fig. 2 "a" and 1 "b" shows that an increase temperature in cryobot by 50 K (from 450 K to 500 K) leads to the fact that at the same time the average bulk cryobot temperature at the initial temperature T = 450 K is substantially lower than the cryobot temperature at the initial value T = 500.

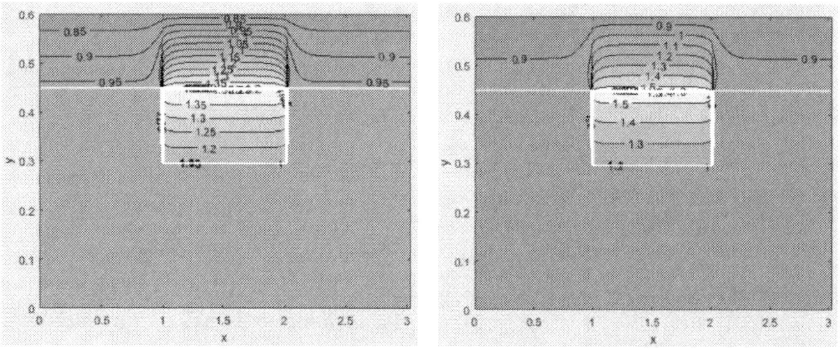

(a) (b)

Fig. 1. (a) The temperature field in the , "ice-cryobot-air" system with a temperature of the last 450 K. (b) The temperature field in the "ice-cryobot" system at a temperature of 500 K.

Fig. 2 "a" and 2 "b" show the tempature fields at the moment of the cryoblot sinking at various initial ice temperatures. The temperature of the ice varied in the range 263K-253K. It can be noted that when the temperature of the ice decreases by 10 K (from 263 to 253), the degree of cooling of the cryobot does not increase. So the average volume temperature of the cryobot has practically not changed. At the same time, practically identical temperature gradients are formed in the cryobote. This is due to the fact that when the cryobot moves in the region of conjugation of the walls of the heat drill and ice, a water membrane is formed, the temperature of which is equal to the phase transition temperature of 273 K. Therefore, the cooling degree of the cryobot remains practically unchanged.

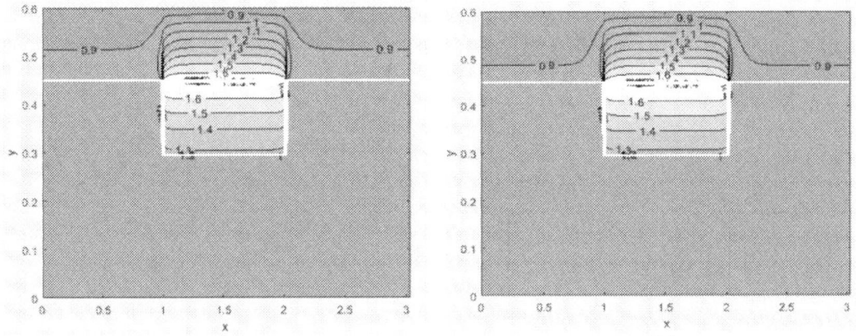

Fig. 2. (a) The temperature field in the "ice-cryobot-air" system at an ice temperature 263 K;
(b) The temperature field in the "ice-cryobot-air" system at an ice temperature 253 K.

5 Conclusion

The article presents a mathematical model of the process of cryobot movement in the conditions of melting of massive ice structures. According to the results of mathematical modeling, it is established that the initial temperature of the cryobot can have a significant effect on the melting characteristics, while the average temperature of the cryobot with an increase in the cryobot temperature from 450 K to 500 K will increase its average bulk temperature at the time of immersion in ice at 383 K. It was also found that a decrease in the initial temperature of ice from 263 to 253 K leads to an insignificant change in the characteristics of heat and mass transfer during the motion of the thermal barrier into a layer of massive ice structures. This is due to the fact that when the cryobot moves in the region of conjugation of the walls of the heat drill and ice, a water membrane is formed, the temperature of which is equal to the phase transition temperature of 273 K. Such membrane creates a kind of barrier that prevents more intensive cooling of the cryobot.

References

1. N. I. Vasiliev, The results of drilling a 5G well at the Russian station "Vostok" and the study of ice cores **10** (2016)
2. N. I. Vasiliev, *Rational technology of drilling wells in ice massifs using an electromechanical projectile on a cable* (St. Petersburg, 2004) [in Russian]
3. P. G. Talalay, Nature **9** (2006)
4. P. Weiss, K. L. Yung, T. C. Ng, Planet. Space Sci. **56**, 1280 (2008)
5. E. N. Chumachenko, D. Dunham, R. R. Nazirov, V. P. Kulagin, Vestnik of the Nizhny Novgorod University named after H. I. Lobakogo **6** (2011)
6. N. I. Vasiliev, Problems of the Arctic and Antarctic **76**, 78 (2007)

7. G. V. Parfenova, E. N. Chumachenko, Mathematics **13** (2010) [in Russian]
8. P. G. Talalay, Nature **1** (2002)
9. O. N. Korolyova, *Dynamic adaptation in the multi-front problems of Stefan* (Abstract, 2006)
10. J. Jonsson, J. Sundqvist, H. Nguyen, Acta Astronautica **79**, 203 (2012)
11. L. Zelenyi, O. Korablev, M. Martynov, Advances in Space Research **48**, 615 (2011)
12. S. V. Syrodoy, N. Y. Gutareva, R. I. Taburchinov, MATEC Web Conf. **72**, 01109 (2016)
13. S. V. Syrodoy, N. Y. Gutareva, K. A. Bugaeva, MATEC Web Conf. **72**, 01035 (2016)

Technique of measuring the emissivity coefficient of solid materials surface

Irina Zharova[1,*], *Valery* Kuznetsov[1], *Ksenia* Perfilieva[1], and *Irina* Zasadny[1]

[1]National Research Tomsk State University, 634050 Tomsk, Russia

Abstract. The option of non-stationary method of measuring the emissivity coefficient of solid materials surface is presented. Results of a model calculating experiment are given.

1 Introduction

Solid materials with low thermal conductivity as heat-shielding materials (HSM) are used in systems of thermal protection of elements of power plants [1-3]. In the operation of HSM at high temperatures in addition to conductive and convective component, radiation of the material surface makes a significant contribution to the total heat exchange. In this regard, there is a need to obtain data on the emission factors of HSM in the high temperatures close to the temperature of thermal destruction. In experimental studies, radiation, calorimetric and non-stationary methods are used to measure the emissivity coefficient of the surface of different materials [4, 5]. Analysis of measurement methods [5] showed that in the high temperature range it is preferable to use radiation or non-stationary methods. In [6, 7] a non-stationary method of determining the emissivity coefficient is proposed. The method is based on the measurement of the sample temperature during its cooling in a vacuum chamber and the interpretation of experimental data using the apparatus of inverse heat transfer problems (IHTP) [7, 8].

In this paper, we propose a modified version of the method [6], which allows to improve the accuracy and efficiency of measurements of the emissivity coefficient. Improved accuracy and efficiency of measurements is achieved through improved methods of thermophysical experiment. To reduce the measurement time, a device is proposed that provides heating of the HSM sample to the required temperature value directly in the vacuum chamber. The accuracy of the emissivity coefficient is provided by increasing the informativity of thermocouple measurements of temperature during the cooling of the sample.

2 Calculation-experimental technique of measuring the emissivity coefficient

* Corresponding author: zharova@niipmm.tsu.ru

The scheme of the device of measuring the emissivity coefficient is shown in Fig. 1. A feature of the device is the ability to heat the sample directly in the chamber. Since the heater is placed in a vacuum chamber, it eliminates the measurement errors of the initial cooling temperature associated with the movement of the heated sample from the thermostat to the vacuum chamber.

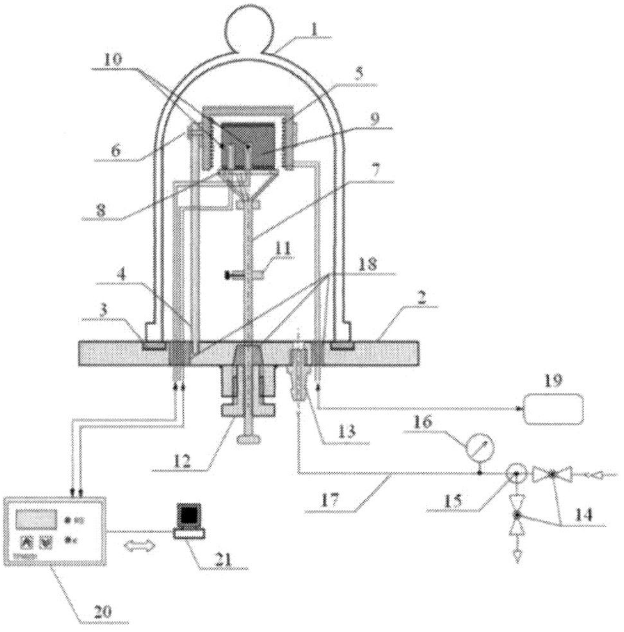

Fig. 1. The scheme of the experimental setup.

The order of measurements is as follows. In the initial position, the cap 1 with blackened walls is removed from the base 2. The movable rod 7 is moved to the lower position until it stops 11. The cylindrical sample 9, the end surfaces of which are covered with a foil with a high reflection coefficient, is coaxially placed on the disc 8 of the movable rod 7. The movable rod 7 is moved to the upper position to place the sample 9 inside the heater 5. To seal the chamber, clamp the seal 12. The glass cap 1 is installed in the groove on the base 2. For complete stabilization of the residual pressure of air in the chamber produces its evacuation. Turn on the voltage source 19 and supply power to the heater 5. The temperature of the sample is controlled by thermocouples 10, connected to the meter-regulator 20. When the required temperature of the sample is reached, the movable rod 7 is moved to the lower position and the heater 5 is switched off. Temperature measurement in the center and on the side surface of the sample during cooling is carried out without turning off the vacuum pump.

The emissivity coefficient is defined from the solution of the inverse heat conduction problem [6–8]:

$$\frac{\partial T(r,t)}{\partial t} = a\left(\frac{\partial^2 T(r,t)}{\partial r^2} + \frac{1}{r}\frac{\partial T(r,t)}{\partial r}\right) \quad (1)$$

with boundary conditions and entry condition:

$$\frac{\partial T(0,t)}{\partial r} = 0\,,\ \lambda\frac{\partial T(R,t)}{\partial r} = -\varepsilon\sigma T^4(R,t)\,, \quad (2)$$

$$T(r,0) = T^{exp}(0,0) + (T^{exp}(R,0) - T^{exp}(0,0))\left(\frac{r}{R}\right)^2 \quad (3)$$

and experimental conditions:

$$T(0,t) = T_1(t)\,,\ T(R,t) = T_2(t)\,, \quad (4)$$

where $T(r, t)$ – temperature dependence on the radial coordinate of the sample and time; t – time; r – radial coordinate of the sample; $a = \lambda/(\rho \cdot c)$ – temperature conductivity coefficient of the sample; ρ, c, λ – density, specific heat, thermal conductivity of the sample; R – the radius of the sample; $\sigma = 5.6687 \cdot 10^{-8}$ Вт/(м^2·K^4) – Stefan-Boltzmann constant; ε – the emissivity coefficient; $T^{exp}(0,0)$, $T^{exp}(R,0)$ – measured temperature values on the axis of symmetry and on the lateral surface of the sample at the start of cooling $t = 0$; $T_1(t)$, $T_2(t)$ – measured temperature dependence time on the axis of symmetry and on the lateral surface of the sample versus time during cooling.

The numerically sought value of the emissivity coefficient is determined by minimizing the functional

$$J(\varepsilon) = \sqrt{\frac{1}{t_k}\sum_{i=1}^{n}\left[\left(T(0,t_i) - T^{exp}(0,t_i)\right)^2 + \left(T(R,t_i) - T^{exp}(R,t_i)\right)^2\right](t_i - t_{i-1})}\,, \quad (5)$$

where t_k – the cooling time of the sample; n – number of points used by time ($i = 1, 2,\ldots, n$); $T(0,t_i)$, $T(R,t_i)$ – calculated values of temperature on the axis of symmetry and on the lateral surface of the sample at the time t_i; $T^{exp}(0,t_i)$, $T^{exp}(R,t_i)$ – the temperature values measured at time t_i on the symmetry axis and on the lateral surface of the sample.

3 Results of a model calculating experiment

The proposed experimental-calculating method was tested by numerical solution of the model problem. Characteristics of the heat-shielding material: density $\rho = 2350$ kg/m^3; specific heat $c = 1000$ J/(kg·K); coefficient of thermal conductivity $\lambda = 1.4$ W/(m·K). The radius of the cylindrical sample $R = 0.015$ m, the initial temperature $T_0 = 293$ K; temperature of thermal degradation 1000 K.

The model calculating experiment consists of two stages.

At the first stage, the temperature field in the sample is determined when heated to the temperature of thermal destruction by a constant heat flow. The temperature of the sample during the heating process is determined from the solution of the heat equation (1) with boundary conditions:

$$\frac{\partial T(0,t)}{\partial r} = 0\,,\ \lambda\frac{\partial T(R,t)}{\partial r} = q \quad (6)$$

and initial condition :

$$T(r,0) = T_0 = \text{const}, \quad (7)$$

where $q = 9.4 \cdot 10^4$ W/m^2 – external heat flow.

It is obtained that on the side surface the set temperature of thermal degradation T_{term} is achieved for $t_k = 86$ c. In Fig. 2 the obtained dependence of temperature on time on the axis of symmetry $T(0, t)$ and on the lateral surface $T(R, t)$ is presented numerically.

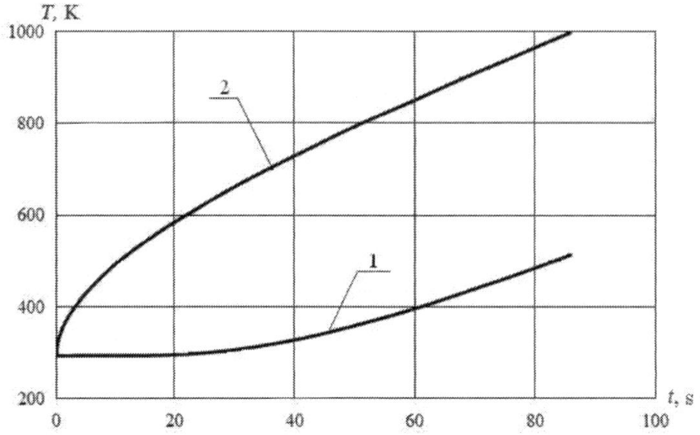

Fig. 2. Temperature dependences time on the axis of symmetry and on the lateral surface of the sample: 1 – $T(0, t)$; 2 – $T(R, t)$.

On the temperature dependences of $T(0, t)$ and $T(R, t)$ temperature distribution along the radius of the sample is determined by the equation [9]:

$$T(r,t) = T(0,t) + (T(R,t) - T(0,t))\left(\frac{r}{R}\right). \tag{8}$$

Temperature profile along the radius of the sample at the time t_k is shown in Fig. 3.

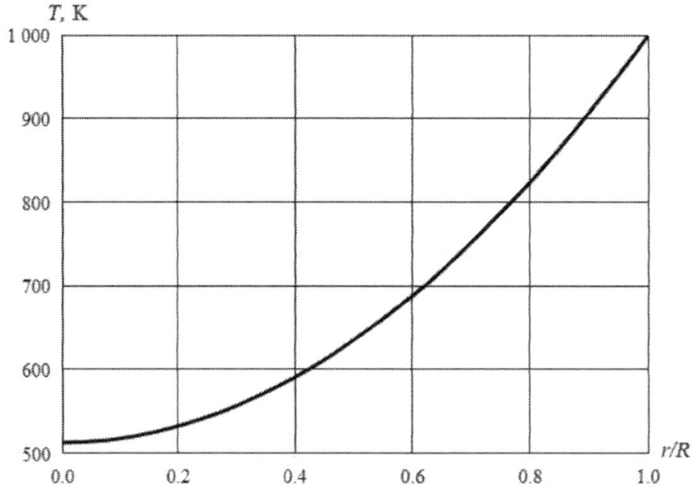

Fig. 3. Temperature profile along the radius of the sample.

At the second stage, the temperature of the sample during cooling under vacuum conditions is calculated on the basis of the solution of the direct problem of heat transfer (DPHT). The mathematical formulation of the DPHT includes equation (1) with boundary conditions (2) and initial condition (3). In the condition (3) as "experimental" temperature values are taken as $T(0, 0) = 513$ K, $T(R, 0) = 1000$ K (Fig. 2) with the introduced perturbations simulating the random error of thermocouple temperature measurements. The initial value of the emissivity coefficient is $\varepsilon_0 = 0.80$.

The sought value of the emissivity coefficient ε is obtained by solving IHTP (1)–(5). In conditions (4) "experimental" dependences $T_1(t)$, $T_2(t)$, are defined as $T(0, t)$, $T(R, t)$ (Fig. 4), that received by solving DPHT while cooling, with the introduced perturbations simulating the random error of thermocouple temperature measurements.

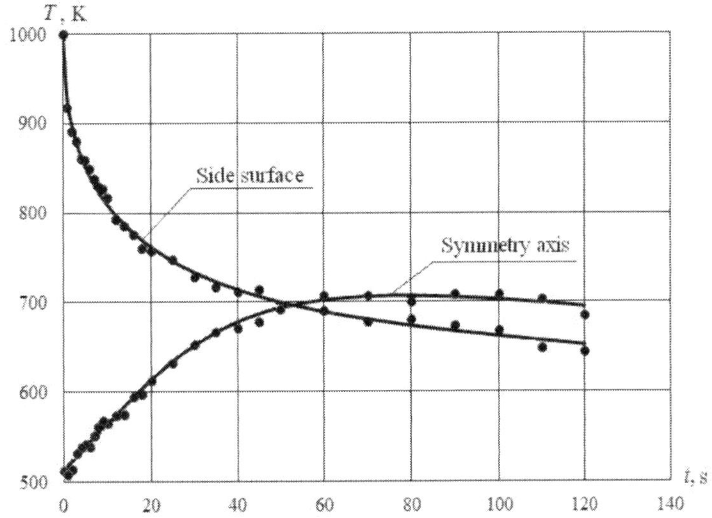

Fig. 4. Temperature dependences on the cooling time of the sample on side surface and on symmetry axis: ● – "experimental"; ▬ – calculated.

By results of calculating experiment we found the specified value ε = 0.805, which is different from the given initial ε_0 less than 1 %.

4 Conclusion

An experimental-calculating technique implementing a modified version of the non-stationary method of measuring of the emissivity coefficient was presented. The technique is approved by carrying out a model calculating experiment. Results of calculating experiment have shown that the offered technique allows to increase the accuracy of determination the emissivity coefficient of solid materials. It is achieved by accounting of the non-uniformity of the temperature field in the sample during cooling, and also due to the increase of the informativity of the measurements due to the placement of an additional thermocouple.

This research was supported by "The Tomsk State University competitiveness improvement programe".

References

1. Yu. V. Polezhaev, F. B. Yurevich, *Thermal protection* (Energy, Moscow, 1976) [in Russian]
2. N. A. Rubtsov, *Radiation heat transfer in continuum medium* (Science, Novosibirsk, 1984) [in Russian]
3. A. G. Blokh, Yu. A. Zhuravlev, L. N. Rizhkov, *Radiation heat exchange: Reference book* (Energoatomizdat, Moscow, 1991) [in Russian]
4. A. Ye. Shtyndlin, *Radiating properties of solid materials. Reference book* (Energy, Moscow, 1974) [in Russian]

5. V. A. Arkhipov, V. D. Gol'din, I. K. Zharova, N. I. Kurilenko, G. Ya. Mamontov, T and A **19** (6), 750 (2012) [in Russian]
6. V. A. Arkhipov, V. D. Gol'din, I. K. Zharova, N. I. Kurilenko, Method for measuring the emissivity coefficient of heat-shielding materials surface Patent RU 2468360 (Int.Cl.: G01N 25/18, G01K 7/02, 2012) [in Russian]
7. A. Arkhipov, V. D. Gol'din, I. K. Zharova, N. I. Kurilenko, Instruments **2** (140), 43 (2012) [in Russian]
8. O. M. Alifanov, *Inverse heat transfer problems* (Mashinostroenie, Moscow, 1988) [in Russian]
9. A. V. Likov, *Heat and mass transfer: Reference book* (Energy, Moscow, 1971) [in Russian]

Influence of dispersion aluminum powder on the burning rate of mixed solid fuel

Nikolay Zolotorev[1,*], *Yana* Dubkova[1], and *Aleksey* Konovalenko[1]

[1]National Research Tomsk State University, 634050 Tomsk, Russia

Abstract. In work results of experimental study of influence dispersion of metal powder on burning rate in mixed solid fuel are provided. As metal fuel used additive of aluminum of different type depending on dispersion of powder. The prepared samples of mixed solid fuel were tested in constant volume combustion chamber. The received results of burning rate show that at reduction of dispersion of aluminum powder the burning rate of mixed solid fuel increases.

1 Introduction

The use of high-energy materials nowadays is important [1]. High-energy materials are now widely used in different areas, in fire-extinguishing systems, orientation of spacecraft, rescue operations etc. [2, 3]. Gas generators which generally represent solid-fuel element which consists of mix components the including fuel and oxidizer.
As is known, the solid-fuel charge of the gas generator has a sufficient amount of oxidizer for combustion in a vacuum. These types of solid fuels are explosive and dangerous during their production, storage and transportation. The danger increases with the size of the charge.

New types of gas generator and engines, operating on metallized mixed high-energy materials, are characterized by a wide range of variation of the regime parameters, for example, the pressure, the relative content of the components, and the metal fuel content and its dispersity. Influence on the main characteristics of the working process must be known in order to achieve a maximum realization of the advantages of metallized mixed high-energy materials. For ensuring increase of power characteristics a combustion catalyst is added to the solid fuel composition or metals powders and their oxides into composition of the mixed solid fuel (MSF) [4]. Preliminary thermodynamic calculation allows defining efficiency of use of metal powders and their oxides entered in composition MSF. However, the thermodynamic analysis does not allow estimating effectiveness influence of dispersion of the entered metal powder. Combustion of metal powder in gas generators turns out to be an extremely complex and multifaceted process.

* Corresponding author: nikzolotorev@mail.ru

2 Object of research

Dependence of burning rate of MSF turns on pressure is one of the important characteristics. For definition of influence dispersion of metal powder on the burning rate of MSF series of experiments on measurement of burning rate has been carried out.

The main components of mixed solid fuel used in gas generators for wide application are an oxidizer and an inert combustible-binder. As model samples MSF used mix from inert combustible-binder 15.8 wt. % butadiene rubber plasticized by transformer oil 20/80, ammonium perchlorate 69.2 wt. % (fraction less than 50 μm and 160–315 μm in the ratio 40 / 60) and 15 wt. % of additive of metal fuel was used. For solidification of mix 0.5 wt. % of hardener of TOH-2 over 100 wt.% were added to it. As metal fuel of the aluminum powder of different dispersion was used [5]. Aluminum is one of the most commonly used metal fuels because of its high oxidation enthalpy, high density, low price and good availability. The use of aluminum powder in samples MSF is promising not only in terms of increasing energy characteristics, but also in terms of reducing the agglomeration on the burning surface of samples, increasing the burning rate, and controlling the exponent in the burning rate law [6].

Table 1. The aluminum powder used in experiments.

Type of aluminum powder	ASD4	ASD6	ASD8	ASD10	ALEX
Size of particles, D_{43}, μm	10.1	6.5	2.3	1.8	0.12

After consecutive mixing of components, the received paste was placed in a mold with the subsequent curing within 24 hours at the room temperature. The studied cylindrical samples of MSF with a diameter of 10 mm and height of 30 mm were produced in the laboratory by extrusion pressing. The samples of MSF were sorted out based on mechanical integrity criteria and on density values. The study uses samples differing on density not exceeding 0.02 g/cm^3.

3 Technique and experimental stand for determination of burning rate

For determination of burning rate different methods can be used [7]. The most widespread method is chamber of constant pressure. In this method combustion of sample of MSF happens with constant pressure and burning rate is defined by usually contact method at combustion of the sample length. This method of determination dependence of burning rate of MSF on pressure is most exact but is very labor-consuming as for each set pressure it is necessary to conduct several experimental tests.

As method measurement of burning rate of samples MSF the burning method in constant volume combustion chamber has been chosen. Use of constant volume combustion chamber allows to determine in one experiment burning rate with the wide range of pressure and to define change of the of burning rate law for the considered range of pressure. The scheme of experimental stand presented in figure 1.

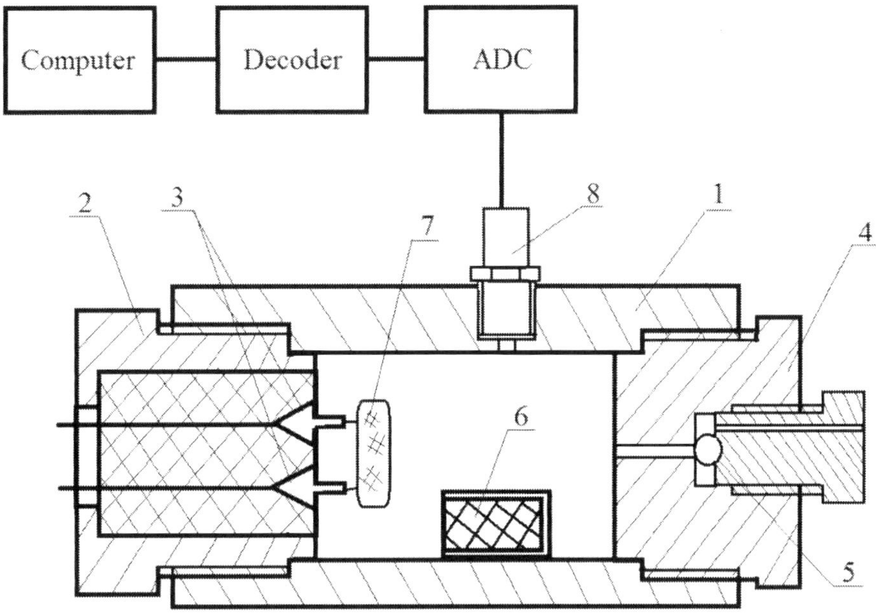

Fig. 1. Scheme of the experimental stand: 1 – chamber casing; 2 – front cover; 3 – insulated wires; 4 – back cover with the pressure release valve; 5 – spherical valve; 6 – MSF sample; 7 – igniter; 8 – pressure sensor.

The volume of constant volume combustion chamber equal to V = 40 cm^3 that at given sizes of MSF sample to provide the rise rate of pressure not less than 10 MPa/s and maximum pressure is not higher than 30 MPa. Measurement of pressure in constant volume combustion chamber was carried out by the pressure sensors tensometric LH-412/400 with use of the digital AT1-8 amplifier and registration of output signal for the computer. The total measurement error does not exceed 1 %.

The ignition of sample was carried out using charge of gunpowder weighing 0.1 g. The influenced of gunpowder combustion products on the test sample weighing more than 1 grams can be neglected. Pressure in constant volume combustion chamber does not exceed 30 MPa, and temperature of combustion products remains close to MSF sample combustion temperature. Thermal-physical properties of gas can be considered constant and equal to thermal-physical properties combustion products of the sample. Thus, law of conservation of energy and law of conservation of mass for combustion chamber internal volume is written as:

$$\frac{dm}{dt} = G_F \tag{1}$$

$$\frac{d(mU)}{dt} = G_F H_F - \frac{dQ}{dt} \tag{2}$$

where m – mass of gas; $G_F = \rho_F u_F S_F$ – formation of gas from combustion of a sample; ρ_F – density of the sample MSF; u_F – burning rate of the sample MSF; S_F – burning surface area; U – internal energy of gas; H_F – enthalpy of the sample MSF; Q – the amount of heat expended on heating the chamber casing; t – time.

Using equations (1) and (2), one can obtain expression for burning rate of the sample MSF:

$$u_F = \frac{V\frac{dp}{dt} + (k-1)\frac{dQ}{dt}}{(k-1)\rho_F S_F H_F - pS_F} \quad (3)$$

where V – combustion chamber internal volume, k – heat capacity ratio for combustion products of the samples MSF.

By the developed technique of determination of burning rate experimental studies of MSF sample combustion have been conducted. The received results of experiments are presented in the form of the degree law of burning rate.

$$u = u_0 p^{\nu} \quad (1)$$

where u – the burning rate, u_0 – burning rate at atmospheric pressure p – pressure in the combustion chamber, ν – is the exponent in the power-law burning rate.

4 Results of experimental study

Investigated model MSF sample comprising ammonium perchlorate, inert combustible-binder on the basis of rubber SKDM-80 and metal fuel of the aluminum powder of different dispersion was used. In table 2 values of burning rate with atmospheric pressure and exponent in the degree law of burning rate for different type of aluminum powder are provided.

Table 1. Results of experimental study.

Type of aluminum powder	ASD4	ASD6	ASD8	ASD10	ALEX
Size of particles, D_{43}, μm	10.1	6.5	2.3	1.8	0.12
u_0, mm/s	4.0	4.8	5.3	5.1	5.2
Exponent ν	0.183	0.13	0.1	0.16	0.16

It is shown that as the particle size of the aluminum powder decreases, the burning rate of MSF increases. For aluminum powder which average diameter of particles equal $D_{43} = 1.8$ μm increase in burning rate in relation to particles size $D_{43} = 10.1$ μm makes about 17 % with pressure of 4.0 MPa. Further reduction of diameter of particles aluminum powder to micron $D_{43} = 0.12$ in comparison with particles aluminum powder $D_{43} = 1.8$ does not lead to significant increase in burning rate and is 2 %. Dependences of burning rate on pressure are presented on figure 2.

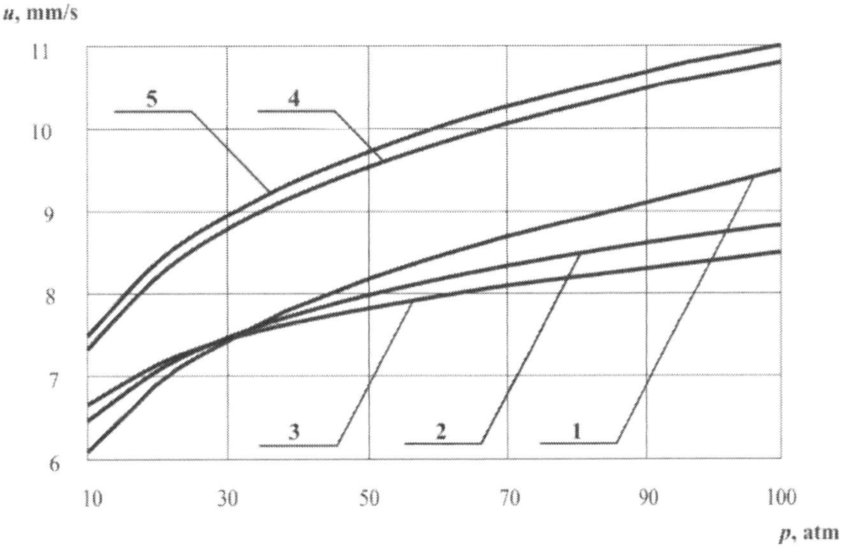

Fig. 2. Dependence of burning rate on pressure: 1 – aluminum powder ASD4, 2 – aluminum powder ASD6, 3 – aluminum powder ASD8, 4 – aluminum powder ASD10, 5 – aluminum powder ALEX.

The experimental results obtained are consistent when compared with known experimental results obtained in chamber of constant pressure. The addition of an aluminum powder of ultrafine size to the composition of the mixed solid fuel allows to increase the density of the samples. The experimental tests carried out in constant volume combustion chamber showed that a reduction of dispersion metal powder used in the composition of fuel (MSF) increases the burning rate of the sample. The burning rate of the mixed solid fuel decreases with increasing particle size of the aluminum powder due to small particle size and small oxide layer thickness, lower temperature of the beginning of rapid oxidation, and the high chemical activity of the Alex ultrafine aluminum powder, which increases the rate of heat transfer into the condensed reaction layer.

Thus, replacement of micro-size aluminum powder by ultra-size aluminum powder can significantly modify the main characteristics of combustion of mixed solid fuels.

This research was supported by The Tomsk State University competitiveness improvement program.

References

1. P. Pokhil, A. Belyaev, Y. Frolov, *Combustion of Powdery Metals in Active Media* (Moscow, Nauka, 1972)
2. S. V. Goldaev, N. A. Babushkin, S. A. Basalaev, EPJ Web Conf. **23**, 01014 (2015)
3. S. Goldayev, S. Basalayev, N. Minkova, MATEC Web Conf. **72**, 01034 (2016)
4. L. T. De Luka, L. Galfetti, F. Severini, Combust. Explos. Shock Waves **41**, 680 (2005)
5. V. A. Arkhipov, S. S. Bondarchuk, A. G. Korotkikh, Gorn. Zh., Special Issue: Tsvetn. Met. **4** (2006)
6. V. A. Arkhipov, A. G. Korotkikh, V. T. Kuznetsov, Khimicheskaya Fizika **23**, 9 (2004) [In Russian]
7. V. Arkhipov, L. Savel'eva, N. Zolotorev, MATEC Web Conf. **23**, 01005 (2015)

Determination of thermophysical and thermokinetic characteristics of forest combustible materials

Alena Zhdanova[1,*], *Svetlana* Kralinova[1], and *Galina* Nyashina[1]

[1]National Research Tomsk Polytechnic University, 634050 Tomsk, Russia

Abstract. The danger of forest fires and their large-scale consequences is facing humanity more and more sharply from year to year. We carried out experimental studies to determine the thermal characteristics (thermal conductivity, heat capacity, thermal diffusivity) and thermokinetic characteristics (activation energy, pre-exponential factor) of typical forest combustible materials and their mixtures for a wide range of temperatures (298–423 K). A generalization of experiments for typical forest combustible materials of the Far Eastern Federal District of Russia was performed. The established experimental values can be used for mathematical modeling of occurrence, propagation and extinguishing of forest fires.

1 Introduction

Deterioration of the environment and air pollution are, of course, a consequence of many factors. Over the past five years, wildfires have claimed the thousands of human lives around the world [1–3]. Rare species and individual animals are dying. In addition to thousands of hectares of burnt forest massifs, numerous cases of suffocation and poisoning of people with combustion products of forest combustible materials (FCMs) should be noted.

There are four relatively independent directions forest fire-fighting: direct assessment of forest fire danger [4], prediction of the combustion front propagation [5], fire extinguishing [6], and localization of combustion source [7]. In recent years, in each of these directions, the number of theoretical studies (for example, [4-7]) is increasing. This is due to the significant progress of the scientific community in understanding the physics of the corresponding heat and mass transfer processes, phase transformations and chemical reactions. It is important to carry out experimental studies of thermophysical and thermokinetic characteristics of FCMs with the aim of forming an information database necessary for the further development of the theory of forest fires.

The purpose of this study is the experimental determination of thermophysical and thermokinetic characteristics of typical forest combustible materials of the Far Eastern Federal District.

* Corresponding author: zhdanovaao@tpu.ru

2 Experimental setup and procedure

In this paper, the laser flash method [8] was chosen to determine the coefficients of thermal diffusivity, thermal conductivity, and specific heat of FCM under laboratory conditions. The main advantages of this method (in comparison with others) are [8, 9]: relative simplicity of measurements, high accuracy and reliability of results, applicability in a wide range of experimental conditions and materials, simultaneously measuring all basic thermophysical characteristics. In addition, the study of thermal conductivity by the laser flash method takes several times less time than the other methods [8, 9].

We used the system "DLF-1200 TA Instruments". The operating principle of the system is based on the evaluation of the propagation velocity of the heat pulse in the sample. Measurements are made at temperatures in the thermostat up to 1500 K. The error in determining the coefficients of thermal diffusivity is ± 2.3%, the heat capacity is ± 4%, and the thermal conductivity is ± 5%.

We used a rotary mill "Pulverisette 14" to grind FCM. The rotor speed was 6000–20000 rpm. The average particle size was about 200 μm. Thus, we prepared samples of powders of the investigated materials with a mass of 0.2–0.25 g. Using a hydraulic press, the samples were pressed into cylindrical specimens (tablets). The dimensions of the tablets corresponded to the requirements of the used measuring system, and did not exceed 12.5 mm in diameter and 3 mm in height. The density of the samples corresponded to the narrow recommended range (ρ=0.71÷0.83 g/cm^3). Powders were stored under ambient humidity at room temperature (18–22 °C). We did not pre-dry the powders before experiments. To increase the degree of absorption of the laser energy and, respectively, to increase the accuracy of measurements, a layer of graphite was applied to the surface of the samples by aerosol spraying. Then the samples were placed in a thermostat using a holder and a specialized trolley (Fig. 1 a). The thermostat working volume was filled with an inert gas (nitrogen).

During the measurements, we used a calibration sample (Pyrex7740), loaded into the thermostat module simultaneously with the test materials. The values of specific heat, thermal conductivity and thermal diffusivity coefficients are determined using the software of DLF-1200 TA Instruments system. We determined mass, height, base diameter and density of each sample before placement in the thermostat. Five measurements were carried out, the results of which were averaged. The height and diameter of the tablets were measured with a micrometer. The systematic measurement error was 0.005 mm. The mass of the samples was determined by the analytical balance Pioneer PA114. The systematic error of the measurements was 0.00005 g. According to the Romanovskii criterion, the geometric characteristics and mass of the test tablets were evaluated. We did not reveal rough errors in the experiments.

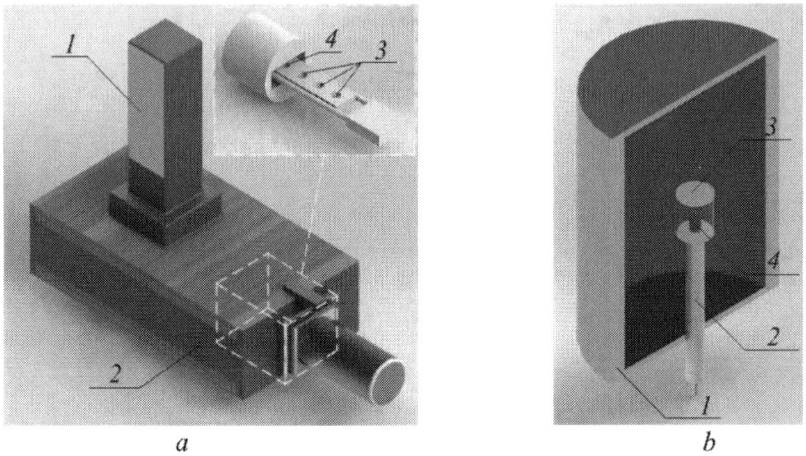

Fig. 1. Scheme of experimental setup for determining the thermophysical characteristics (*a*): *1* – laser pulse generation unit, *2* – thermostat unit, *3* – test FCM samples, *4* – calibration (reference) sample and conditional scheme of the thermo-system for thermal analysis (*b*): *1* – furnace, *2* – holder, *3* – sample place; *4* – temperature sensor.

In the present work we used synchronous thermal analyzer NETZSCH STA 449F3 for experimental determination of thermokinetic characteristics of the studied materials. The device has a wider temperature range than [10] and the possibility of performing the analysis in an inert and oxidizing atmosphere. The conditional scheme of the experiment is shown in Fig. 1 b. The thermal analysis was carried out under the following conditions, corresponding to the experiments [11]: the sample mass was 25 mg; heating from 300 K to 1270 K at a rate of 10 K/min in argon and oxygen with a gas flow rate of 100 ml/min. The kinetic parameters of pyrolysis were calculated by the method [12].

3 Results and Discussion

Fig. 2 illustrates the established values of thermophysical characteristics of FCMs, typical for the Far Eastern Federal District, in the temperature range 298–423 K. The choice of the temperature range is due to the real conditions under which the forest leaf litter is thermally decomposed and the subsequent combustion is initiated. If we consider in general the conditions under which the forest fire begins, then in the selected temperature range, intense heating of the material and the yield of bound water occur [13]. An intensive material decomposition is realized in the temperature range 350–450 K [6]. It is inadvisable to carry out experiments in the temperature range up to 500–600 K because the acceleration of thermal decomposition processes will lead to a change in the sample density and a corresponding increase in the error of determining the thermal conductivity, heat capacity, and thermal diffusivity.

The established results (Fig. 2) contribute to the significant development of modern concepts about dependencies (type, nature, variation ranges, scale of growth) of thermophysical characteristics of typical FCMs on temperature.

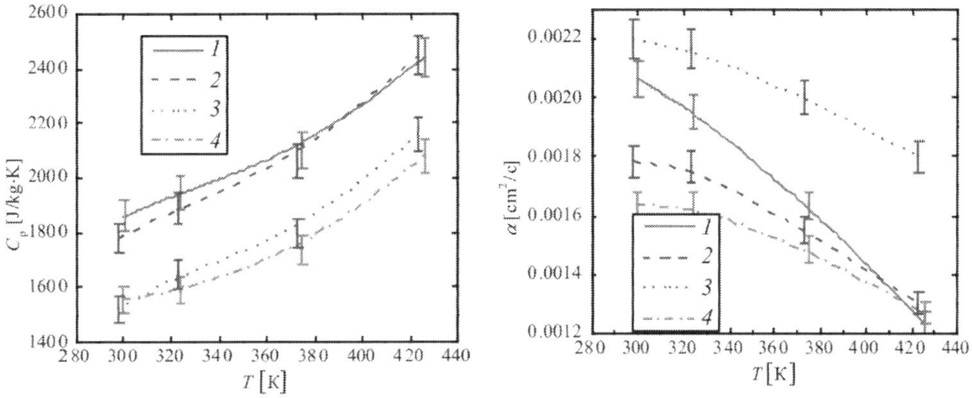

Fig. 2. The experimental values of heat capacity (*a*) and thermal diffusivity (*b*) of forest combustible materials: *1* – birch bark, *2* – birch branches, *3* – mixture of branches, *4* – grass.

Tables 1 and 2 show, respectively, the established thermokinetic parameters of the FCM organic mass decomposition and the formation of volatile substances in an inert and oxidizing atmosphere for all the studied boreal materials in three temperature sections.

Table 1. Thermokinetic characteristics of FCM pyrolysis in an inert atmosphere.

FCM	***E*, kJ/mol**	***k*, 1/s**	***E*, kJ/mol**	***k*, 1/s**	***E*, kJ/mol**	***k*, 1/s**
	Section 1 (430-540 K)		Section 2 (540-630 K)		Section 3 (630-840 K)	
Birch branches	77.27	$4.83 \cdot 10^6$	93.49	$2.95 \cdot 10^7$	13.46	$1.21 \cdot 10^0$
Birch bark	139.58	$1.86 \cdot 10^{13}$	93.50	$7.27 \cdot 10^7$	14.16	$1.09 \cdot 10^0$
A mixture of branches (60% of spruce, 40% of birch)	70.34	$1.46 \cdot 10^6$	97.82	$9.86 \cdot 10^7$	20.57	$4.28 \cdot 10^0$
Grass	–	–	61.54	$9.78 \cdot 10^4$	28.45	$1.73 \cdot 10^1$

The kinetic characteristics were obtained for three temperature ranges (Table 1): 430–540 K (*first*), 540–630 K (*second*), and 630–840 K (*third*). The temperature ranges were varied by no more than 10 K (in accordance with the error of the thermocouples) from experiment to experiment.

It was established that the thermophysical characteristics of FCMs differ significantly in different temperature ranges (Fig. 2, Tabs. 1 and 2). We can conclude that this can significantly affect the theoretically calculated times of warm-up, drying, thermal decomposition and burning out of FCMs, as well as localization and suppression of the corresponding processes.

Table 2. Thermokinetic characteristics of FCM pyrolysis in an oxidizing atmosphere (430–840 K).

FCM	***E*, kJ/mol**	***k*, 1/s**
Birch branches	99.93	$2.32 \cdot 10^7$
Birch bark	176.22	$6.11 \cdot 10^{13}$
A mixture of branches (60% of spruce, 40% of birch)	85.89	$1.09 \cdot 10^6$
Grass	70.17	$1.08 \cdot 10^5$

4 Conclusion

Temperatures close to the values corresponding to the beginning of thermal decomposition (350–450 K) are of greatest interest for applied calculations (for example, in problems of gasification, pyrolysis, fire extinguishing). The results of a comparative analysis of the heat capacity and thermal diffusivity of FCM mixtures and their individual components have shown that, in the first approximation, it is possible to fairly accurately determine the average (it is considered to be "effective") values of heat capacity, thermal conductivity, and thermal diffusivity of a mixture of materials by analogous parameters of individual components, taking into account their relative mass fractions in mixture.

The research was funded by the grant of the President of the Russian Federation (project MK-1684.2017.8)

References

1. A. Dimitrakopoulos, C. Gogi, G. Stamatelos, I. Mitsopoulos, Pol. J. Environ. Stud. **20**, 327 (2011)
2. F. X. Catry, F. C. Rego, F. Moreira, F. Bacao, 1st International Conference on Modelling, Monitoring and Management of Forest Fires **119**, 213 (2008) DOI: 10.2495/FIVA080221
3. D. H. Klyde, D. J. Alvarez, P. C. Schulze, T. H. Cox, M. Dickerson, AIAA Atmospheric Flight Mechanics Conference, 97625 (2010)
4. N. V. Baranovskiy, G. V. Kuznetsov *Prognoz vozniknoveniya lesnykh pozharov i ikh ekologicheskikh posledstviy* [Forecast of occurrence of forest fires and their ecological consequences] (Siberian Branch of the RAS, Novosibirsk, 2009)
5. M. S. Vdovenko, G. A. Dorrer, P. S. Shatalov, Computational tech. **18**, 3 (2013)
6. A. O. Zhdanova, R. S. Volkov, I. S. Voytkov, K. Y. Osipov, G. V. Kuznetsov. Int. J. Heat Mass Transf. **126**, 703 (2018), DOI: 10.1016/j.ijheatmasstransfer.2018.05.085
7. N. P. Kopylov, A. L. Chibisov, A. L. Dushkin, E. A. Kudryavtsev, Fire safety **4**, 45 (2008)
8. W. Hohenauer, L. Voza¨r, High Temperatures – High Pressures **35/36**, 253 (2003/2004)
9. W. J. Parker, R. J. Jenkins, C. P. Butler, G. L. Abbott, J. Appl. Phys. **32** (1968)
10. M. B. Gonchikzhapov, A. A. Paletskiy, O. P. Korobeinichev, III All-Russian Conference with International Participation "Combustion of Solid Fuels", November 13-16, 2012. Institute of Thermophysics named after. S. S. Kutateladze [in Russian]
11. P. A. Maryandyshev, A. A. Chernov, N. V. Shkayeva, V. K. Lyubov, Int. J. Exp. Educ.**11-1**, 71 (2013)
12. T. V. Bukharkina, N. G. Digurov, *Osnovy postroyeniya kineticheskikh modeley: uchebnoye posobiye* [Fundamentals of the construction of kinetic models: a textbook]. (RKHTU im. DI. Mendeleyev, Moscow, 1999)
13. A. M. Grishin, *Matematicheskiye modeli lesnykh pozharov* [Mathematical models of forest fires] (Izd-vo TSU, Tomsk, 1981)

The main reasons of rebound, coagulation, and explosive disintegration of the liquid drops in gas-vapor-droplet streams

Ivan Voytkov[1], *Olga* Vysokomornaya[1,*], *Maksim* Piskunov[1], and *Nikita* Shlegel[1]

[1]National Research Tomsk Polytechnic University, 634050 Tomsk, Russia

Abstract. We have reviewed the known data about mechanisms, conditions, reasons, and characteristics of rebound, coagulation, and explosive disintegration of drops in gas-vapor-droplet streams. The three main factors of altering a direction of motion, a velocity, a size and a concentration of droplets in a gaseous stream have been highlighted. Among of them are the thermal factor (heating and evaporation of drops), the aerodynamic factor (acceleration, reverse motion, and deceleration of drops due to their entrainment by gaseous stream), the dynamic factor (change in size, velocity, and direction of motion of drops after their collision). We have generalized the findings obtained by various authors in experiments with two drops, little group of drops, and aerosol. In addition, we have reviewed the published results about the fields of velocities and motion trajectories of drops in a flow, about the change in size and concentration of drops due to rebound, coagulation, disintegration (*i.e.* the complete destruction of the parent drops and the detachment of the liquid fragments of different size and volume from the surface).

1 Introduction

Gas-vapor-droplet technologies present a wide group of practical applications based on the interaction between the flows of sprayed liquid and gas. A considerable part of such technologies includes heating or evaporation of a liquid in the medium of hot gases. Among such applications are the contact heat exchangers [1], the drier installations [2], the technologies of the polydisperse fire suppression [3], the waste-heat recovery by a formation of gas-vapor heat carriers [4], the thermal purification of wastewater [5], etc. In addition to the mentioned heat and power technologies, as well as firefighting solutions, it is extremely important to consider structures and compositions of gas-vapor-droplet flows when treating of different surfaces by atomization (protective and paint coatings), in the fuel-feed systems, in the meteorology, medicine [6, 7], etc.

Currently, the three main groups of factors of altering a direction of motion, a velocity, a size and a concentration of droplets in a gaseous stream are known. Among them are the thermal factor (heating and evaporation of drops) [8–10], the aerodynamic factor

* Corresponding author: vysokomornaja@tpu.ru

(acceleration, reverse motion, and deceleration of drops due to their entrainment by gaseous stream) [8, 11], the dynamic factor (change in size, velocity, and direction of motion of drops after their collision) [12–17].

The purpose of the study is to review the state-of-art concepts about the three main mechanisms of varying the drop concentration in the gas-vapor-droplet flows.

2 Thermal Factor

As known, liquid evaporation occurs even at relatively low temperatures. However, with an increase in temperature, an evaporation rate nonlinearly grows. Homogeneous drops usually evaporate monotonously from a free surface. This process is accompanied by a step-down of their size [18]. Numerous studies focus on an analysis of particularities and characteristics of the liquid evaporation in a dropwise state. However, to date, an adequate prediction of the mass rate of evaporation is possible in a limited range of temperature only, up to 600 K. A more tremendous challenge is to describe a behavior of the evaporating heterogeneous drops because a presence of the interfaces causes the inhomogeneous fields of temperatures and concentrations, the vortices, and even the vapor bubbles that can result in disintegrating a drop.

A controlled use of a disintegration of the heterogeneous drops when heated can be applied in not only conventional heat and power technologies [1–5], but also in the present-day technologies of formation and combustion of the composite fuels based on wastes of the coal conversion and the oil refining [19].

The study of mechanisms, modes, and characteristics of heat and mass transfer in the case of the drop disintegration when heated [18] has enabled us to highlight the main governing factors. Among them are nucleate points (solids, bubbles of dissolved gases, etc.), differences in thermos-physical properties of components of the heterogeneous drops, a dominant heating mechanism, thermal radiation transmittance and accumulation, etc.

In addition, studies [20, 21] state that the intermolecular interactions and the influence of the drop components against each other on a change in their thermos-physical properties during heating can play a defining role. In particular, when heating the 0.1% solution of surfactant in water becomes less transparent in comparison with pure water. As a result, an aggregation of the amphiphilic molecules occurs. Based on experiments [20, 21], a correlation between temperatures of the water-based solutions and a level of turbidity has been found. An analysis of thermocouple measurements and experimental data obtained using highly precise microscopes confirms the conclusions drawn in studies [20, 21].

3 Aerodynamic Factor

One of the main problems of predicting a behavior of the gas-vapor-droplet streams is a difficulty to control an entrainment of drops by gases [22, 23]. Researches [8, 11] provide data about the main characteristics of phase transformations and deformation of drops of the polydisperse water flow under conditions close to the technologies with gas-vapor-droplet media [1–7]. Studies [9, 24] generalize, in a dimensionless form, the experimental data presented in [22, 23]. The results presented in such a manner characterize the conditions for deceleration and possible subsequent entrainment of drops by high-temperature gases.

An analysis of the experimental results presented in [10, 25] allows us to deduce that the drops of water and the water-based emulsions keep an integrity during motion through the gaseous medium in the conditions of varying their size and velocity in a limited range. It is assumed that water drops disintegrate in the gaseous flow when the Weber numbers are

10–12 [25, 26]. Study [27] reports that during motion of water drops through high-temperature combustion products the limit Weber numbers are 9–11. In addition, research [27] considers an influence of velocity and size of drops on characteristics of their deformation and an influence of the temperatures of gaseous medium and drops on characteristics of these processes. Moreover, the authors of study [27] show an effect of the aerodynamic factor at a variation of velocity and size of drops. Consequently, the links between the temperature gradients at the liquid / gas interface and the characteristics of deformation of water drops moving through a gaseous medium have been found.

Research [28] reports that the limit Weber numbers for the drops of mercury, alcohol, and kerosene are about 10. Study [29] describes that for the drop moving through the gaseous flow the critical Reynolds number characterizing a transition from the laminar flow regime to the turbulent one is about $3 \cdot 10^5$.

4 Drop Interaction Factor

Another factor considerably influencing the gas-vapor-droplet stream characteristics is a constant collision of the liquid drops. Currently, in the published studies researchers have try to find regimes of the drops collision and probabilities of consequences corresponding to these regimes [12–17].

To consider such factors as size of the colliding drops, their thermos-physical properties and velocity, the dimensionless criterion (the Weber number (We) can be used. Studies [12–17] provides the following scenarios of collisions: (1) coagulation; (2) rebound; (3) short-term coagulation with subsequent disintegration and formation of the secondary drops, which are similar in size to the primary ones; (4) collision scenario when the drop passes through another one; (5) explosive disintegration of drops with a formation of great number of small sub-droplets. The Weber numbers ranges have been obtained based on experiments for each collision scenario.

Note that studies [12–17] use the similar experimental methodologies. However, the results on the Weber numbers ranges considerably differ. In particular, the scenarios of the explosive disintegration and the passing the drop through another one in [12] are characterized by We > 25 and We > 50, respectively. Whereas, in [13, 14] the scenario of passing the drop through another one occurs at We > 50, but the explosive disintegration scenario – at We > 100. Moreover, the authors of researches [12–14] show zero probabilities of occurrence of any other consequences of collisions in each range mentioned. At the same time, studies [16, 17] depict the insignificant but not zero probability of occurrence of each scenario in any Weber numbers range.

The authors of studies [13–16] make several assumptions about the possible reasons for occurring each scenario of the drops collision consequences. In particular, the reason for coagulation of drops at small Weber numbers can be the drops surface oscillation of and the diminished pressure in the airspace between them. The rebound of drops stems from a presence of the gas space between them. The possible reason for the short-term coagulation with the subsequent disintegration is a rotational movement of drops due to a tangential collision. Lastly, the explosive disintegration of drops arises from cavitation at high kinetic energy during collisions.

5 Current Issues

The most challenging task of studies considering the thermal, aerodynamic, and dynamic factors is a regulation of the possible ranges of increase in surface area of liquid in the gaseous medium during the drop disintegration. To date, there is no comparative assessment describing a manifold increase in the surface area of liquid in the gaseous medium by using one of the factors considered in the article. Such an assessment is of particular importance for heat-exchange technologies to attain an enhancement of phase transitions [30, 31]. Most likely, the maximum efficiency can be reached when using a combined effect of thermal, aerodynamic, and dynamic factors, during the collisions of the heterogeneous drops heated up to high temperatures. These conditions are often fulfilled in multiphase and multi-component gas-vapor-droplet streams, but while they are underexplored. A detailed analysis of change in structure and composition of such streams is of importance for understanding their main parameters and the considerable resource saving in the appropriate technologies.

6 Conclusion

The analysis of literature devoted to the study of characteristics of the gas-vapor-droplet streams allows us to deduce that over a period of several decades results of such investigations remain in high demand. During this time, researchers have obtained numerous experimental and analytical findings enabling the predictions of the main parameters of the gas-vapor-droplet stream, their dynamics depending on many factors, namely, gas temperature, velocities of drops and a gaseous medium, droplet size distribution, etc. However, a difference in the results obtained by various authors is most likely due to an influence of the conditions, which have not been considered yet up to date in studies, namely, shapes of drops during collisions, properties of a gaseous medium, a presence of the vapor film around the evaporating drops, etc. The most complete consideration of all the particularities of fluid dynamics and heat mass transfer will considerably expand a theory of the formation of the gas-vapor-droplet streams and provide a basis for the development of the adequate methods for predicting their characteristics.

Research was supported by the Russian Science Foundation (project 18–71–10002).

References

1. R. M. Schirmer, R. L. Eson, J. Petrol. Technol. **37** (10), 1903 (1985)
2. M. G. Lee, Jajaei P. Mansouri, S. Mansouri, K. Hapgood, M. W. Woo, Dry. Technol. **36** (2), 178 (2018)
3. Z. Tang, Z. Fang, J. P.Yuan, B. Merci, Fire Saf. J. **55**, 35 (2013)
4. M. Ren, S. Wang, L. Qian, Y. Li, Appl. Mech. Mater. **577**, 523 (2014)
5. P. R. von Rohr, K. Prikopsky, T. Rothenfluh, Strojnicky Casopis **59** (2), 91 (2008)
6. D. Tarlet, C. Josset, J. Bellettre, Int. J. Heat Mass Transfer. **95**, 689 (2016)
7. K. Sollohub, K. Cal, J. Pharmaceut. Sci. **99** (2), 587 (2010)
8. G. V. Kuznetsov, P. A. Strizhak, High Temp. **52** (4), 568 (2014)
9. R. S. Volkov, G. V. Kuznetsov, P. A. Strizhak, J. Eng. Phys. Thermophys. 88 (4), 937 (2015)
10. R. S. Volkov, O. V. Vysokomornaya, G. V. Kuznetsov, P. A. Strizhak, Adv. Mech. Eng. **2014**, 865856 (2014)
11. G. V. Kuznetsov, P. A. Strizhak, Techn. Phys. Lett. **40** (6), 499 (2014)
12. D. G. Pazhi, V. S. Galustov, *Basic techniques of spraying liquids* [*Osnovy techniki raspylivaniya zhidkostey*] (Chimiya Publ., Moscow, 1984) [in Russian]

13. V. A. Arkhipov, V. F. Trofimov, J. Appl. Mech. Tech. Phys. **46** (1), 42 (2005)
14. V. A. Arkhipov, G. S. Ratanov, V. F. Trofimov, J. Appl. Mech. Tech. Phys. **19** (2), 201 (1978)
15. V. A. Arkhipov, I. M. Vasenin, V. F. Trofimov, J. Appl. Mech. Tech. Phys. **24** (3), 371 (1983)
16. D. V. Antonov, R. S. Volkov, G. V. Kuznetsov, P. A. Strizhak, J. Eng. Phys. Thermophys. 89 (1), 100 (2016)
17. R. S. Volkov, G. V. Kuznetsov, P. A. Strizhak, Techn. Phys. Lett. **41** (9), 840 (2015)
18. O. V. Vysokomornaya, I. S. Voytkov, G. V. Kuznetsov, A. V. Abramova, Int. J. Heat Mass Transf. **126**, 1043 (2018)
19. D. O. Glushkov, P. A. Strizhak, M. Yu. Chernetskii, Therm. Eng. **63** (10), 707 (2016)
20. A. V. Kim, N. N. Medvedev, A. Geiger, 2014. J. Mol. Liq. Spec. issue: Fluid phase associations **189**, 74 **(**2014**)**
21. V. P. Voloshin, A. V. Kim, N. N. Medvedev, R. Winter, A. Geiger, Biophys. Chem. **192**, 1 (2014)
22. R. S. Volkov, G. V. Kuznetsov, P.A. Strizhak, Int. J. Therm. Sci. **88**, 193 (2015)
23. A. Maatar, S. Chikh, M. Ait Saada, Int. J. Heat Mass Transf. **86**, 212 (2015)
24. Yu. V. Alekhanov, M. V. Bliznetsov, Yu. A. Vlasov, V. I. Dudin, A. E. Levushov, A. I. Logvinov, S. A. Lomtev, E. E. Meshkov, Techn. Phys. Lett. **29** (3), 218 (2003)
25. J. Eggers, E. Villermaux, Rep. Progr. Phys. **71**, 036601 (2008)
26. J. E. Sprittles, Y. D. Shikhmurzaev, Phys. Fluids **24**, 122105 (2012)
27. R. S. Volkov, A. O. Zhdanova, G. V. Kuznetsov, P. A. Strizhak, Techn. Phys. **60** (10), 1443 (2015)
28. V. V. Dubrovskiy, V. V. Podvysotskiy, A. A. Shrayber, Inzhenerno-fizicheskiy Zhurnal **58** (5), 804 (1990) [in Russian]
29. E. H. Trinh, R. G. Holt, D. B. Thiessen, Phys. Fluids **8** (1), 43 (1996)
30. O. V. Vysokomornaya, M. V. Piskunov, P. A. Strizhak, App. Therm. Eng. **127**, 1340 (2017)
31. P. A. Strizhak, M. V. Piskunov, R. S. Volkov, J. C. Legros, Chem. Eng. Res. Des. **127**, 72 (2017)

AUTHOR INDEX

AUTHOR INDEX

EDP Sciences – Web of Conferences
17, Avenue du Hoggar
Parc d'Activité de Courtabouf
BP 112
F-91944 Les Ulis Cedex A
France

ISBN 978-1-5108-7004-8

CURRAN ASSOCIATES INC.
proceedings.com